After the Boom in Tombstone and Jerome, Arizona

T0141397

WILBUR S. SHEPPERSON SERIES IN HISTORY AND HUMANITIES

Eric L. Clements

After the Boom in Tombstone and Jerome, Arizona

DECLINE IN WESTERN RESOURCE TOWNS

▲▲

UNIVERSITY OF NEVADA PRESS / RENO & LAS VEGAS

Wilbur S. Shepperson Series in History and Humanities

Series Editor: Jerome E. Edwards

University of Nevada Press, Reno, Nevada 89557 USA

Copyright © 2003 by University of Nevada Press

Photographs copyright © by the author unless otherwise noted

All rights reserved

Manufactured in the United States of America

Design by Carrie House

Library of Congress Cataloging-in-Publication Data

Clements, Eric L., 1958–

After the boom in Tombstone and Jerome, Arizona : decline in western resource
towns / Eric L. Clements

p. cm. — (Wilbur S. Shepperson series in history and humanities)

Includes bibliographical references (p.) and index.

ISBN-10: 0-87417-571-2 (hardcover : alk. paper)

ISBN-13: 978-0-87417-571-4 (hardcover : alk. paper)

1. Tombstone (Ariz.)—History. 2. Jerome (Ariz.)—History. 3. Tombstone (Ariz.)—
Economic conditions. 4. Jerome (Ariz.)—Economic conditions. 5. Frontier and
pioneer life—Arizona—Tombstone. 6. Frontier and pioneer life—Arizona—
Jerome. 7. Company towns—Arizona—Case studies. 8. Resource-based
communities—Arizona—Case studies. 9. Tourism—Arizona—Tombstone—
History. 10. Tourism—Arizona—Jerome—History. I. Title. II. Series.

F819.T6 C5 2003

979.1'53—dc21   2003008000

University of Nevada Press Paperback Edition, 2014

23   22   21   20   19   18   17   16   15   14

5   4   3   2   1

ISBN-13: 978-0-87417-958-3 (pbk. : alk. paper)

For my parents,

Anna and George Clements,

exemplars of the life of the mind

CONTENTS

ILLUSTRATIONS

Figures

Maps

Tables

ACKNOWLEDGMENTS

While hiking the long trail that ends at a book, one becomes beholden
to many people. First among these, in my case, is my wife, Barbara.
She has lived with this project for a decade and earned most of the
money that sustained it. My interest in the mining history of the
American West goes back twenty years; hers goes back four generations to Leadville, Colorado, in the 1880s. She has explored dozens
of mining towns and made numberless side trips to the middle of
nowhere in support of my odd enthusiasms. Then there are all the
sacrifices that she has made so that I could research and write. Thank
you for everything, Barbara, however inadequately I may express it.

My parents, Anna and George Clements, and my uncle Tom and
aunt Yoko Clements have also contributed to this project morally and
financially. My parents have supported this long journey with sympathy, sage council, and quite a bit of money. My uncle and aunt invested
thousands of dollars at two crucial points in the process, which kept
me relatively free from debt and permitted me to concentrate on
completing the initial stages of this work. My other parents, Alan and
Vera Patrick, have been understanding, hospitable, and supportive
over the twenty-five years that I have known them. Thanks to you and

the rest of my family for making this goal possible and the journey meaningful.

I've also been fortunate to have had excellent mentors to guide my work. Peter Iverson of Arizona State University managed this manuscript through its original development. He has shared with me much of his wisdom, both about writing a book and about life in the history profession. He has also rescued you from many shallow thoughts and much bad writing. Duane A. Smith, one of America's foremost mining historians, has let me draw upon his extensive knowledge of and enthusiasm for the subject, provided me with much good advice about the history business, and become another valued friend. Historians Robert Trennert and Philip VanderMeer both carefully examined and criticized this manuscript and provided me with pertinent advice and valuable support. A special word of thanks to Jerome Edwards, editor of the Wilbur S. Shepperson Series in History and Humanities of the University of Nevada Press, whose original enthusiasm for this project sustained it in its early days in the publication process.

Anna Clements—a doctoral candidate in English and a copyholder at Doubleday before she was my mother—proofed the entire manuscript with care and in detail. Sandy Crooms, former Assistant Director of the University of Nevada Press, guided this manuscript through its initial evolution into a book and showed boundless good humor and patience with an often ignorant and occasionally grumpy author. Managing Editor Sara Vélez Mallea ably oversaw the final stages of this book's production and Michelle Filippini served as in-house editor. Design and Production Manager Carrie House designed this book, Assistant Production Manager Jeanette Nakada managed its illustrations, and Bill Nelson created its maps. All of these people have greatly improved the quality and presentation of this book. The mistakes that remain are mine alone.

Of course, without the expertise and dedication of archivists, there would be little evidence of Tombstone's and Jerome's past left to research. My sincere thanks to the professionals at these facilities for their knowledge, assistance, and courtesy: the Hayden Library, Arizona State University, Tempe; the State of Arizona's Department of Library and Archives, Phoenix; the Arizona Historical Foundation and the University of Arizona Library's Special Collections, Tucson; Tombstone Courthouse State Historic Park, Tombstone; the Cochise County Recorder's Office, Bisbee; the University of Northern Arizona Library's Special Collections, Flagstaff; Jerome State Historic Park, Jerome; the Sharlot Hall Museum's Archives, Prescott; and the Yavapai County Recorder's Office, Prescott.

I would also like to mention that the chapter titles of my book are quotations from the historical newspaper articles I cite within the chapters. And, my thanks to the editors at the *Journal of the West* for allowing me to reprint segments of this book's text, which appeared in a different format in the article "Bust and Bust in the Mining West" in vol. 35 no. 4 (October 1996). Copyright © 1996 by Journal of the West, Inc. Reprinted with permission of *Journal of the West*, 1531 Yuma, Manhattan, KS 66502 USA.

Last, two special acknowledgments: First, thanks to all of my comrades in the Mining History Association, who are too numerous to list here. If you have any interest at all in mining history, I recommend that you join this organization immediately. The association is a wonderful and enthusiastic amalgam of scholars, preservationists, industry professionals, and enthusiasts. The association's annual conference, held in a mining district, is always one of the high points of my year. Who knew that you could learn so much and have so much fun at the same time? Second, an insufficient word of thanks to the scholars I've met along the trail in the past fifteen years who have become my friends. You are in this, too.

After the Boom in Tombstone and Jerome, Arizona

Introduction

Gunfire crackled in the hills above Tombstone, Arizona, in the early morning darkness of 9 August 1884. The gun battle involved perhaps fifty men behind a woodpile, shooting at seven men holed up in a building. But this was not a fight between lawmen and desperadoes, the sort of thing from which modern Tombstone claims its fame and makes its living. This time labor battled management, for the men behind the woodpile were striking miners, the men in the building were guards protecting the hoisting works of the Grand Central Mine, and the contest resulted from a wage reduction in a mining boomtown gone bust.[1]

No books have been written or movies made about this battle, and one is hard put today to find much information about this long-forgotten skirmish. Thus, it may serve as an analogy for bust itself. Bust is the gunfight that everyone forgets. Hundreds of books have been written about mining boomtowns and regions, but mining bust, which claims all of them sooner or later, has received very little notice. Standard practice when writing the history of a particular mining camp is to mention its bust as an epilogue. It is the termination of the story, but not a story in itself.[2]

It turns out that there is an excellent reason for this practice. Bust is difficult to research. One is frustrated time and again by sources that go as dry as an Arizona arroyo during a town's declining years. People move away and take their diaries with them. Interest turns to other subjects as a once busy camp wastes away. Yesterday's news. Even the sources that survive can betray the researcher. One purpose of a nineteenth-century newspaper was to boost its community, and editors had no qualms about ignoring negative facts that got in the way of a positive story. "On the contrary," as a Tucson editor remarked of journalism in Tombstone, "there was woe in store for the man or newspaper who dared to question the ramification, extent or permanency of the ... [town's] mines, the beauties of its climate or ladies, or the gallantry and rapidity of its men and horses."[3]

One expects that sort of rampant boosterism in a nineteenth-century bonanza camp, but even the mid-twentieth-century industrial town of Jerome advertised itself on highway signs and postcards as "The Most Unique Little Town in America," and proclaimed itself "The Billion Dollar Copper Camp," although its actual production never justified such a boast. Thus in Jerome as well, one has to blast through solid boosterism to get at the truth beneath. Given these problems with sources, one must sometimes live by inference and intuition when dealing with bust. The reader can therefore understand why some of what follows can be only speculation, written without the assurance one could have about the same subject and location during its boom days.[4]

Difficulties notwithstanding, this is an important subject. Economic decline and urban abandonment are certainly not unique to the American West, as the history of the Rust Belt demonstrates. One urban historian tallied 2,205 abandoned settlements in Iowa alone. Nor is the ghost town solely a product of the mining industry. Colorado has more than three hundred mining ghost towns—and over two hundred agricultural ghost towns. But the resource-extractive West has been renowned for the habituality of its boom-and-bust cycles, and it is the mining ghost town that fascinates people. Patricia Limerick, historian of the American West, notes that one would have trouble finding a popular guide to the ghost towns of Massachusetts or Ohio. She might have added that it is almost as difficult to purchase a guide to the agricultural or lumbering ghost towns of the Old West.[5]

A list of busts that have caused economic damage to towns, cities, and even whole regions of the North American West would be long and varied. The West is, of course, almost as famous for its gold and silver busts as it is for its gold and silver booms. One travels a deeply rutted road when observing that for every Leadville or Virginia City or Lead,

Fig. o.1. This view of modern Jerome, "the Billion Dollar Copper Camp," shows the main part of the town, with the United Verde Hospital above the town on the left, Cleopatra Hill in the center, and United Verde's open-pit copper mine to the right. The town also promoted itself on billboards and postcards as "America's Most Unique City."

there were dozens of places such as Safford, Nevada, or La Plata, Utah, or Elkhorn, Montana, which had an active life of a few years at most before being swallowed by obscurity.[6]

Although the gold and silver rushes produced the most famous ghost towns, bust has struck every mined metal at some time and place. Copper has boomed and busted in the West for more than a century. The state of Montana lost a congressional representative after the census of 1990 because of a declining population resulting from the curtailment of its copper industry. The more exotic minerals have also had their ups and downs. The decline in demand for tungsten after World War I effectively terminated the mining and smelting industries in Nederland, Colorado, yet that town has fared better than Tungstonia, Nevada, which ceased to exist. Pinchi Lake, British Columbia, had the largest mercury mine in the world during World War II, but the town died after 1947, when the mine closed due to low postwar demand. The uranium frenzy in the four-corners region of the American Southwest during the 1950s and 1960s has long since collapsed. Moab, Utah, ground zero in that uranium boom, returned to being a quiet farm town for a time, though it later surged anew when discovered by relocating Californians. The slump in the price of molybdenum in the 1980s caused yet another suspension of operations at the giant Climax Mine above Leadville, Colorado, seriously damaging that town's vitality.[7]

It is not just the metals, of course. Residents of Bonanza, Wyoming; Median, Kansas; and Mentone, Texas, learned all about the nomadic nature of the oil business. Many little towns like Walsenburg, Colorado; Thurber, Texas; and Van Houten, New Mexico, which used to mine the coal that stoked railroads and industries, have dwindled or disappeared. The abandoned charcoal, coke, or lime kilns at places such as Gardiner, New Mexico; Nicholia, Idaho; and Electric, Montana, now serve as the headstones for those settlements. The diminution of CF&I Steel at Pueblo, Colorado, brought about the abandonment of regional lime, coal, and coking operations; ended iron mining at Sunrise, Wyoming; and for many years seriously injured Pueblo itself.[8]

Farming and ranching have also experienced economic disasters in the West, and to such a degree that it is difficult to know where to begin when discussing the subject. Some of the more well known of these include the destruction of the open-range cattle business during the severe winters of the late 1880s. Perhaps half of all open-range cattle died, and many of the companies that had entered this latest bonanza industry went under. The cattle trade busted in towns such as Kimball, Texas, because the cattle trails shifted westward or at Elgin, Kansas, or Doan's Crossing, Texas, because the railhead relocated closer to the range. The greatly expanded production of wheat on the Great Plains during and after World War I contributed to the collapse of wheat prices during the Great Depression, and, some believe, produced the Dust Bowl, which killed towns like Foss, Oklahoma, and McAllaster, Kansas. Most of the larger prairie towns along northern Colorado's front range saw a sugar beet mill close in the 1970s, and Montana also had a sugar beet bust.[9]

In the Pacific Northwest, the timber industry, up and down for over a century, has lately been down again, with mills shut down and mill hands unemployed. In earlier days, the exhaustion of trees sealed the fate of logging towns such as Lumberton or Kitchener, British Columbia. Rumors have circulated in the Pacific Northwest's deep-sea commercial fishing industry that it, too, may be closed down in a few years—run out of business by the farm fisheries of other countries.[10]

The long and dreary list of busts in the North American West has not been confined to the extractive industries. Transportation systems have made and broken communities. When stage lines or wagon trails shifted or declined, places like Diamond Springs, Kansas, and Pinery, Texas, found themselves without a means of subsistence. Houston, Idaho, was described as "relatively prosperous" before the railroad bypassed it in 1901. The construction of the Canadian Pacific and Union Pacific Railroads made temporary boomtowns of Emory City, British Columbia, and Bear River City, Wyoming. But these communities quickly declined after the tracks had been laid. Railroad construction sounded a death knell for steamboat landings such as Port Sullivan, Texas, and Polhamus Landing, Arizona, and stage stops like San Antonio, Nevada, and Kennekuk, Kansas. A new stage road bypassed and doomed Cottonwood, California, in favor of Henley in 1854, but Henley's glory proved short-lived. In 1887 the railroad passed through Hornbrook instead of Henley, and Henley became another California ghost town. Such mortal transportation rivalries often took place between communities in the West. The coming of the railroad secured the future of Wichita, Kansas, while it obliterated nearby Park City. Antelope, Oregon, declined as a nearby railroad town prospered, while Otero, New Mexico, which had the railroad, lost the division point—and most of its population—when the rails pushed north five miles to Raton.[11]

Changes in markets or technologies affected the fortunes of railroad communities in their turn. Steins, New Mexico, lost its helper station; Montello, Nevada, its roundhouse and shops; and Pumpville, Texas, its depot and water stop when steam locomotives were phased out, a fate sociologist W. F. Cottrell terms "death by dieselization." Other small communities, like Raso, Arizona, were wiped out when automatic signaling or mechanized track maintenance ended the railroad functions for which they had been created. Even if a town survived, the loss of its railroad or its depot could deliver a hard blow to its economic and psychological vitality. Railroad market towns such as Perico, Texas, and Mingo, Kansas, also died when improved roads, automobiles, and trucks rendered them superfluous.[12]

Steins experienced the double misfortune of losing its railroad facilities and then, within a few years, being bypassed by the new interstate highway, another fate suffered by numerous towns. Highway engineers designed the interstate system to bypass small towns, an idea that increased highway speed and capacity, but doomed many small-town business districts. The automobile and rural free delivery ended the careers of many rural villages by making their post offices and general stores redundant. The farm town of Richmond, Oregon, busted when the automobile reduced the trip to its larger neighboring community from a daylong wagon ride to a half-hour drive. The mechanization and consolidation of agriculture in the twentieth century considerably reduced farm populations and doomed many rural communities like Dubuque, Kansas.[13]

Sometimes bust goes beyond economics to politics. Contests over the location of the county seat were often a question of victory or death for the communities involved. Hiko, Nevada, and Rayner, Texas, both disappeared after losing the county seat, though Rayner did not go quietly, subjecting Stonewall County to eight years of litigation on the subject. Occasionally, these questions went beyond the law. An attempt by citizens of Ingalls, Kansas, to seize and remove the Gray County records from Cimarron in 1889, after disputed elections, resulted in a fatal gun battle. Similar incidents took place in several other county seat disputes in Kansas.[14]

One of the most vindictive of these wars erupted between Ravanna and Emenence for the seat of Garfield County, Kansas. In their enthusiasm to gain the county seat, the citizens of Ravanna liberalized their suffrage in 1887, permitting sixty deceased persons to vote. When Ravanna won the election, Emenence went to court and gained the county seat. Ravanna appealed, but in 1889 the state supreme court upheld Emenence. Unable to gain the victory, Ravanna returned to court with the ultimately successful claim that the area in question was too small to be a county under Kansas statutes. Ravanna's revenge caused Garfield County to be absorbed into Finney County in 1893. This last bit of spite effectively doomed both Ravanna and Emenence.[15]

More recently, the end of the cold war added to the chronology of bust in the West by curtailing the California and Washington aircraft and electronics industries and injuring many western communities beholden to nearby military bases for their economic vitality. The region has seen this before. When the army deemed Fort Belknap in Texas and Fort Wallace in Kansas no longer necessary, their namesake communities were damaged or destroyed. That was more than a century ago. When Columbus, New Mexico, served as the headquarters for General Pershing's punitive expedition into Mexico in 1916, it billeted almost five thousand soldiers, but seven years later only sixty. When World War I ended, so did the need for Army City, Kansas, created in 1917 to serve Fort Riley and Camp Funston.[16]

Some towns have had the hardihood to survive a series of boom-and-bust cycles. Aspen, Colorado, boomed in silver a lifetime before the town boomed again with skiing. Breckenridge, Colorado, had three kinds of gold mining in different periods: placering, lode mining, and dredging. Samfordyce, Texas, boomed when the railroad arrived in 1905, had a revival after 1910 as a military supply town during the border troubles of the Mexican Revolution, and rebounded again during an oil boom in the 1930s. Samfordyce busted for good when the oil dried up after World War II. Lenado, Colorado, may have the record for number and types of boom-bust cycles over the past century, with a silver-lead boom that ended in 1893, a short silver-lead encore after 1900, a lead-zinc boom during World War I, a timber boom in the 1930s, and a counterculture boom in the 1970s.[17]

Historians and social commentators, with feelings ranging from indignation to nostalgia, have long cited these boom-and-bust cycles as central to the history of the American West. If we are to accord the busttown such significance in western history and thought, we should make a better attempt to understand it. Although there has been plenty of writing to date about bust as symbolic of this or that, there has been next to no attempt made to study the phenomenon on a case basis.

The case-study work that has been done on communities in disaster is mostly the work of sociologists, although they, too, have been much more interested in boom than bust. There is even a subfield of sociology known as "disaster studies." The U.S. Strategic Bombing Survey served as the cornerstone of disaster studies. The survey, conducted in Germany and Japan after World War II, attempted to determine the effectiveness of Allied strategic bombing in destroying the Axis economies and morale. The field of disaster studies subsequently developed during the 1950s and 1960s through numerous monographs and articles.[18]

Its practitioners studied the reactions of town officials, institutions, and residents to physical disasters—such as tornados, floods, or atomic attacks—visited upon their communities. The field examined the psychological behavior of individuals in disaster situations, as well as the sociological behaviors of unorganized masses of people and of formal organizations in disasters. It also sought to discover more rational responses that might help communities better withstand such cataclysms, atomic or otherwise. The shadow of the cold war loomed over disaster studies. As one of its leading authors notes, "[S]ince 1945 [the field] has been supported mainly by government agencies attempting to plan civil defense against atomic weapons."[19]

Although certain elements of sociological theory derived from disaster studies are useful in understanding bust, generally these studies of physical disasters are of no more help in understanding a community's economic collapse than is the study of traumatic injury helpful in understanding disease. More helpful are sociological examinations of mass unemployment. This line of inquiry began with several studies made during the Great Depression. As with disaster studies, the research was driven as much by an effort to find a way out of present difficulties as it was by academic curiosity. In consequence, the field went into eclipse during the boom times of the 1940s and 1950s. It revived during the 1960s and 1970s. A renewed interest in social history and the readjustment of the U.S. economy away from heavy manufacturing, with its attendant dislocation, produced another wave of unemployment studies. The collapse of the basic steel industry received particular attention.[20]

The decline of the small town also gained some attention in this era, with particular attention being devoted to fading midwestern farming communities. First noted in the 1950s, the small town's lost relevance and decay in the face of postwar urbanization and suburbanization received the attention of social scientists in the 1960s and 1970s. These small, independent "island communities" dried up and died as the twentieth century progressed, losing their purpose because of changes in transportation systems, living styles and patterns, farming techniques, and an increasingly nationalized economy.[21]

Again, although elements of these fields are helpful to an examination of bust in the American West, there are limits to their utility. Unemployment studies have, reasonably enough, focused on large groups of unemployed persons and sometimes upon the industrial context that produced their unemployment. Often, in quest of solutions, these studies examine the efficacy of social service agencies charged with assisting the unemployed. Seldom have these studies dealt with the effects of mass unemployment on whole communities. Small midwestern farm towns—though closer in scale and purpose to the subjects considered here—do little to help establish the significance of boom and bust in and to the American West. As with disaster studies, most of the case examples from which these fields draw their conclusions are from the modern era of an established and extensive welfare bureaucracy. These studies were created in and concern, therefore, a social and governmental context quite dissimilar to that which existed throughout most of the history of the West.

Similarly unhelpful in the study of bust are most community and urban histories, with their focus on growth and development rather than their antitheses. Popular community histories are often written by interested parties, and are essentially biographies of the communities under study. As such, their utility for making more general statements

is limited to reading a number of them and extracting their common-
alities. More analytical studies in community history tend to be meta-
physical examinations of the nature of community, an intellectual
exercise mostly outside the bounds of this study.[22]

Scholarly urban history has, like the sociological studies discussed
above, produced some theories useful to us, even if urban historians
have not been inclined to take mining communities too seriously.
Mining towns, argues one, "were for the most part too ephemeral to
leave any enduring mark in our history. Colorful and dramatic, their
importance was peripheral to the story of the town. The same could be
said for most cattle and lumbering towns, although there are of course
important exceptions." It need hardly be said that these kinds of towns
were not peripheral to the story of the town in the West. They are central to the history of bust in the West. The weakness of urban history,
for our purposes, is its orientation toward large eastern cities.[23]

The little historical work that has been done on bust deals mostly
with the period after World War II. The best of the bust histories is
William Robbins's *Hard Times in Paradise*, a study of the timber industry in Coos Bay, Oregon. This book presents perhaps the most systematic examination of bust in western history. Using extensive personal
interviews and sociological evidence, Robbins deals primarily with the
effects of lumbering shutdowns in the Coos Bay area upon the mill and
camp workers left suddenly to fend for themselves during the Great
Depression and in the early 1980s. He uses his study as a platform for
an extended discussion of the nature of industrial capitalism and its
injurious effects on the working class and the environment.

The best works considering bust in mining history include *Treasure
Hill: Portrait of a Silver Mining Camp*, by W. Turrentine Jackson. Jackson
studies the history of the White Pine District in eastern Nevada, rather
than bust per se, but in his account Jackson makes sure to describe the
fall as well as the rise. He understands that bust was by far the more
probable outcome for a mining settlement than everlasting fame and

glory, but also recognizes the handicap under which he works. "The relative silence about the failures in the mining kingdom," he writes, "has resulted from a lack of records. Men have a tendency to forget rather than record disappointment and failure."[24] But what he lacks in sources, Jackson makes up for by his intrepid willingness to pioneer into the unmapped territory of bust.

Since Jackson's work appeared, several other studies have been published that have examined the complete life cycle of particular mining communities. One of the best of these is Duane Smith's *Silver Saga: The Story of Caribou, Colorado.* Caribou, high in the mountains west of Boulder, showed much promise in the 1870s and 1880s, but disappeared after 1900. Smith uses the example of Caribou's brief career to examine not only the lives of the people involved, but also the society, economy, motives, and expectations that molded their experiences. Ideas and questions Jackson and Smith raised in *Treasure Hill* and *Silver Saga* inspired the work before you.[25]

An important recent publication is *The Roar and the Silence,* by Ronald James. James, in his comprehensive treatment of the history of Nevada's Comstock Lode, illustrates that "Virginia City alternated between boom and decline, again and again." The Comstock boomed during its discovery period in the late 1850s, again after the district's consolidation and development by the Bank of California syndicate in the latter half of the 1860s, and once more during the "Big Bonanza" from 1873 to 1879. After each of these booms came bust, with its mass exodus and predictions of the district's demise. James believes that "a single portrait cannot capture the nineteenth-century community . . . since each period of prosperity and depression assumed its own distinct form."[26] Like Jackson and Smith, James devotes as much thought and emphasis to bust as he does to boom.

Ralph Mann's *After the Gold Rush: Society in Grass Valley and Nevada City, California, 1849–1870* is another good and important work, being a rare examination of two California mining camps after the first blush of success had passed. Based largely on census data from 1850, 1860, and 1870, Mann's book is mainly concerned with social and ethnic relations in the two "maturing" towns of Grass Valley and Nevada City, rather than with towns plunging into bust. There was a serious economic depression in the period he examines, however, and he makes many observations that are pertinent to this study.

Perhaps the work that comes closest to the aims of this study is Andrew Gulliford's *Boomtown Blues: Colorado Oil Shale, 1885–1985.* Gulliford sees the oil-shale bust of the 1980s, which followed Colorado's western-slope oil-shale boom of the 1970s, largely as a question of "capital versus community." He believes that "if there are lessons to be learned from boom-and-bust cycles in the West, the oil shale saga is more than just an excellent case study. It is *the* case study."[27]

Gulliford ardently argues that his case study illuminates the corporate exploitation and victimization of the West. The villain of the piece is the Exxon Corporation, which created the boom by planning to carry out an unmitigated ecological disaster upon Colorado's western slope, then causing the bust by not committing that ecological debauchery in the face of low oil prices. Gulliford maintains that "the recent oil shale boom in the Colorado River Valley was blatantly manipulated from the start without adequate consideration for the people whose lives would be most affected." The small-town West, he believes, suffers from corporate colonialism and economic dependence.

Gulliford's book seems to lie more in the realm of public policy than history, however. Indeed, he chides historians for their wait-and-see attitude, holding that "the opportunity to reflect and gain perspective is *now*—not at some distant point in the future when the cycle begins to repeat itself."[28] He also resided in one of the towns struck down when the oil-shale bubble burst, and classifies himself as a "participant-observer," rather than a disinterested scholar.

Although all of the works mentioned above are valuable, the fact remains that no one has yet made a systematic study of the nature of bust itself. All have considered bust, to the extent that they have, as a means to some other end. This work will examine bust itself. As the topic has not been addressed for its own sake from a historical perspective before, the purpose of this study will be largely descriptive—endeavoring to discover and disclose what actually happens. This work should probably not be classified as community history, urban history, labor history, or even mining history, although each of these will play its part and have its influence. The object of this study is to use the histories of two towns, and of particular subjects within them, to examine bust. Hence, the only questions it seeks to answer about women, or fraternal organizations, or politics, or boxing matches are: How does bust influence X? Or the reverse: How does X influence bust?

The two subjects of this study are Tombstone and Jerome, Arizona. The Tombstone mines, discovered in the late 1870s, supported the second-largest town in the Arizona Territory by 1882, but that success could not be sustained. Flooding in the mines and the falling price of silver eroded Tombstone's economy. By 1890, the town had been reduced to about one-third of its peak population, and even with a modest revival of mining after 1900, its bonanza days were over.

The original Jerome claims were located a few years before those at Tombstone. In this case, copper attracted the mining men. Although copper is not as valuable as silver, because of their slow development and enormous size Jerome's deposits outlasted Tombstone's by decades. The Jerome ore deposits eventually proved to be some of the great discoveries of high-grade copper ore in human history, but even such a vast resource could not protect Jerome from economic misfortune. Dependent on a good price for copper, the town experienced local depressions when the price of the metal did not justify mining. The most serious of these occurred during the Great Depression, when, for several years, the market price of the metal fell below its cost of production. The consequent closure of the principal mine in the district, combined with the exhaustion of the second mine, produced at least a 50 percent reduction in Jerome's population between 1930 and 1940.

Both of these communities have the necessary characteristics that a candidate for a case study of bust must possess. As we have seen, many towns in the West have suffered economic reverses, but unless the cause of the downturn is fairly obvious, the date of its occurrence reasonably easy to determine, and the downturn itself fairly sharp, it will not lend itself to systematic analysis. Over the past century, Rico, Colorado, has been reduced from a center of mining and smelting and a railroad division point to an isolated mountain hamlet, but that decline took place over the course of several decades, as the town lost its mines and smelters, and finally its railroad. Its decline was so diverse and so gradual that there is little chance to examine it critically. Both Tombstone and Jerome satisfy the criteria outlined above.

Two objections could be made against selecting Tombstone and Jerome for this study. The first is that to compare Tombstone, a nineteenth-century frontier bonanza silver camp that boomed and busted within a decade, to Jerome, a mid-twentieth-century base metal–processing industrial town of seventy-five years' duration, is to compare apples and oranges. But one simply must compare apples and oranges if one wishes to understand fruit. Bust is not some quaint nineteenth-century phenomenon caused by goldbugs and vanished with the buffalo. Boom and bust in western mining have lasted from the frontier days right through the twentieth century. A serious consideration of bust as a topic, therefore, must embrace not only the bonanza camps, but also modern industrial mining, base as well as precious metals, and mining communities of long duration as well as those long gone.

A more serious objection is that both Tombstone and Jerome are atypically large and renowned, and thus do not represent the "average" mining community. This is perhaps true, at least in regard to nineteenth-century mining. There is no gainsaying the fact that

Leadville, Colorado, and Tombstone, Arizona, were the most famous bonanza silver camps of the early 1880s or that the Jerome deposits were among the greatest mineral discoveries in North America. Certainly, no one could seriously claim that Jerome is representative of Quijotoa, Arizona; Cleary, Alaska; Tuolumne, California; or hundreds of other towns in the mining West that sparkled briefly then quickly disappeared.[29]

But the problem of being oversized is not insurmountable; in some aspects it is helpful. Small towns, here and gone in a season or two, are not particularly informative about the process of decline. To return to the medical analogy, death, in itself, is not terribly instructive. Obviously, in a town quickly denuded of its entire population, every aspect of town life, institutions, and population is affected 100 percent. Of interest here is the process of dying, and the larger towns, which took longer to decline, can reveal that process more clearly.

Beyond the theoretical lies the practical. Any systematic study of bust will favor larger communities simply in quest of enough material with which to work. The towns chosen have to be large enough and survive long enough to show up in census tables as discrete communities for more than one census, to sustain a newspaper for a number of years, and to establish recognizable patterns of community before their collapse. It is worth noting that Jerome lost its newspaper in the middle of the period under study and that both Tombstone and Jerome declined to such small sizes at different times that they ceased to receive attention in published census tabulations. Thus, even when dealing with the great camps, one suffers from a lack of information. As the size and duration of a subject community decrease, the problem of the dearth of sources increases until it is soon insufferable.

Further, our stereotypical perception of western mining camps as here today and gone tomorrow may be in need of some revision. True, many of the gold-placering settlements were never more than short-lived, rude camps, but the other metal-mining processes were heavily

industrialized, even in the nineteenth century. These industrial processes required large capital investments, sophisticated physical plants,
and substantial labor forces. These, in turn, fostered complex urban
communities to support them. Numerous mining towns founded in
the 1870s or 1880s survived well into the twentieth century. A considerable number produced for over a half century and havé lasted in
some other form for almost a century and a half. One guide to Arizona's ghost towns mentions 117 communities that had a post office
with known opening and closing dates. The mean life of their post
offices was twenty-one years, and the median was seventeen years.[30]
The careers of these communities lack the stereotypical evanescence,
and the same is true in other states as well.

The following four chapters will recount the developments and declines of Tombstone and Jerome. The fifth chapter considers changes
in the mining industry during the declines of the two communities.
This chapter examines the industrial situation in a social and, to a
limited extent, economic and technological sense. Does bust change
the nature of ownership or operation of mines? Does it necessarily
produce labor-management conflict?

The sixth chapter concerns itself with the economy of a mining
town beyond its mining industry. This economic community includes
town merchants and business leaders, of course, but also encompasses
the professions, trades, occupations, and less savory ways of making
a living known to occur in mining towns. How was the economic community injured or transformed by bust? Did some types of businesses
suffer more than others? Did businesspeople suffer disproportionately because they lacked the physical mobility of capital and labor?

The seventh chapter examines busttown demography. Did bust
change the demographic composition of these communities in such
identifiable categories as race, ethnicity, sex, or age? Did bust have
a greater effect among minority groups? Did it necessarily produce
ethnic tension?

The eighth chapter addresses the fate of busttown society. The vitality of these societies will be examined by studying such formal institutions as churches and fraternal organizations, as well as less formal social expressions like community celebrations and sporting activities. What changes, if any, were wrought by bust? Did a breakdown of the social order occur under the stresses of bust, manifested by such things as more crime and violence or altered rates of marriage and divorce?

The ninth chapter considers the changes bust produced in the physical realities of mining towns. How did bust, or the reaction to it, alter the community physically? What sorts of changes were produced in its infrastructure? Did bust change patterns of residency or commercial activity? In a more general sense, what were the processes involved in creating a ghost town?

The tenth chapter discusses community leadership. Did bust produce political transformations in local government, like new leaders or new political organizations? What changes, if any, did bust create in such government functions as education or city services? What did people want or expect from local government in the face of bust?

The eleventh chapter examines the mitigation of bust. What did people do, individually or collectively, to defend themselves and their community against economic reversal? What did they do to relieve the suffering of the unemployed and to bring about an economic revival?

The concluding chapter briefly reviews what happened to these two towns in the periods after the study and then discusses the larger significance of boom and bust in western history and historiography. In sum, this book will attempt to provide a better understanding of bust, the responses to it, and its place in an accurate understanding of the history of the American West.[31]

CHAPTER ONE

Future Growth and Prosperity Is Assured

Nineteenth-century mining camps are famous for being located in forbidding places. The Tombstone District was certainly one of them. John Gray, who as a young man made the stage trip from the railhead a few miles east of Tucson to Tombstone in June 1880, wrote many years later:

> That day's stage ride will always live in my memory—but not for its beauty spots. . . . Leaving Pantano, creeping much of the way, letting the horses walk, through miles of alkali dust that the wheels rolled up in thick clouds of which we received the full benefit, we couldn't then see much romance in the old stage method of traveling. But the driver said that was *his* daily job which made us ashamed of our weakness. . . . If it had not been for the long stretches when the horses had to walk, enabling most of us to get out and "foot it" as a relaxation, it seems as if we could never have survived the trip.[1]

Another stage passenger who arrived at about the same time, Adolphus Noon, wrote to the *Chicago Tribune* that those destined for Tombstone "can now travel for $4 where they recently had to pay $10. . . . The

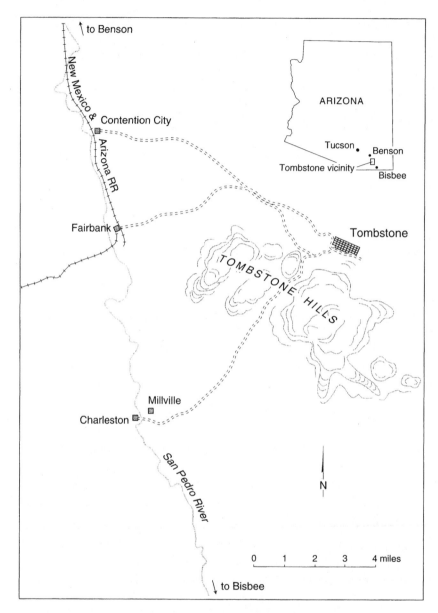

to Benson

New Mexico &

Arizona RR

Contention City

Fairbank

Tombstone

ARIZONA

Tucson
Benson

Tombstone vicinity
Bisbee

TOMBSTONE HILLS

Millville

Charleston

San Pedro River

N

0   1   2   3   4 miles

to Bisbee

Map 1.1. Southern Arizona and the Tombstone District, c. 1883, showing Tucson, Tombstone, Charleston, Fairbank, and Contention City.

seventy-five miles are made in about eleven hours—all the glorious effect of free trade and competition, for they formerly took about twenty-four." Noon seemed as unimpressed by the destination as by the trip. "Tombstone is not a very attractive place," he recorded. "It is windy and dusty, scantily supplied with water, with no trees to break the landscape ... yet it will probably be found healthy ... and good mines and appropriate position will build a town more quickly than sylvan scenes."[2]

Gray and Noon had come a long way to see this elephant. Located in the arid, hot, and desolate San Pedro River valley in far southeastern Arizona, Tombstone lay about thirty miles north of the Mexican border and deep inside Apache territory. It was the sort of country in which—

Fig. 1.1. Tombstone in the early 1880s. This was the scene that greeted new arrivals after their long and dusty stage ride from Tucson. (Arizona Historical Society, Tucson, #44688)

as one prospector had been warned—a fellow was as likely to find his tombstone as his fortune.

In spite of the dangers, that professional prospector, Edward Schieffelin. persisted with his explorations. He discovered a series of outcroppings of horn silver so rich, it was said, that one could make an impression in the soft metal with a knife or a coin. After making his initial location in September 1877, Schieffelin recruited his brother Al to help prospect and a friend named Richard Gird who agreed to assay the ore samples in return for a one-third interest in the venture. The three men returned to the area in the spring of 1878 and made a series of valuable locations, which they then recorded at Tucson.[3]

The recording of the locations ended the secrecy that the three had done their best to maintain, and the stampede was on. By the time Gray took his dusty stagecoach ride two years later, Tombstone had a population of more than two thousand, two newspapers, several stage lines and hotels, and many stores and restaurants. By 1882 the population

Fig. 1.2. Tombstone at the height of its boom in 1882. Some of the town's mines are visible on the hills to the right. (Arizona Historical Society, Tucson, #8295)

had increased another 200 percent and Tombstone had become the county seat of newly created Cochise County and the preeminent mining camp of the Southwest.

By then the mines had passed out of the control of their original locators. The district included thousands of claims by that time, but its heart consisted of three great mining concerns. The first and arguably foremost of these was the Tombstone Mill and Mining Company, formed out of the original Schieffelins-Gird holdings. Having the claims but needing the capital to develop them, the discoverers brought in a series of investors, including former territorial governor A. P. K. Safford and the Corbin family of Philadelphia.[4]

In October 1878 the Schieffelins, Gird, and the Safford group reached an agreement, granting the Safford group a one-quarter share in a series of claims in exchange for their construction of a wagon road to the San Pedro River and a ten-stamp mill at the river to process the ores from those claims. This combination of investors organized as the Tombstone Gold and Silver Mill and Mining Company, and their stamp mill, the first in the district, crushed its first ores and shipped its first silver in June 1879. By August the Tombstone Company had produced over ninety thousand dollars in bullion and that month announced its first dividend after fewer than nine months in operation.[5]

In early 1879 essentially the same parties organized the Corbin Mill and Mining Company out of another set of Schieffelins-Gird claims. In January 1880 the stamps of the Corbin Mill dropped for the first time, and in May of that year the Schieffelin brothers sold their share of the two companies, which were then consolidated as the Tombstone Mill and Mining Company of Hartford, Connecticut. Richard Gird sold out in March 1881. The company had produced more than one million dollars and paid almost a half-million dollars in dividends by that time. By the end of 1881 the company had paid over one million dollars in dividends.[6]

The Contention claim produced the second great mining concern in the district, nearly the equal of the Tombstone Mill and Mining Company. The Contention claim was originally incorporated as the Western Mining Company in 1880. The Contention had its own mill on order from San Francisco by August 1879 and produced twelve hundred pounds of bullion per week from that mill by the following March. By the end of 1880 production closed in on one million dollars, the company had paid out more than a half-million dollars in dividends, and plans called for a larger mill to handle the mine's abundant ores. The Western Mining Company combined with several other important properties and was reorganized as the Contention Consolidated Mining Company at the end of 1881.[7]

The Grand Central Mining Company of Youngstown, Ohio, may be considered the Tombstone District's third great mining enterprise. This property developed more slowly than either the Tombstone or the Contention. In May 1879 eastern investors purchased the Grand Central properties, clearing the way for "active work on the mine at an early day." It did not take the Grand Central long to catch up. By the end of 1881 the company had produced over one million dollars in metals and paid out around three hundred thousand dollars in dividends.[8]

Among them, these three great properties attained almost eleven million dollars in production by the end of 1882. Including the output of the lesser mines of the district, like the Girard, the Head Center, the Vizina, the Bob Ingersoll, and several others, production for the Tombstone District in its first three and one-half years totaled more than twelve and one-half million dollars. Nor did things seem to be slowing down. One source reported that the monthly production average for the district rose in the early part of 1883 over that of the previous year.[9]

Merchants numbered among the first boomers into the new district, following closely in the wake of prospectors and speculators. They supplied the material needs of this latest bonanza. As early as September 1878, a correspondent to the *Arizona Weekly Citizen* in Tucson reported that "the new townsite is progressing finely; two stores, one butcher shop and a restaurant under way already." As the boom days rolled along, Tombstone's business district grew and evolved. That sole restaurant soon had plenty of competition. Somewhat stunted in growth by a lumber shortage in its earliest days, Tombstone nonetheless saw dozens of business houses being erected toward the end of 1879.[10]

By mid-1880, the tumultuous bonanza camp had turned into a prosperous town with a bustling business district. "It is a query," wrote a reporter for the *Weekly Citizen*,

> if there is not a business man to every working miner in the
> camp. How they all live is a source of wonder to the stranger, and
> I am not certain that it is not to many of the traders themselves.
> Of course, some of the merchants here do a princely business,
> and the town itself is growing rapidly; but so is the business
> community. On all sides you see new stores in various stages
> of progress toward completion, and some of the buildings are
> permanent and handsome structures.[11]

At the beginning of 1881, a correspondent reported over one hundred business houses in Tombstone, including at least six "doing a very large and profitable business in general merchandise," along with several stables, two newspapers, and two thriving banks.[12]

By then business establishments had begun to stratify and diversify to cater to various interests and economic classes. While numerous boardinghouses provided room and board for miners, teamsters, and laborers, those with greater means could eat at the Elite or Maison Doree restaurants, or bed down at the Occidental, Cosmopolitan, or Grand hotels. Specialty stores began to replace the general merchandise houses. Rudolph Cohen turned his general store into a furniture shop, and B. Leventhal turned his into a men's clothing store. Tasker,

Pridham, and Co., general merchants, apparently evolved into the Cochise Hardware and Trading Company. The town also had a foundry to serve its mining industry, and the lumber shortage had been erased by the operation of five sawmills in the mountains surrounding the district.[13]

Such business opportunities were not the exclusive domain of American-born men. Irish and German names appeared among the proprietors of hotels and saloons, while some Chinese were involved in the restaurant, laundry, and mercantile trades and others made a living supplying vegetables to the district from gardens along the San Pedro River. Nellie Cashman and Kate O'Hara opened the Nevada Boot and Shoe Store in early 1880, and Cashman and her sister Mrs. Cunningham leased the Russ House Hotel three years later.[14]

Fig. 1.3. Allen Street, Tombstone's main business thoroughfare, in the early 1880s. Nellie Cashman and Kate O'Hara's Nevada Boot and Shoe Store—"Gents Furnishing Goods"—is visible on the right side of the street. (Arizona Historical Society, Tucson, #14835)

Other women joined Cashman and her partners in operating Tomb-stone businesses. Some ran small businesses or, like Mary Tack, the Swedish lodging-house keeper, performed domestic services for their mining clientele. Many other women worked at night. A few of these prostitutes have since become part of the legend of Tombstone, but many more lived and worked anonymously in the red-light district along Allen Street. Prostitution, too, was segregated by class, with the most desirable women working at the Bird Cage Theater, a middle class of prostitutes working in the dance halls and saloons, and the lower class working out of cribs in the red-light district. They did not lack for clients among the footloose men who inhabited boomtown Tombstone.[15]

Those men also had plenty of opportunities to drown their isola-tion in drink. A correspondent estimated at the end of 1880 that 75 of Tombstone's 105 business houses were saloons. The following year the town granted 110 liquor licenses, although not all of these were issued to saloons, as liquor could be purchased in some stores. Arthur Laing reported to a Tucson paper in 1881 that "the saloons are in many instances elegantly fitted up, and, in truth, beat anything I have seen in Tucson." The Alhambra and Oriental saloons were the most luxu-rious in Tombstone, but other houses catered to working men or eth-nic groups.[16]

In addition to drinking and womanizing, a miner could always squander his earnings at one of the town's fourteen faro banks or on numerous other games of chance. Describing the saloon trade in nearby Charleston, James Wolf probably spoke for Tombstone as well. "Every saloon," he recalled, "was sure to have as part of its regular equipment one or more roulette wheels, besides faro and poker tables. These places were open day and night, Sundays and holidays. A few were respectably conducted but the rest were conducted under decid-edly more or less flexible codes of ethics. In these places all the games were as crooked as they dared to be."[17]

Most of the gambling and much of the solicitation occurred in the saloons, which some observers held to be Tombstone's most important business activity. When fire demolished the business district in May 1882, "the saloonists" resumed business first. The *Weekly Citizen* reported that "on the day following the fire some of the most enterprising of them opened out over their smoking ruins. . . . It is reported that one man took in over five hundred dollars the first day, and many others did proportionally well."[18] As long as money flowed in the bonanza camp, the sin businesses would continue to do proportionally well.

Learned professionals headed for the new bonanza along with the gamblers and prostitutes. At the beginning of 1880 a man identified as Retlef reported six or seven doctors and ten lawyers in Tombstone "all trying to eke out a precarious living, your correspondent among the number." Their ranks included George Goodfellow, formerly an army surgeon at Fort Whipple, in Prescott, Arizona. In September 1880 he resigned his post at Fort Whipple to establish a private practice in Tombstone. The lawyers engaged mostly in mining litigation and real estate deals until the following year, when Cochise County was created and the county seat located at Tombstone.[19]

True to form in the mining West, a real estate bonanza quickly followed the mineral one. "Town property has a real value now," a correspondent reported in September 1879, not eighteen months after the initial locations. "Lots are selling from $150 to $250, and corner lots at fancy prices, from $400. The real estate business during the past month has been very brisk." Almost five months later, another witness confirmed that "lots are held at fabulous prices."[20]

The fabulous real estate prices meant high rents. An *Arizona Quarterly Illustrated* article of July 1880 noted that "the great difficulty here, is to get places to do business in. Small houses are also very scarce and in demand, for small families, and prove profitable investments, as a small plainly furnished two-roomed house rents from $25 up to $40 per month. Board can be had at $8 per week, and good meals at

50 cents each." More than a year later, the *Weekly Citizen* reported that "there is not a vacant residence in the city. As soon as the foundations of a new house are laid there are a dozen applicants for it." Prices for goods were also high.[21]

Good wages somewhat offset isolated Tombstone's high rents and prices. The *Arizona Quarterly Illustrated* article noted that artisans made six dollars a day, miners made four dollars a day, laborers made three dollars a day, and that "house servants [were] very scarce, at high wages." In this same period, teamsters made two dollars a day and board, and clerks up to three dollars a day and board. As a result, despite its isolation, the Tombstone District prospered. In describing the town at the beginning of 1881, Arthur Laing wrote that "one sees many well-dressed ladies on the streets. Two-horse buggies are not scarce, and there is an amount of well-to-doedness visible all over the town, which one would hardly expect to find in a two-year-old town in Arizona."[22]

Tombstone's growing wealth encompassed more than extravagance or display; in only two years miners and merchants had turned desert into market. In July 1880, not three years after the original locations, the county sheriff reported the assessed valuation of Tombstone's real and personal property as $212,910. Two years later, at the height of the boom, that figure had increased 663 percent, to $1,411,919.86.[23] In mid-1883 an unidentified correspondent to the *Arizona Weekly Citizen* summed up the economic community of boomtown Tombstone:

> Every branch of business is represented in Tombstone. You can buy anything here, from a jews-harp up to a piano, from cotton cloth to the finest silks; from a pony glass to a barrel of whiskey, or from a claw-hammer to a quartz mill. The nicest groceries for your family or very ordinary goods with which to pay off Mexican "doby makers." . . . Nothing is too good for the Tombstone market. . . . In fact there is nothing that an old-timer needs that he cannot buy in Tombstone, except "lost opportunities."[24]

That passing reference to Mexican adobe makers told outsiders what would have been obvious to anyone walking the streets of the town: Americans were not the only ones who had come to Tombstone. What had been a lonely patch of desert in 1878, 2,173 people called home in 1880, according to the federal census. In 1882 a county census determined that the population had more than doubled to 5,300, and it might have been even greater.[25]

These censuses listed natives of most continents and many countries, but there were four significant minority groups in 1880s Tombstone. One, women, made up only a small percentage of the population, and the male residents of Tombstone greeted their arrival with enthusiasm. In August 1878 a milling engineer and his wife visited the district, she being "the first lady who has favored our district with a visit. We extend a hearty welcome to this female pioneer and hope the day is not far distant when many more ladies will enjoy the beautiful scenery and delightful climate of the Tombstone [District]." The writer engaged in a little false advertising to attract more women—Tombstone does not have a delightful climate in August—but he did not have long to wait for their arrival. Five months later another correspondent to the *Weekly Citizen* reported "four American ladies in Tombstone now," and when Endicott Peabody arrived to establish a church in Tombstone at the beginning of 1882, he found "a ladies' society of 30 females—fancy that when I had expected to get alon[g] among the men."[26]

Women and children had begun to arrive in some numbers by the end of 1879. The federal census of 1880 indicated that women constituted 10 percent and women and children together 20 percent of Tombstone's population. Sex was not listed as a category in the county census of 1882, but children under sixteen had increased from 10 to 13 percent of the population in those two years, and we may safely assume a corresponding or even greater increase in the percentage of women as well.[27]

Enough Irish and Irish Americans had come to Tombstone to organize a chapter of the Irish Land League in time to sponsor a St. Patrick's Day ball in 1881. The census of 1880 listed 6 percent of Tombstone residents as Irish. Two years later those with Irish nativity constituted 10 percent of Tombstone's population, about 550 people, making the Irish the most numerous ethnic minority in Tombstone during its boom years.[28]

Latinos were the second-largest ethnic group in the district. Unlike the Irish, Latinos largely segregated into their own quarter on the southwest side of Tombstone. The census of 1880 indicated that 5 percent of the population came from Mexico, mostly from Sonora. By 1882 that percentage had risen to almost 8 percent. Quite probably these numbers are too low, as they take into account only Mexican nativity and do not include Latino citizens of the United States.[29]

Chinese began to enter Arizona in the 1870s and became Tombstone's fourth significant ethnic group. By mid-1880 they represented 2 percent of the town's population and had already begun to draw the ire of white residents. By 1882 Chinese made up 4 percent of the population and were largely segregated into "Hoptown"—"hop" being slang for opium—a two-square-block area bordered by Second, Third, Fremont, and Toughnut Streets. In its prime, Hoptown mirrored Tombstone, with gambling halls, restaurants, stores, brothels, and even a Masonic lodge. The residents of Hoptown made their livings by providing various trades and services to fellow Chinese or other customers. However they subsisted, the much maligned Chinese, segregated by race, language, and culture, faced the longest odds of any group in Tombstone.[30]

Women, Irish, Latinos, and Chinese ranked as the four most important minority groups in Tombstone, but others also migrated to the town. Every nationality from Argentineans to New Zealanders appeared in tumultuous Tombstone in the early 1880s. Germans pioneered in the district in significant numbers, as manifested by the Turnverein

Hall they built in 1880. The *Tombstone Epitaph* reported at the end of the year that the Turnverein society "is now in a flourishing condition, and bids fair to become one of the largest and most influential in the camp. Our German fellow citizens do nothing by halves." When the Turnverein threw their first ball, in the spring of 1881, the *Epitaph* reported "the huge attendance" by "the very best of Tombstone's people" and concluded that "it was generally conceded that the affair was the greatest social success ever witnessed in Tombstone."[31]

Ethnic organizations such as the Turnverein, the Irish Land League, and an Italian band were joined by other social organizations, including the Grand Army of the Republic, the Red Men, the Good Templars, and the Knights of Columbus. Other notable Tombstone fraternal organizations included King Solomon Lodge no. 5, Free and Accepted Masons, which met to organize in March 1881 and held its first regular meeting in June of that year. By 1883 the organization had eighty members. Close upon the heels of the Masons, Arizona Lodge no. 4, Knights of Pythias, organized at the end of June 1881 with eight charter members. By 1883 their membership had risen to sixty-five.[32]

Cochise Lodge no. 5, Independent Order of Odd Fellows (IOOF), originated in the summer of 1881 with six members. The lodge did so well that Unity Degree Lodge no. 3, IOOF, Daughters of Rebecca, was instituted in March 1883. Its twenty-two members also testify to the growing presence of women in the district by that date. Another Tombstone order, with a working-class orientation, was Tombstone Lodge no. 3, Ancient Order of United Workmen, founded in January 1882. This death-benefits fraternity had twenty-eight charter members and forty-three in 1883. By then the prospering society held its meetings on Thursday nights at Schieffelin Hall.[33]

Perhaps people on the frontier needed the diversions of society and recreation especially keenly. In any case, they pursued these activities with astonishing enthusiasm, given the urgent nature of economic life in bonanza Tombstone. "We are pleased to note," commented the

Tombstone Nugget in January 1880, less than a year after Tombstone was platted, "that some of the young gentlemen of our town are organizing a home dramatic association, in which some of the ladies of our city will participate."[34] Given the camp's isolation from professional entertainment in its early days, such amateur groups assumed special importance.

The first amateur theatrical in Tombstone apparently took place in November 1880. The Tombstone amateurs had a fondness for Gilbert and Sullivan, performing *H.M.S. Pinafore* in 1882 and *Pirates of Penzance* the following year. The amateurs often donated their proceeds to charities, and sometimes they reinforced the professional companies that soon began to play to Tombstone audiences. The first of these, the Nellie Boyd Troupe, performed at Ritchie's Hall to a packed house in December 1880. By 1882 Tombstone had become a regular stop on the Pacific slope theater circuit and had, besides several smaller venues, two large theaters to accommodate traveling shows.[35]

Respectable undertakings, like Shakespeare's plays or political debates, were staged at Schieffelin Hall, the community auditorium completed with Schieffelin money in mid-1881. Entertainments of a less refined nature occurred at the Bird Cage Theater, a vaudeville house that encouraged drinking and permitted womanizing. Since Tombstone's proper women did not enter the Bird Cage, its acts were occasionally sanitized and presented at Schieffelin Hall between visits by traveling companies.[36]

Other social and recreational outlets included the Tombstone Literary and Debating Club, formed at the end of 1881; a dance club; a glee club established in July 1880; and a nine-piece city band organized by 1883. Residents also enjoyed picnics and calling upon friends, and the town had its first wedding ceremony before the end of 1880.[37]

Athletic events had great appeal in a frontier camp full of young men. When the young missionary Endicott Peabody brought the Word to Tombstone in 1882, he shrewdly "organized a baseball nine which created a great deal of interest and brought me in touch with the younger people." Some idea of the primacy of sport in the district can be gained by reading the schedule of events for Christmas Day of 1880. "Christmas will be a lively day here," a correspondent reported to the *Weekly Citizen*. "There are to be races at Charleston . . . [that] promise to be most exciting, as some of the best horses in the Territory are entered. . . . At the Tombstone course there will be a foot race of 100 yards dash between Messrs. Whicher and Helyer, for $150 a side. There will also be a turkey shooting match, and all the best shots are entered."[38] Presumably, there were also religious services that day, though these do not seem to have attracted as much attention.

Many diversions existed for the sporting man in Tombstone. Shooting matches, hunting, and races drew interest. Cockfights took place on Sunday afternoons at the cockpit on Allen Street behind Walsh's Saloon. Boxing and wrestling matches occurred in the Bird Cage Theater, with betting assuredly much a part of these contests. Gamblers also favored the Tombstone District's frequent horse races, although they faced the same risks at the track as they did at the faro tables. One perhaps disgruntled witness wrote that "fairness is a quality wanting in most of our races and today's race did not prove an exception. We hope that in the future things will be done more on the square." Those robust sportsmen who preferred participation to observation organized the Tombstone Athletic Club in October 1883, the club "having for its object athletic exercise and muscular development."[39]

Baseball was probably the most popular local pastime. The Reverend Endicott Peabody's team was not the first, and certainly not the last, to compete in Tombstone. In May 1882 Tucson and Tombstone nines engaged in a "championship contest," with a rematch in December of that year. Tombstone fans showed considerable loyalty to their town team. When the team traveled twenty miles through the Arizona desert in August 1883 to beat a nine from Fort Huachuca, many Tombstoners attended. Not all of the baseball played in Tombstone matched the town team against other town teams. That same month featured a contest between two mine teams, the Tombstones prevailing over the Grand Centrals twenty-nine to eight.[40]

Occasionally, Tombstone's social interactions went beyond the acceptable to become criminal. Much of that now celebrated lawlessness occurred beyond town limits and is thus irrelevant to this study, but even the lawlessness of Allen Street has perhaps been exaggerated. Newspaper editors, then as now, paid particular attention to incidents of violence, but other witnesses, writing then and later, felt that the town's reputation for criminal behavior in its bonanza days had been exaggerated. Although since celebrated for one notorious series of incidents, Tombstone hosted about as much crime and violence as its contemporaries on the mining frontier.[41]

Pious people outnumbered those with criminal intent. Even on the far frontier, individuals strove early to establish religious institutions in their community. James Wolf, a resident of the district by mid-1883, remembered "quite a procession of buckboards and buggies to the Catholic and Protestant churches in Tombstone" on Sundays. Things did not begin quite so grandly, however. The tent occupied by Sam Danner's saloon provided the setting for the first religious service in the new community in 1879. In February 1880 George Parsons heard a sermon in a tent, sitting on boards upon boxes, with, he wrote in his diary, "good attendance considering." By the end of 1881 the town had Methodist, Catholic, Episcopal, and Presbyterian churches and a Hebrew Association.[42]

Tombstone had several other religious groups, though none was large enough to build a church. Congregationalists started services in 1880, making use of the Presbyterian church, and occasionally its minister, until they built their own chapel many years later. A Baptist organization apparently existed by 1881, and the Hebrew Association observed Yom Kippur in 1881, held a celebration at the Turnverein Hall in 1882, and organized its own cemetery.[43]

The building of churches contributed to the transformation of Tombstone from rude tent camp to substantial frontier boomtown. Tombstone had been preceded by other towns located near the mines on the hill. As a reliable water supply posed a problem in this south-western desert, the original settlement had been established in September 1878 at Watervale, several miles from the riches, but at the site of a small spring. Another early town, Richmond, was located a little closer to the mines, but as the district's population increased it soon became apparent that the mesa directly below the mines, known as Goose Flats, offered the best place to locate a town of any size. One correspondent claimed that this was the natural location for the town, on level ground with easy access to the mines.[44]

The town of Tombstone was formally located by a town-site company on 5 March 1879, and development began immediately, in many instances with buildings relocated from the other town sites. A resident reported two days later that "we have plenty of buildings for present purposes, but until lumber can be had, visitors must refrain from remarks upon our architecture." By the summer of 1879 another correspondent indicated that new houses were being started almost every day "and seem to grow like mushrooms," and that fall still another wrote that "the temporary canvas houses are fast giving way to structures of more substantial material."[45]

By the time Tombstone had been incorporated and awarded the county seat of the new Cochise County in 1881, the town was generously described as a "large and constantly growing city, with well laid-out streets, kept clean and orderly, and buildings such as would do credit to a town of much longer growth and greater pretensions in any part of the older states." At that same time Arthur Laing described the town's principal streets as "lined with good, substantial frame buildings, with here and there an adobe edifice."[46]

A rapidly growing controversy between the rights of the town-site company and those of the residents paralleled Tombstone's rapid growth. The town site lay atop several significant mining claims, which complicated the matter of who owned town lots, and a dispute developed over the contested surface ground. Whatever the merits of the various claimants, the town-site company did at least lay out the pattern of streets to start the settlement of Goose Flats in good order. Although the company platted almost one hundred square blocks on the mesa, the town itself never occupied the entire area. The streets that ran in a north-northeasterly direction were numbered, while those running east-southeasterly were named. The town site—company directors had assumed that Fremont Street would be the center of business activity and thus made it wider than the others; however, merchants ignored the plan and instead lined Allen Street, centering on Fifth.[47]

The property owners of Tombstone faced a bigger threat than disputes with the town-site company. Like all other mining camps built of wood, and especially because of its arid location, Tombstone faced constant danger from fire. Disaster came quickly. On 22 June 1881 a fire roared through the business district, and in the absence of an adequate water supply, townspeople had to fight the blaze with blasting powder. The fire consumed sixty-six buildings and their contents, worth $175,000. Residents quickly rebuilt the town. A Tombstone correspondent to the *Arizona Weekly Citizen* reported two months later that very few of the damaged businesses had not been reconstructed.[48]

They rebuilt too quickly, perhaps, using the same combustible materials, and only eleven months later faced another ordeal by fire. On the afternoon of 25 May 1882, a fire started on Fifth Street between Toughnut and Allen Streets. Once again, citizens lacked enough water to fight the fire, and they fared even worse than the preceding year. About 116 business houses, almost the entire business district, were destroyed, including the office of the town's original newspaper, the *Tombstone Nugget*, which burned, and the post office, which had to be blasted. Total damage ran to about one-half million dollars.[49]

Once again residents rebuilt. A month after the fire the *Epitaph* reported reconstruction nearly complete and asked: "Is it not a convincing proof that they who know Tombstone best have unlimited confidence in its resources and prosperity?" By 1883 "not a vestige of the burnt district" remained, and this time Tombstone's residents did not repeat their previous mistake. "Since those fires the buildings erected have been nearly all of adobe, and, except in the suburbs, the city is mostly built of that material," one source reported. In the aftermath of the second fire, community leaders also came to the unsurprising conclusion that they needed to enhance their water supply, and embarked on an aggressive campaign to address that situation.[50]

The shortage of water and concern over the fire hazard had been with Tombstone from its earliest days. In the spring of 1880 the town elders awarded a contract for a water service from the Dragoon Mountains, about fourteen miles to the east. That system supplied basic needs but did not yield enough water to fight fires. By the following year the residents of Tombstone looked forward to another system to pipe in a more reliable source of water from the Huachuca Mountains. While the waterworks of the Huachuca Water Company were not completed in time to keep the town from suffering its second great fire, by 1883 even the *Weekly Citizen* in rival Tucson believed that "Tombstone lays claim to the best water works on the Pacific slope."[51]

The waterworks offered only part of the town's defense against fire. A month after the first inferno citizens initiated a subscription campaign in support of fire companies. Only two months later the town's new hose cart was delivered. Volunteers originally organized Engine Company no. 1 and the Rescue Hose and Ladder Company, which, by October 1883, listed a combined total of 107 members. The town organized another hose company later in the decade. These fire companies were community social organizations as well, with many leading citizens among their members.[52]

Other amenities of civilization came to the desert boom camp with surprising celerity. In October 1879 a petition circulated demanding daily mail service between Tucson and Tombstone, with the contract for same awarded in February of the following year. The year 1881 witnessed the completion of a telegraph line to the outside world and a telephone system between the Grand Central's mine office in Tombstone and mill in Contention City. Only two years later, in 1883, boosters and citizens had further evidence of progress with the opening of the Tombstone gasworks. That company quickly signed a contract with the city, and soon illuminated Tombstone's streets and city buildings with gaslight.[53]

The most noticeable improvement probably came in transportation. The first stage line to Tombstone debuted at the end of 1878, and barely a month later another began offering service. Within a year the fare and travel time from Tucson to Tombstone had been reduced more than half, thanks to the competition. By 1881, Tombstone had two daily stage lines north to the railhead at Benson, as well as regular service to Charleston to the west and the copper mining town of Bisbee to the south.[54]

Reliable roads and a lively stage competition were good for the town, of course, but everyone agreed with the *Weekly Citizen* that "what we now need most for rapid development in this country is the extension of the Southern Pacific Railroad to Charleston, on the San Pedro River . . . where the Lucky Cuss and Tough Nut Mines have their mills." The railroad rumors began almost as soon as the Southern Pacific began building across southern Arizona in 1880. When a rail line reached the towns along the San Pedro River, only nine miles to the east, at the beginning of 1882, many people assumed that a branch line would soon be pushed east to Tombstone. That assumption seemed confirmed in May 1883 when the newspaper reported that the Santa Fe had men grading between Fairbank and Tombstone. The paper stated its belief "that trains will be running into Tombstone inside of sixty days," but this development did not come to pass.[55]

The San Pedro River towns served by the railroad were the mill towns for the ores from Tombstone's mines. Wagons hauled ores from the mines to Charleston and Contention City, where they were concentrated and then shipped out of the district. The southernmost of these towns, Charleston, had been established in 1878 on the east bank of the river and originally named Millville. The main settlement removed to the west bank of the river in 1880 and was renamed. Charleston contained the mill site and headquarters for the Tombstone Mill and Mining Company and had perhaps seven hundred residents and more than thirty business houses at the height of the boom. The next town to the north, Fairbank, had been established as a mill town and railhead for Tombstone by 1882, with a post office opening the following May. The northernmost town on the San Pedro, Contention City, had been founded in 1879 and became home to the thirty-stamp ore-reduction mill of the Contention Mill and Mining

Fig. 1.4. The southernmost of Tombstone's milling towns, Millville (superseded by Charleston), in the early 1880s, with the Tombstone Mill and Mining Company's mill and office to the left. (Arizona Historical Society, Tucson, #10237)

Company, reportedly the largest mill in the territory. Contention City, the largest of the San Pedro towns, boasted over a thousand residents at its height. In addition to the three towns, several other mill sites with their tiny settlements dotted the San Pedro.[56]

But none of the San Pedro towns could rival Tombstone. The Pima County Board of Supervisors incorporated Tombstone as a village at the end of 1879, with a mayor and common council sworn in in January 1880. The following year the community gained the status of a town in, and became the county seat of, Cochise County. By 1883 Tombstone's government totaled nineteen employees. Elected officials included a mayor, four councilmen (one from each ward), an assessor, treasurer, tax collector, recorder, chief of police, and attorney. In 1882 Tombstone had an assessed valuation of real and personal property of more than $1.4 million, from which its government derived almost $28,500 in revenue. The town had a positive balance of $4,595.36 as of 1 May 1883 and had been able to pay off $3,000 of indebtedness accrued for the city hall's construction from surplus city revenue.[57]

Contrary to the general perception of participants in mining rushes, Tombstone's residents seem to have taken some interest in local politics. Cochise County and its government were heavily Democratic, while the town of Tombstone itself was Republican. This fact seems to have contributed to some of the now legendary factionalism of the early 1880s and kept interest in politics much higher than in most frontier mining camps. Voters cast a total of 947 votes in the town's primary elections in September 1880.[58]

Citizens also took great and early interest in a school system for Tombstone. The demands for a school began in the fall of 1879, with twenty-six children reported to be living in the town. Residents started a schoolhouse fund, the town-site company donated a lot, and in February 1880 the public school opened under the charge of Miss Lucas, recently arrived from Tucson. At the height of the boom in 1883, one publication reported that Tombstone had "a well graded public school with five teachers, and an attendance of 275 pupils [that] offers very good educational advantages." Although the Tombstone School District still owed $3,000 on its original school building, there were enough students in the six grades that the three-and-one-half-year-old school district had to rent two other buildings, including Turn-verein Hall, for classrooms. In spite of the district's debt, the same publication reported that $15,000 worth of bonds had been issued for the construction of a new building.[59]

That seemed like a good investment in 1883; the future of Tombstone looked bright. After the second fire, in May 1882, the *Tombstone Epitaph* assured its readers that, while this was a serious setback, the district would be quickly rebuilt, because "so long as the marvelous mines of this camp continue to send forth their treasure, so long will the town of Tombstone exist and flourish. . . . The sun of our future prosperity shines as bright as ever. . . . The resources which have built up the most flourishing camp in the territory are still here, and will yet make this town of Tombstone the Virginia City of the Southwest."[60]

While the *Epitaph*'s editor might be accused of a little hometown bias, other disinterested parties had reached the same conclusion. One gazetteer called Tombstone "one of the most active towns on the Pacific Coast" in 1881, informed its readers that Tombstone's "future growth and prosperity is assured," and opined that the district would eventually rival the Comstock. The encouraging word even came from one of Tombstone's rivals. When Mayor Strauss of Tucson visited Tombstone in May 1883, he predicted in a speech before the Tombstone City Council that Tombstone would overtake Tucson, then the largest city in the territory, in less than two years. The following month, one of the *Arizona Weekly Citizen*'s correspondents wrote to the paper that "notwithstanding the doubts of some people, who are ever on the lookout for the collapse of a mining camp or town, they will look in vain for it in this camp."[61]

CHAPTER TWO

Apartments or Houses
Are Impossible to Obtain

Residents of Tombstone could find some solace in the belief that even if its mines gave out, the town could survive as a ranching and border supply center and county seat. People who lived in the great copper mining town of Jerome could find no such comfort in their situation. Dug in high on the side of Cleopatra Hill at the foot of Mingus Mountain, about thirty miles northeast of Prescott, Jerome was vulnerable, both to the continuing existence of commercial ore in the area and to the market price for copper.

In his Christmas message of 1928 the editor of Jerome's newspaper, the *Verde Copper News*, wrote that "the welfare and success of every business or professional man, every executive, every employee and every property owner in the Verde District hinges on exactly the same thing—the production and marketing of valuable minerals. Without this enterprise there would be no excuse for the existence of such a district."[1]

Morbid thoughts about the Verde District's future hardly seemed appropriate at the beginning of 1929, however. Although mineral development had been going on on Cleopatra Hill for more fifty years, 1929 was arguably Jerome's greatest year to date. In 1929 the district's great United Verde Mine had been, once again, the leading copper-producing mine in the leading copper-producing state in the nation. The United Verde also led the state in gold and silver production that year. The company had been the leader in all three categories every year since 1924. The United Verde Extension, the district's second great mine, was the state's fifth-leading producer of gold and silver and was sixth in copper in 1929. Even better, another mine, the Verde Central, started up its own mill in January and was well on its way, residents assumed, to becoming the third great mine in the district.[2]

Jerome's successes constituted only a part of Arizona's good fortune. The copper mining industry boomed throughout the state in 1929, with a record production of more than 415,314 tons of copper—

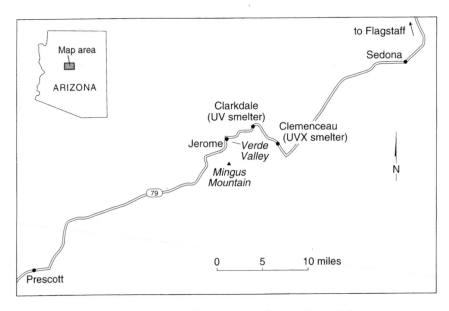

Map 2.1. Northern Arizona and the Verde District, showing Flagstaff, Prescott, Jerome, Clarkdale, and Clemenceau-Cottonwood.

40 percent of U.S. and 19 percent of world output. That product was worth $146,190,600, a record for value exceeded only during the great boom years of World War I. Copper prices, which had held at around fourteen cents a pound since 1923, soared to twenty-one cents in the first quarter of 1929, the highest price for the metal since World War I. The price then receded to eighteen cents a pound but held there solidly for the remainder of the year. Consequently, both the mines and the smelters of Jerome's two great mining companies remained in operation continuously in 1929.[3]

In the town of Jerome itself, the three major mining companies employed almost two thousand men. These men received their highest wages to date, and as a result Jerome's consumer economy boomed. Five licensed automobile dealers did business, and over thirteen hundred automobiles were registered in the small town on the hillside in 1929. At midyear local merchant Charles Robinson announced he would open radio station KCRJ in Jerome—only the seventh radio station in the state. At the beginning of 1929, the *Verde Copper News* reported that "nearly a dozen stores in the Verde District have unofficially announced proposed improvements and a general program of expansion for the early part of 1929. Business has been good during the year just past," the paper announced, and prophesied: "It will be better in the coming year."[4]

This was not Jerome's first boom, nor had bust been previously unknown to the district. The first claims in what came to be called the Verde District had been located in 1876. Territorial governor F. A. Tritle and his partner, Frederick F. Thomas, a mining engineer from San Francisco, purchased these claims in 1880. Tritle and Thomas in turn solicited financial support in the East, where they gained the backing of James A. MacDonald and Eugene Jerome of New York City, among others. Together these men purchased eleven claims on Cleopatra Hill, which they organized into the United Verde Copper Company in February 1883. MacDonald became president of the company, and Jerome, for whom the little settlement adjoining United Verde's properties was named, became its secretary.[5]

The new company set to work with some energy, constructing a small smelter in Jerome and establishing a regular water supply. It built wagon roads to Prescott, the Verde Valley, and sixty miles over the mountains to Ash Fork, and a connection with the Atlantic and Pacific Railroad. But these modest improvements could not compensate for decreasing ore values and a 50 percent decline in the market price of copper. In December 1884 United Verde suspended operations, and the rude little settlement on Cleopatra Hill experienced its first bust.[6]

Jerome might have died then, busted after a brief period of promise like so many other camps in the West, save for one of those happy accidents of fate that enliven mining history. Montana mining magnate William Andrews Clark happened to see some samples of United Verde ore at a mining exposition, spent three weeks underground investigating the property, then purchased it from the MacDonald group in March 1888. Clark—already a multimillionaire through his Montana commercial and mining ventures—had taken the trouble to receive formal training in mining and metallurgy. Here was the man with the money, the expertise, and the determination to break Jerome's transportation and smelting bottlenecks. Within five years he had the smelter enlarged and undertook construction of a narrow-gauge railroad to replace the tortuous wagon roads. Jerome's first boom days had arrived.[7]

The town's population, only 250 in 1890, had increased more than ten times by 1900. The railroad into Jerome, completed in January 1895, brought in families as well as smelting supplies, and Jerome began its transformation from crude frontier camp into twentieth-century industrial town. By the turn of the century the United Verde Mine had become the leading producer in the territory, employing around 800 men, and in the opinion of a reporter from the *Los Angeles Mining Review*, "the property is conceded to be one of the greatest and most valuable in the known world." The town of Jerome had been incorporated, and it had at least the rudiments of municipal services, schools, churches, fraternal organizations, and a thriving business district with brick buildings, electric lights, and a telephone . system.[8]

Thus Jerome remained, with minor ups and downs, for over a decade; however, the town's dependence on one mining company ceased at the end of 1914 due to the discovery of a second bonanza by the United Verde Extension Mining Company. That company, known locally as UVX, had been formed in 1899. Its mine developed fitfully until the company was taken over at the end of 1911 by a group headed by James S. Douglas, the son of one of the most prominent mining men in the Southwest. The younger Douglas thereupon made quite a name for himself in mining by winning a stupendous gamble.[9]

Fig. 2.1. Jerome, ca. 1900. The United Verde Mine and smelter complex are to the right. (Arizona State Library, Archives, and Public Records, Archives Division, Phoenix, #97-1452)

A geologic fault runs through the Verde District. Some mining men believed that the top of the United Verde ore body might have been sheared off along the fault line and relocated hundreds of feet north and down toward the valley below. The Douglas group intended to search for that theoretical ore body. For the next four years they chased what seemed to be a dream. On more than one occasion they contemplated abandonment, or used personal funds to keep the operation going. By the end of 1914 the group had invested around four hundred thousand dollars in uvx, and seemed to be at the end of their resources, but in December 1914 they hit one of the great bonanzas in U.S. mining history. For their investment the group had struck an ore body that ultimately yielded more than $125 million in ore and over $50 million in dividends.[10]

Their timing could not have been better. World War I greatly excited the demand for copper and produced Jerome's second, and greatest, boomtown period. As the price of copper soared in the World War I buying frenzy, some individual or company claimed just about every piece of ground not previously located, and investment capital poured into the district. In 1913 six major mining companies operated in the Verde District. By 1916 thirty companies had claims in Jerome, twenty-two of them in actual development. In those three years the district's mine payroll had increased from two thousand to three thousand men. One estimate, based upon school enrollments, put Jerome's population at seven thousand by 1916 and perhaps ten thousand in 1917.[11]

The town experienced a severe housing shortage during the war boom, with many people sleeping in the park or sharing beds. One resident reported to the Yavapai County Chamber of Commerce at the end of 1916 that "hotel accommodations are at a premium and no houses [are] available for rent." When Chris Thomas arrived in 1917, he recalled later, "you couldn't get a house for love or money. I moved into an old shack with bedbugs." While accommodations might have been unsatisfactory, Thomas, a veteran miner, had no trouble finding work. The mining boom precipitated a construction boom that never quite caught up with demand during the war years.[12]

In the days leading up to the boom period United Verde had been busy greatly expanding its own operations. In 1912 the company began constructing a great modern smelter complex and an associated company town named Clarkdale in the Verde Valley below Jerome. The Clarkdale smelter complex and a standard-gauge railroad connecting it to the United Verde Mine commenced operations in 1915. The company thereupon removed Jerome's smelter and began open-pit operations at that site in 1919. In 1918 the United Verde Extension began operating its own smelter in the Verde Valley, near Cottonwood. That company's town site was christened Clemenceau in 1920, the same year that Jerome's original narrow-gauge railroad was abandoned upon completion of the standard-gauge line into town from the Verde Valley.[13]

Fig. 2.2. Smoke drifts over Clarkdale from the stack of the United Verde's smelter in 1920. After 1915 the Verde District included the valley smelting towns of Clarkdale and Clemenceau, as well as the mining town of Jerome. Unlike Jerome, Clarkdale was a company town, as is suggested by its uniform architecture. (Sharlot Hall Museum Photo, Prescott, Arizona, #M372pe)

The war years produced the most contentious labor relations in Jerome's relatively peaceful labor history. There had been labor activity in 1884, when United Verde could not pay wages, and again in 1907, in quest of the eight-hour day and higher wages, which the miners won. Though he could never be classed an egalitarian, W. A. Clark had started his career in mining camps, and he knew that good wages and amenities attracted good workers. Jimmy Douglas of uvx, who had to answer to stockholders, could not match Clark's expenditures on employees. But he, too, had spent some time in the ranks of the mining industry, understood the issues, and tried to keep his employees contented. Between them, Clark and Douglas presided over a mining town generally free of labor unrest.[14]

When World War I generated rising profits and prices, without a corresponding rise in wages, however, Jerome's Local 101 of the International Union of Mine, Mill, and Smelter Workers struck in May 1917. That strike, marred by some violence, ended in compromise after eleven days. That compromise proved unsatisfactory to the local chapter of the Industrial Workers of the World (iww), which tried to persuade more militant miners to renew the strike. Though only a small minority of miners chose to participate in the iww action, they did cause some disruption of mining operations, pleasing neither the companies nor the other union miners. After a rock-throwing contest between the two union factions, about seventy-five members of the iww local were detained by local authorities and armed citizens and deported from the district in freight cars on 10 July 1917. That action effectively ended the town's wartime labor agitation.[15]

Although the 1920s would see the apex of Jerome's development, the decade began poorly for the town and the copper industry. The copper market was glutted at the end of World War I, and the war boom ended with a crash. The postwar copper surplus produced the second of Jerome's major busts. By the end of 1920 United Verde was selling no copper at all, and the company laid off all but 275 of its 2,200 men in 1921. United Verde and uvx tried to soften the blow by keeping men on construction and development work and by providing some money for poor relief. Several hundred of Jerome's Mexican residents returned to their native country in search of work or succor.[16]

But even in 1921 United Verde proceeded with plans to reconfigure its surface plant and to build more company housing in Jerome in expectation of better times. By 1922 the market had begun to turn around, and both companies had resumed operations. The rest of the decade was an era of good prices for copper, sustained growth and profitability for the district's mining companies, and rising affluence for the residents of Jerome.[17]

By the beginning of 1926 the *Verde Copper News* reported that copper was "in the strongest statistical position of any year since the end of the war, and certain years prior to the war." The next month the paper informed gratified readers of a "record peace time production of copper" in Arizona. The state's 1925 copper production equaled that of the nine other western mining states combined, and the total value of all metals produced in Arizona also far exceeded that of its nearest competitor. The year 1926 produced similar figures. Although the industry suffered a slight slump in 1927, the state retained first place in both copper production and total value of all metals produced, and both United Verde and uvx paid dividends. After the pause in 1927, the following two years were a heady time indeed. By the third quarter of 1928 the price of copper had passed fifteen cents a pound, and it went roaring aloft from there. The price continued to rise steadily for the rest of 1928 and into 1929, reaching a peak of twenty-four cents at the end of March 1929.[18]

Wages, tied to the price of copper, climbed proportionately. In April 1929, Arizona miners received their fourth pay increase in six months and the highest wages yet paid in Arizona mines. Although the price of copper retreated to eighteen cents by the middle of April, and the local editor might complain about market instability, he reported record sales as late as the first week of September and asserted that an advance in the price of copper "seems inevitable." His confidence stemmed from the strong and steady growth of the mining industry in Arizona during the previous seven years. He had observed a sustained boom rather than the frenzied one that had occurred during the war years, a boom that must have been very satisfying to residents, merchants, and mining company officials alike.[19]

Developments in the district, which residents beheld firsthand, had been as substantial and sustained as those throughout the rest of

Fig. 2.3. Jerome cityscape, ca. 1924. The central business district, dominated by the four-story T. F. Miller Company Building, is at the center, with company housing on the hill above, and the "Mexican Colony" in the foreground along Rich, Juarez, and Diaz Streets, below. Compare this to the modern photograph on page 269. (SPC 94: 2.74, Forrest Doucette Collection, Arizona Collection, Arizona State University Libraries)

Arizona. In March 1924 an area magazine reported "that there are today more men on the mine payrolls than ever before in the history of the 'Billion Dollar Copper Camp.'" Several of the mines, like the Copper Chief and the Jerome Verde, active in the frenetic war years, had ceased operations since, but activity at the United Verde and the UVX more than offset these losses. The big mines both produced at or near capacity from 1923 through the end of the decade, and the Verde Central also showed great promise.[20]

United Verde began its open-pit program in 1917. The company tore down the old smelter and relocated the mine's surface plant downhill in 1920. It completed removing eight million cubic yards of waste rock from the pit in 1927 and drew ore from both the pit and underground from that date. In mid-1929 the *Verde Copper News* stated that United Verde had increased its production in the first five months of 1929 half again over the same period in 1928. The district's other sure bet, the United Verde Extension, had experienced quite a decade as well. That decade was crowned by the year 1929, in which UVX dug almost three and one-half miles of new works, sold over 28,391 tons of copper, and paid $3.75 per share in dividends. The company reported to stockholders at the end of October that production had been so great because of the improved copper market.[21]

Observers watched the Verde Central, the district's up-and-coming mining property, with special interest in the latter half of the 1920s. Organized in 1916, the Verde Central's main shaft reached almost 2,000 feet deep by 1929. In that year the mine produced more than 2,167 tons of copper and almost 20,000 ounces of silver, and employed 173 men. But the most promising news from the Verde Central came with the opening of its new 300-ton-per-day concentrating mill on 1 January 1929. The *Verde Copper News* called the opening of the mill "an event of the utmost importance," and predicted that it would add an eighty thousand–dollar annual payroll to the district and support two hundred families.[22]

Before leaving this examination of the industry's progress in Jerome during the 1920s, we must reflect briefly upon the subject of mining company "paternalism," which also reached its zenith in the Verde District during that decade. Jerome was never a company town, thanks in part to the presence of two large mining companies. The town also boasted a diverse business district, largely outside the control of the mining companies, and, unlike the company smelter towns of Clark-dale and Clemenceau, Jerome was an incorporated municipality with a reasonably independent city government.[23]

That said, it would be hard to overstate the influence of the mining companies—particularly United Verde—upon business, governmental, and social activities in Jerome. In some cases, such as the T. F. Miller Company mercantile house and the *Verde Copper News*, United Verde held a financial interest. During World War I one of that newspaper's brief competitors derided it as the "United Verde Copper News," but even given complete editorial freedom, it is hard to imagine the paper having a much different opinion of the copper industry. Residents of Jerome knew where their economic interests lay.[24]

Mining company officials tried to keep their workers content. United Verde president Robert Tally explained in an April 1930 maga-zine article that to create good labor relations, "confidence must be established. The employee must be satisfied of the good intentions of his employer." After the labor unrest of 1917, company officials had established a large and efficient organization and were loath to see it reduced through employee dissatisfaction or any other cause. Dur-ing the depression of 1920, then general manager C. W. Clark told a newspaper reporter that "we would break [the organization] up only with the greatest reluctance. . . . We would much prefer to run along as at present for several months, even at a loss."[25]

In the 1920s management continued to sweeten the deal for its employees in order to keep labor peace and retain the best workers. In that decade United Verde built a baseball field, tennis courts, and swimming pools for its workers and remodeled its former hospital building into an employee clubhouse. Besides backing the Jerome and Clarkdale town baseball teams, United Verde also supported games between different departments within the company. On more serious matters, the company created a group disability and life insurance plan for its employees, with whom it split the costs, and wrote checks to cover half the value of employees' losses after the Bank of Jerome failed in 1925. These efforts to improve the lives of employees apparently paid dividends. United Verde officials reported at the end of the decade that labor turnover and absenteeism had been greatly reduced and that labor-management relations were cordial "to a degree which 15 years [before] would have seemed impossible."[26]

To be sure, the company pursued its self-interest through its benevolent behavior, but self-interest aside, both United Verde and UVX involved themselves in civic affairs to a remarkable degree during the 1920s. Both companies made donations to the Jerome Public Library, quartered in the United Verde clubhouse. United Verde built a public park and playground in 1927 and, in cooperation with the U.S. Forest Service, developed a picnic site atop Mingus Mountain. When Jerome's Methodists decided to build a new church, United Verde provided financial support, and UVX donated the land. The following year United Verde cast a five hundred–pound church bell for the new Episcopal church. UVX officials solved the problem of school overcrowding in the late 1920s by donating the company's former hospital building to the school district. When the Business and Professional Women's Club—an organization to which United Verde owed very little—held monthly meetings, member Helen Droubay remembered that "the mining company paid a very good cook to prepare the dinner and it was held in the clubhouse the company provided." Many residents later fondly recalled the companies' involvement in the community; such participation would be sorely missed only a few years later.[27]

The business life of the community began soon after the mines opened in the 1870s, and it considerably antedated the municipality of Jerome. George W. Hull established a general store in the rough young camp in 1880, and by the time United Verde took over in 1883, the settlement also had a restaurant, a saloon, several lodging houses, a blacksmith, and a barber. Thomas Miller founded the T. F. Miller Company in 1891, and his four-story mercantile building, constructed in 1898, became the keystone of Jerome's business district. This enterprise—which had some financial backing from Clark and at which his miners cashed their checks—was generally regarded as "the company store," but Jerome's consumers had many other options.[28]

The merchants, too, suffered in the depression of 1920, but a diverse business district survived. Among those enterprises were two banks, three substantial mercantile establishments besides the T. F. Miller Company, five hotels, three drugstores, three automobile dealerships, a movie theater, a brokerage house, a lumber company, and numerous boardinghouses, pool halls, cafes, groceries, artisans, and tradesmen. The town also had a flourishing trade in those adult entertainments common to mining towns. Things only got better for business as the twenties rolled on.[29]

The suspension of the Bank of Jerome caused a temporary depression in 1925, but at the end of 1927 the *Verde Copper News* boasted that "Jerome merchants generally reported Christmas business this year to have surpassed that of any previous season, which is a good indicator of general prosperity throughout the community." As the 1920s moved to a close, optimism seemed to be only confirmed by events.[30]

The Charles Robinson Jewelry Store, which opened at the end of 1926, was an important addition to the business community, and Robinson brought radio to the Verde District three years later. In 1928 the Popular Store added to its franchises in Prescott and Flagstaff by purchasing the D. J. Shea Company—one of Jerome's earliest business houses—on Main Street. In March 1928 the Popular Store signed a lease on the Main Street property through 1 April 1933. That same month the Shea family leased another of their properties up the street to E. K. Reese and Harry Amster, who opened the New State Motor Company. Reese and Amster entered into competition with about a dozen gas stations and four other auto dealers, including the Liberty Garage of Jerome, owned and operated by A. R. Tipton.[31]

The prosperity of the late 1920s was not confined to the town's white Americans. In mid-1927 Manuel Gutierrez announced the opening of his new $35,000 Gutierrez Building, which featured his Victory Grocery Store—complete with marble counters—and apartments on the second floor. Most Latino businesses could not afford such embellishments as marble counters, but many of them enjoyed prosperity in the 1920s. Most were small establishments, many of which catered mainly to Latinos. These included business houses like that of Carlota

Fig. 2.4. Jerome's central business district on Main Street in the early 1920s, with the Miller Building at left. Motorized transportation has replaced horse drayage—almost. (Arizona Historical Society, Tucson, #29047)

Ruiz, Francisco Madrid's lodging house and mercantile establishment, and John Perez's pool hall. Some Latino establishments, like Marie Vasquez's hairstyling salon and Dan Gonzales's hotel and apartment house, were located outside of the "Mexican Colony" and served customers from throughout Jerome.[32]

Other businesses made important, though less wholesome, contributions to the community. The backstreet businesses of drinking, gambling, and prostitution had been part of town life since the very early days. Gambling and prostitution were nominally illegal activities in early-day Jerome. A movement against prostitution produced a segregated red-light district on the western end of Main Street during the 1890s, but this highly visible vice district proved unsatisfactory to its opponents, and it was relocated down the hill to Hull Avenue. At the insistence of the army, the state health officer closed down the red-light district during World War I. The adoption of state prohibition in 1915 also drove the liquor trade underground.[33]

Prostitution recovered somewhat during the 1920s, with the return of the red light to the northern end of Hull Avenue. Several madams operated houses in the area, and individual women worked out of cribs or boardinghouses. An illicit distilling industry, centered in Deception Gulch on the east end of town, produced the wine to go with the women. Jerome reached the summit of its brazen disregard for the Volsted Act in October 1927, when a resident attempted to lubricate the inmates of the town jail in the middle of the night, using a jar of hooch and a rubber tube. The drunken laughter of the prisoners betrayed the scheme, and its perpetrator quickly found himself, as the *Verde Copper News* put it dryly, "on the wrong end of the hose."[34]

Most of the white mule seems to have issued from Jerome's pool halls, which probably also accounted for some of the prostitution, and formed the center of Jerome's illegal gambling trade. These pool halls were frequently cited for possession of liquor or illegal gambling and occasionally threatened with dire consequences, but the latter seldom came to pass. Jerome's authorities usually took the live-and-let-live attitude toward the vices adopted by most mining communities, preferring regulation to prohibition.

The demands of justice could generally be satisfied by arresting the flagrant drunkards and making occasional sweeps against gambling and prostitution. Even that guardian of good standards, the *Verde Copper News*, admitted that "it must be recognized that a mining camp, with hundreds of single men, must of necessity be run on lines not so closely drawn as in more settled communities.... This may be a utilitarian way of looking at the matter, but it has the merit of being common sense." The editor believed city officials should draw the line "between reasonable liberty and unreasonable license," a compromise satisfactory to most during the 1920s, if the sporadic enforcement of the prohibitions against these activities is any indication.[35]

Although the area's original claims had been located by nearby white ranchers moonlighting as prospectors, Jerome very quickly became an ethnically mixed place. The railroad brought different groups from east and west. By the turn of the century Jerome had Chinese, Irish, and Italian communities, with people of Slavic origin constituting the largest minority element in the town up to World War I. By then, Mexican nationals began to arrive in large numbers and to settle below the main part of town to the north and east. In 1920 the United Verde Copper Company employed men of twenty-three different nationalities.[36]

The 1920s produced a significant shift in Jerome's demographic composition. United Verde reported that Latinos represented roughly 20 percent of the company's workforce in 1920. The Latino community bore a heavy share of the depression of the early 1920s; perhaps four hundred people returned to Mexico from Jerome with the assistance of the Mexican government in 1920 and 1921. But once copper mining resumed, the Latino community recovered and experienced relative prosperity. The Chinese, Irish, Italians, and Slavs declined in proportion and significance as the years advanced.[37]

The census of 1930 listed Latinos as 57 percent of Jerome's population, and the many Latinos who resided outside of the corporate
limits of the town would not have been included in that calculation.
The size and significance of Jerome's Latino community encouraged
both United Verde and UVX to cease mining operations during the Fiestas Patrias (Mexican Independence Day) celebration of 1927. Jerome
schools also dismissed for the occasion. In 1928 both mines shut down
for Cinco de Mayo, but neither did for Memorial Day. The economic
importance of Latinos to the town's merchants can be seen in the *Verde
Copper News*'s Spanish-language page, "El Departmento Español," inaugurated in May 1927 and continued until 1930, when it was replaced
by a Spanish-language newspaper.[38]

The prosperity of Jerome's boom days reached the Latino residents
of the city and served to attract others to the town. Charles Mann, who
sold appliances door-to-door in the Latino quarter during the 1920s,
reported that the average Mexican national in Jerome had resided in
the community for five to ten years' duration and liked "American
high standards of living but [had] no immediate ambition to become
a citizen." These foreign-born residents enjoyed the American con-
sumerism of the 1920s. Mann reported that "phonographs, radios
and sewing machines are great attractions to these Mexicans, and they
will buy readily. Almost every family owns a sewing machine and a
phonograph or radio." Latinos, Mann discovered, were as enthusiastic about the credit plan as the town's other residents. He claimed that
he had no trouble making sales to Latinos, that "the time payment plan
is the ultimate sure winner argument to which there is no debate." Of
course, there were limits for persons at the lower end of the economic
order. Mann believed that Mexican residents were "more budget-
minded than the average American in Jerome and [were] not as likely
to exceed their incomes."[39]

The Latino community's partial inclusion in the boom economy
of the 1920s paralleled its partial integration into the larger society of

Jerome. The nature of Latino segregation was complex. Some community organizations completely segregated, while others did not. The high school sports teams were integrated, apparently without difficulties, while some of the numerous town baseball teams were segregated. Sometimes these segregated teams played each other. Jerome certainly had "Mexican" neighborhoods, but these were not rigidly defined, being inhabited by Latinos, other ethnic groups, and native whites, while other Latinos lived outside of these neighborhoods. In this small town many people made friends and communicated across cultures. John Krznarich, son of a Yugoslavian immigrant who lived near the uvx Mine, recalled, "My father would take a gallon of wine over to a Mexican family, and in return he'd get maybe two dozen tamales or enchiladas or something like that."[40]

Latino social life proved to be no more sharply isolated than Latino neighborhoods. The local press lauded the achievements of prominent Latinos and occasionally announced marriages between Latinos and whites. The big Mexican holidays were celebrated uptown as well as in the Latino quarter. On those occasions a grandstand would be constructed in the town plaza, from which music and patriotic speeches would issue—in Spanish and English—while overhead fluttered the flags of the United States and Mexico. The program would usually conclude "with a dance in Miller Hall and on Main Street which [would] be roped off for the light footed." "They'd really celebrate," remembered Krznarich, who added, "Oh yeah, by all means, we'd join in." Sometimes non-Latinos helped to organize these occasions, as when Matt Shea and Milton Scott directed the sports activities for a Mexican Independence Day celebration.[41]

Still, Latino inclusion in the general community remained only partial, a situation apparently approved of by most people in Jerome. United Verde built a segregated clubhouse and pool in the Latino quarter at the request of the Latino community. While some Latinos—among them the Mexican consul to Arizona—regarded the "Americanization" school, founded by United Verde in 1925, and its English-language training as a good thing, others preferred to keep their distance. Jerome's police often had trouble making arrests for crimes committed between Latinos; the suspect usually disappeared into the community, among witnesses reluctant to speak to the authorities.[42]

Some of Jerome's many social organizations, like the Alianza Hispano Americana (AHA) or the Croatian Fraternal Union, were essentially ethnic organizations, but many others were not. The fourth floor of the T. F. Miller Company building housed the Masonic Hall, which also served as the meeting place for many other fraternal organizations. At various times in its history Jerome had chapters of the Grand Army of the Republic, the Red Men, the Ancient Order of United Workmen, the Knights of Pythias, and the Woodmen of the World. A Moose Lodge was constructed in 1914, and chapters of the Odd Fellows and the Elks were organized during World War I. The AHA made its appearance in April 1899, and the local chapter of the Croatian Fraternal Union was organized two years later. The array of other fraternal and social organizations that appeared in the town at one time or another is too numerous to list.[43]

Fraternal organizations continued to flourish in the 1920s. The Latino community formed a "Mexican" Boy Scout troop and added Masonic Lodge Gloria a Juarez no. 20 to the ranks of Jerome's fraternal orders. The Masons joined Alianza Hispano Americana Logia 13, and Lodge 59 of the Woodmen of the World, as the major Latino fraternal orders in Jerome. In the summer of 1927 Jerome's leading Latino entrepreneur, Manuel Gutierrez, announced plans to build a clubhouse next to his new Gutierrez Building to host the Latino fraternities.[44]

Several organizations increased the ranks of the "American" fraternal orders in the 1920s. The Jerome Rotary Club organized in 1922, and the Knights of Columbus established a chapter in town the following year. The twenties also saw the founding of the Alexander Moisa Post of the American Legion, an organization destined to take a very important place among the fraternal orders and in the community. By 1929 the Rotary, the Elks, and the Business and Professional Women all held their meetings in the United Verde clubhouse, which also hosted Christian Science services.[45]

Jerome's churches, if not as important to the average miner as its saloons, constituted an early and significant part of the community. In 1896 Baptists erected the first church in the town, to be followed the next year by a Catholic church. A three-story brick church, constructed in 1900, replaced the original Catholic church, destroyed by fire in 1898. Local Methodists organized and built a church in 1900, due in large part to money raised by the church's Ladies Aid Society. Jerome's Episcopalians acquired the former Baptist church building in 1904. Given Jerome's large Latino and Eastern European populations, the Catholic church had the biggest congregation and played the most influential role. During the 1920s the Holy Family Roman Catholic Church held three masses on Sundays, as well as two Sunday schools and two weekday church schools, one each in English and Spanish. In the summer of 1927 both the Methodist and the Episcopal congregations dedicated new church buildings in Jerome.[46]

Social activities in the copper camp often lacked the formality of lodge or church events. The simplest of these might be a gathering downtown to follow the progress of a World Series game being displayed at the Post Office Cigar Store or to listen to a boxing match at one of the businesses that had a radio. More formal activities included attending the occasional fund-raising entertainments put on by one of Jerome's fraternal or religious organizations or the concerts provided by the town's bands. During the 1920s the town supported two large bands, the Miners' Band and the Reception Band, the latter so-called because it practiced at the Reception Pool Hall. Both bands attracted "an interested audience of both Americans and Mexicans who gather[ed] to witness the art of thirty or forty musicians" at Sunday and holiday concerts. The Kopper Kids Orchestra, Chub Burner's

Paramount Players, and other small groups played at the various dances and events held throughout the year. Perhaps one of these bands entertained the Hillside Dance Club, a group of twenty couples who organized in 1928. Those who could not dance could always take refuge in the town library, which in just five years after its founding in 1924 had grown from a dream into the federal depository for northern Arizona.[47]

Almost everyone participated in Jerome's major holidays, which were Independence Day and Christmas in the early days, with Cinco de Mayo and Mexican Independence Day added as the proportion of Latinos increased. The programs organized to celebrate these occasions constituted the most elaborate social events of the year. The newspaper announced in 1928 that "bathing beauty contests, children's sports, band concerts, prize awards, aquatic meets, baseball games, [a] fireworks display and a free dance constitute the Fourth of July celebration in Jerome." Main Street served as the seat for most of this activity and for the speeches and miners' contests held on warm-weather holidays. The Christmas festivities usually began in early December with a fund drive to raise money for a community Christmas tree, erected in the plaza in front of the Miller Building. The rest of the proceeds would be used to buy presents and candy for the town's needy children. Santa would distribute these items during a party held Christmas week in Miller Hall.[48]

Like any other mining town, with its overabundance of young males, sporting activities always enjoyed great favor in Jerome. As a *Verde Copper News* editorial explained: "We need baseball, or something of the sort, here perhaps more than other places do, for we are comparatively isolated and must, perforce, go without some of the amusement advantages that other places have; hence the need for anything we can get in the way of wholesome diversion and amusement." Whatever the reason, Jerome witnessed a constant parade of formal and informal sporting events into the 1920s. Boxing and wrestling matches were held a floor below the lodge meetings in the Miller Building. During the 1920s Jerome had horseshoe pits and a gun club. Locals followed the Muckers, the high school football and basketball teams, with great interest.[49]

As elsewhere in the United States, baseball ranked as the most popular sport by far. Every spring saw the Jerome Miners, or Mineros, the town team, take the field to represent Jerome in the Northern Arizona League. "Whoopee," the paper crowed in April 1927, "it's baseball time in Jerome! Sunday's the time, Jerome's the place and Prescott's the victim." The rivalry between Jerome and Clarkdale became so corrosive by the mid-1920s that United Verde suspended its support for separate town teams, choosing for a time to back a district team instead. Within Jerome, American Legion baseball, junior baseball sponsored by the lodges, baseball between company teams, and good old-fashioned pickup games added to the choices available.[50]

Prosperity meant more than celebration and sport to Jerome; it meant money for physical improvements in the town as well. In mid-1927 the newspaper reported an "epidemic of painting prevailing uptown." At the beginning of the decade a magazine article characterized Jerome as a town "where fine residences overlook or are alongside of old wooden buildings, where there is much incongruity to the architectural development." The town had not lost its eclectic architecture by the end of the decade, but it had undergone numerous improvements.[51]

The copper companies contributed hospitals and houses, as well as recreational facilities. When the new United Verde Hospital opened at the beginning of 1927, the *Verde Copper News* boasted that it was the "last word in facilities, equipment and every possible modern convenience." Perhaps, but not for long. In 1929 the company announced that a fourth floor would be added to the building to accommodate a maternity ward and other facilities. The hospital served patients from all over the Verde Valley. The need for more hospital space paralleled an increasingly serious need for more homes; by the end of the 1920s housing was again in short supply in Jerome.[52]

The company originally owned a few houses in Jerome, which it rented to its supervisors and foremen, but during the last four years of the 1920s, United Verde built 117 houses in the town and included workers among its tenants. These houses supplemented a company dormitory for single employees and a company-owned apartment building. A United Verde official estimated at the beginning of 1930 that 330 mine employees, over one-quarter of the company's force, were housed in company properties. Only a week before the stock market crash, United Verde announced plans to build another 39 houses in Jerome, part of "an extensive building campaign started," the company promised, "to relieve [the] congested housing situation."[53]

The copper companies were not the only local agencies making physical improvements. Jerome's municipal government and its services had evolved considerably over the forty-plus years from the creation of the United Verde Copper Company to the 1920s. Government had arrived in the form of a post office in 1883, but municipal government remained almost twenty years away. Influential property owners resisted establishing a local government, with its taxes and regulations, but events literally engulfed those opposing municipal incorporation. Fires ravaged the rapidly growing community in 1897, 1898, and 1899. Because Jerome lacked municipal water and fire services, the town could not fight the blazes that roared through its wooden buildings.[54]

Jerome finally incorporated on 9 May 1899, and the new municipal government worked to reduce the fire danger. It quickly passed a building code and established a town water system and volunteer fire department, with water pipes and fireplugs installed at the end of 1900. Jerome had a good fire system by the boom years of the 1910s. In 1917 residents voted to fund construction of a town sewer system and begin municipal garbage collection. While the newspaper editor might decry the condition of Main Street as "a disgrace to the town" during World War I, plans called for paving the streets and building concrete sidewalks, and the town had certainly come a long way from the log and tent camp of the 1890s.[55]

The boomtown prosperity of the 1920s led to higher property valuations and more revenue for the local and county governments, which took advantage of the situation to make further improvements. The county and state completed the road between Prescott and the Verde Valley in 1920. The town of Jerome concurrently upgraded its own roads to match the quality of the new road, a program that included both surfacing and sidewalks for the major streets of the town. By the middle of the decade the paper noted improvements to culverts, drainage, retaining walls, and stairways between the different levels. The gravel streets were regularly maintained, and a paving program began in 1928 when the county surfaced the Jerome-to-Clarkdale road. At the end of the year the town answered when a special bond election approved paving all the main streets in Jerome. Those improvements continued into 1929.[56]

City officials ordered municipal improvements besides street construction. They purchased new fire trucks in 1926 and 1929 and changed the town's streetlights from arc to incandescent in 1926, providing better lighting at lower cost. The November 1928 bond election also approved monies used to construct a new combined city hall and fire station, which the town occupied in September 1929.[57]

Expenditures notwithstanding, the *Verde Copper News* bragged about the town's excellent financial condition at the time of the 1928 bond election, reporting that "Jerome has a smaller outstanding indebtedness per capita than any other town in Arizona." Economy or "efficiency," to hear the *News* tell it, was the object of governments, national, state, and local. "We all want good roads—and a thousand other things," its editor opined in 1927, "but we must look to the taxable values and the probable revenues, rather than to our aspirations."[58]

Most of the partisanship and interest expressed by Jerome's news-paper and by its electorate centered on national political campaigns, with state politics receiving somewhat less attention. In local elections a "Citizens' Ticket," generally composed of leading merchants, repre-sented the establishment. The Citizens' Ticket usually ran unopposed, and even on those occasions during the 1920s when an opposition ticket formed, the Citizens' Ticket won handily. In the city election of 1926, no member of the Citizens' Ticket received fewer than 426 votes, while no member of the opposition received more than 265. There was little interest in those elections in which the Citizens' Ticket ran unop-posed. For the presidential election in November 1928 Jerome regis-tered 1,360 voters, 831 Democrats and 529 Republicans, but only 65 residents had cast votes in the municipal primary election held the pre-vious spring. Jerome's editor could not contain his sarcasm over local voter apathy after the municipal election of 1930, reporting "the huge total of 35 votes being massed up. All during the day the police were in evidence to quell disorders. Following the counting of the votes [elec-tion officials] were removed to their homes in a state of exhaustion."[59]

The Jerome School District showed an evolution similar to that of the municipal government. Organized in 1884 and designated Yavapai County School District no. 9 in 1889, the district began classes in a rented one-room house, replaced by a four-room frame school build-ing in 1896. By 1910 a four-year high school curriculum and building had been added to the elementary system. The boom period of the 1910s resulted in a burgeoning school population, which tripled between 1910 and 1920. Churches served as classrooms for over a year, and though the district built another large primary school building and seven freestanding classrooms to accommodate the rush, the school system remained seriously overcrowded throughout the boom days of the 1910s.[60]

The district sought to rectify that situation during the 1920s with an extensive school construction program, completing construction of a new and impressive high school complex on the hogback below and to the northeast of town in 1923. During the rest of the decade the town grew out to and around the new school buildings. The following year District 9 opened another elementary school in Deception Gulch, and the year after that, 1925, completely rebuilt the Clark Street School, the original primary school. These improvements barely met the demand. The district had almost two thousand students during the 1929–1930 school year, almost double the 1919–1920 school population. By the end of the 1920s the Jerome School District consisted of three elementary schools and the modern high school complex.[61]

The booming economy that caused those overcrowded schools in the 1920s certainly boded well for the coming decade. At the end of the 1920s the *Verde Copper News* gloated over record copper production tonnages and values. While the stock market crash in October 1929 dampened some of the enthusiasm, New York seemed a long way from Arizona, and stock market speculation seemed far removed from Jerome's basic industrial economy. In November 1929 the paper reported that "Jerome, standing on the threshold of its most prosperous winter, presents a picture that is the envy of virtually every town in Arizona. So rapid has been [the] influx of new residents that all hotels are crowded and apartments or houses are impossible to obtain. Throughout the year business accounts and savings deposits in the Bank of Arizona have shown extraordinary increases." A continued copper price of eighteen cents a pound, according to the paper's sources, "will assure the Verde Valley the most prosperous winter in its history." Earlier in the year the newspaper's editor had opined that mining "'slumps' and failures are becoming increasingly rare." Who could argue with that?[62]

CHAPTER THREE

This Ill-Omened City

Tombstone's problem at first appeared to be its salvation. When miners struck water at 520 feet in the Sulphuret Mine at the end of March 1881, the *Tombstone Epitaph* rejoiced, "Here's Richness." Tombstone's groundwater could be pumped to the surface and used in local stamp mills, which would end having to haul ore nine miles to the San Pedro River for milling, an operation that cost the mining companies around $3.50 a ton.[1]

Each in its turn, other mines in the district sank shafts below the 500-foot level and struck water. Early predictions about the beneficial nature of this water seemed correct. The Girard company quickly erected a twenty-stamp mill in Tombstone that used mine water to treat its ores, and 1882 was Tombstone's most productive year. But by the spring of 1883, it began to become apparent that Tombstone's mines suffered from too much of a good thing. At the beginning of May 1883, the Grand Central's management announced that the mine would suspend operations because of a renewed flow of water.[2]

In order for development to proceed, mining companies would have to undertake large-scale pumping of the underground reservoir beneath Tombstone. This scenario did not seem particularly ominous at first because rumor had it that the ores below the water table were richer than those already developed above. If true, that would mean bonanza indeed, one that mining companies would spare no expense to unearth.

In the spring of 1883 the Grand Central, the first mine to attack the problem, installed steam pumps capable of lifting five hundred thousand gallons of water a day. These succeeded in lowering the water level somewhat, not only in the Grand Central but also in adjacent mines. After witnessing this success, directors of the Contention spent $150,000 installing twelve-inch Cornish pumps capable of lifting one million gallons of water a day. A Tombstone correspondent to

Fig. 3.1. Workers load sacks of ore into mule-drawn wagons at a Tombstone mine in the early 1880s. Tombstone's dry location and its lack of a railroad meant that ore had to be hauled by wagon nine miles to the San Pedro River. Throughout its early history, boosters of the district sought to lower its production costs by acquiring a railroad. (Arizona Historical Society, Tucson, #44692)

the *Arizona Weekly Citizen* informed its readers in June 1883 that there could be no doubt that the pumps of the Grand Central and the Contention would remove all of the water from the mines in the vicinity.[3]

Much to the chagrin of boosters and investors, these newest pumps also failed to solve the problem. By mid-February 1884 the *Tombstone Republican* admitted that "the pumps in the Tombstone mines have not yet overcome the water, and it may be that the moving capacity will have to be increased before the water question is fully settled." Neither the *Republican*'s editor nor the mines' managers would yet concede that this was anything but a temporary setback. The Grand Central proceeded to spend another $200,000 installing a set of fourteen-inch Cornish pumps with a capacity of one and one-half

Fig. 3.2. This giant walking-beam pump engine was installed at the Grand Central Mine in 1885 to try to remove the water encountered in the district below the five hundred–foot level. (Arizona Historical Society, Tucson, #7151)

million gallons a day. The *Republican* reported in March 1884 that $500,000 would be invested in pumps in the district in the next few months.[4]

Editors and mining men assured the public that each improvement in pumping capacity solved the water problem, but the deeper shafting undertaken after each improvement revealed that the problem remained. Although the *Republican*'s editor had claimed at the end of 1883 that "the crisis has passed, and Tombstone as a permanent mining center feels confident of a great future," outsiders not as jealous of Tombstone's reputation as the bonanza camp of the Southwest were less certain of the town's prospects. A writer for the *Arizona Mining Index* of Tucson concluded at the same time that "if rich ore is found below the water level . . . the permanence of the district will be established. If not so found, it is estimated that the ore above that point will be worked out within a couple of years."[5]

Whichever writer's opinion would be borne out in the future, the situation in Tombstone had changed. Companies that had previously paid out dividends freely suddenly found themselves without the capital reserves necessary to meet their new higher costs of operation. The Contention suspended dividend payments in June 1883 in order to meet the expense of the new pumps, and that company was not the only one facing tougher financial circumstances.[6]

Although the Tombstone Mill and Mining Company cleared a profit of almost $120,000 in 1883, its president, George Burnham, reported to his stockholders that their profit could just as easily have been a loss had the company not undertaken significant economies. Burnham's report showed decreasing ore values in the years 1881 to 1883, declining ore production after March 1882, and exhaustion of ores in the Goodenough and Toughnut claims, the company's two most important holdings. He also mentioned that the company was reworking its own tailings, another sign of tougher times. Clearly, Tombstone's bonanza days of high-grade ores, high wages, and plump dividends had departed by the spring of 1884. In the face of heavy expenses, retrenchment was in order.[7]

Rumors began to circulate in Tombstone at the end of April 1884 that the major mining companies intended to reduce miners' daily wages to three dollars from the then prevailing four dollars per day. In reaction to these rumors miners organized the Tombstone Miners' Union, and in reaction to that the Grand Central, the Contention, and the principal property of the Tombstone Mill and Mining Company threatened to close. The Grand Central shut down on 1 May and the Contention and Toughnut shortly thereafter. Superintendent E. B. Gage of the Grand Central indicated that his company had lost $25,000 during the previous month, that its pumps could not hold back the water, and that unless miners agreed to work for three dollars a day the mine might cease production for up to six months while new pumps were installed.[8]

On 5 May a parade of four hundred miners visited Gage and John A. Church, mine superintendent for the Tombstone Mill and Mining Company, but the two sides could not resolve the dispute. Observers disagreed considerably over the value of the ores that remained above the water level, hence the profitability of the mines. There is some cause to believe both sides in this contest; the miners knew the quality of the ores that they had until recently been working, while the superintendents better understood the cash reserves of their companies and realized the enormous expenditures for pumping equipment that lay just ahead. In the middle of May a citizens' committee met with the union in the hope of reaching some sort of compromise, but neither side was willing to concede anything at that early date, so the work stoppage continued, and up to four hundred miners remained idle.[9]

Some Tombstone residents were concerned from the earliest days of the lockout that the miners might express their frustration through vandalism or violence. Rumors of an attack on the Grand Central began to circulate on 7 May. The company posted guards, and George

Parsons, a resident who might best be described as a member of the "law and order" faction of Tombstone, predicted "lively times ahead probably" in his diary. He stood his first guard duty on the night of 9 May. "Not much sleep these times," Parsons recorded at the end of that day, but added hopefully, "miners cooling off I think."[10]

But the situation became grim enough by the third week in May that mine superintendents Gage and Church found it prudent to leave the district. Area newspapers reported threats to destroy mine works and that the sheriff had organized, armed, and deployed one hundred deputies to protect the mines, with the ringing of the bell at the Vizina Mill as their call to arms. Tombstone remained calm for the time being, but then came news from the Grand Central that its directors had decided not to resume work except at the wage rate of three dollars a day. Two days later Tombstone's residents learned that the district's mines would not employ union men even at the three-dollar rate. Union miners began to leave the camp in large numbers.[11]

By the last week in May things began to look very bad for the union. One witness believed "that the miners are playing a losing hand. The town will surely become a $3 camp. . . . The mines can afford to lie idle; the miners cannot. The latter will be forced by idleness to seek employment elsewhere, and there will eventually be no trouble in putting three-dollar men to work." This prediction seemed to be borne out by another observer who reported in the second week of June that only the Contention remained in operation, reworking tailings; that about two hundred miners were idle; and that those who had money were leaving the district. Although it was certainly an uneven game, the union still had a few cards to play.[12]

Parsons noted at the end of May that about fifty chloriders—small-scale contract miners—had been compelled to stop working by the miners' union. The *Arizona Daily Star* of Tucson reported at about the same time that many miners were willing to work for three dollars but were afraid of the union.[13] The pinnacle of union intimidation came in

the first week of June in a letter to the *Tombstone Epitaph* signed by
S. D. Stevens, president of the Tombstone Miners' Union. This read
in part:

> It has lately been brought to my notice that there are several
> persons in this city who are continually speaking disrespectfully
> of the Miners' union. Now, I wish to notify these gentlemen to
> be careful. I find a strong feeling in the union to uphold our
> rights. We have tried our best to be law-abiding citizen[s], and
> have not as yet taken any advantage of our strength: but I am fully
> convinced that there is a feeling in our union to remove from our
> midst those persons who are trying their best to "down" the
> union. . . . I am of [the] opinion that harsh measures will be meted
> out to such persons if they persist in using such language. It is well
> known to the people of Tombstone that we have done nothing
> wrong so far; but I assure you that some of the miners of this camp
> are of the opinion that it is time to pursue some other line of
> action. . . . I hope that in the future parties who are interested
> will be careful how they molest a quiet but determined union of
> miners.[14]

Whether they intended simple bluff or serious threat, union leaders made an egregious tactical error by authoring this letter. When
the home of Grand Central foreman Charlie Leach "mysteriously
burned itself up" within two weeks, blame naturally attached to the
union. The mining companies then upped the ante at the beginning of
July, announcing that they would not resume operations as long as the
union remained in existence. Two weeks later, advertisements began
to appear in the *Epitaph* for miners willing to work for three dollars a
day, and a week after that, at the beginning of August, the Head Center
and Emerald Mines resumed work with three-dollar men.[15]

The *Epitaph* reported in the second week of August that things were
looking up, that mining and business were coming around, and opined

that there would be no trouble from the union. But the crisis was rapidly building. At the beginning of August, Stevens called for a town meeting at Schieffelin Hall, at which he informed those attending that he had lost control of the union and asked for a citizens' committee to intervene in the dispute. That meeting and a visit by union men to the Stonewall Mine, both in an attempt to stop the use of three-dollar labor, ended without resolution.[16]

The situation remained quiet, but the *Arizona Daily Star* reported on 6 August that "there is a manifest uneasiness throughout the community that indicates the apprehension of the people that danger may still be lurking in their midst." Events reached a head, and the union played its final card, after the Grand Central resumed operations in the first week of August. By its second day of operations, the company had more than 100 nonunion miners employed and had announced its intention to employ 250 men as soon as they could be hired.[17]

Fig. 3.3. The Grand Central Mining and Milling Company's mine, ca. 1884, the year of the shootout at the mine. This photo, which appeared in the Arizona governor's report of 1896, identified the home on the left as belonging to the mine's manager, E. B. Gage. (Arizona Historical Society, Tucson, #28307)

The union's final desperate act, the attack on the Grand Central, occurred early on the morning of 9 August. The trouble began with an altercation ending in gunplay between a union miner and a nonunion miner on the evening of 8 August. Neither man was hurt, but a mob appeared ready to take action against the nonunion miner. Law officers dispersed that crowd, but just after three the next morning guards spotted a group of between fifty and seventy men heading up the ore road toward the Grand Central. Foreman Leach readied his seven guards to resist, and within a few minutes a union man came up to ask if Leach intended to continue to work at the three-dollar rate. After Leach took the man prisoner, the group outside began firing into the hoist house from behind a woodpile. The guards returned fire, the prisoner escaped, and then the men outside retreated as whistles from the mines alerted the town. Around one hundred shots were exchanged, but apparently only one man was injured, helped away with those who retreated.[18]

The companies reacted to the assault on the Grand Central by publishing a notice in that evening's *Tombstone Epitaph* that demanded that the group's leaders be handed over to the sheriff and that the union be dissolved within twenty-four hours. The five companies that signed the notice promised to permanently blacklist every union miner unless those conditions were met.[19]

The crisis quickly drew to its conclusion. On the morning of 12 August, Company C, First Infantry, arrived from Fort Huachuca and took up positions on the hills above Tombstone where the major mines were located. Within a month the superintendents had returned to town and the major mines had resumed their activities at three dollars a day or had announced plans to do so. What was left of the union had disbanded, and its leaders had scattered. On 10 September, Company C retired to Fort Huachuca, "it being considered that no further necessity existed for their presence."[20]

When the Grand Central reopened, the *Tombstone Epitaph* reported that "business men are once more wearing smiling countenances, and the prospects of better times than ever are very flattering," but the woes continued in Tombstone after mining resumed. The work stoppage and flooding problems reduced district output from over $5 million in 1882 to under $1.5 million in 1884, and Tombstone never regained its stride after the shutdown.[21]

The mining companies began paying wages into the district again, which townspeople welcomed, yet these same companies still faced debt and still had an ocean of water to move. The Grand Central began operating a new larger set of pumps in early July 1885, which seemed at first to remove the water and revitalize the camp. Both the Grand Central and the Contention began sinking shafts again in the autumn of 1885, and the Tombstone Mill and Mining Company was also active. In December the newspapers reported that the Contention and the Grand Central were pumping more than two million gallons of water a day and that the superintendents agreed that they had solved the water problem.[22]

A look behind the cheerful accounts in the local newspapers reveals a less optimistic situation among Tombstone's major mining companies. In September 1886 the *Engineering and Mining Journal* reported on the Tombstone Mill and Mining Company's operations. The *Journal* found a company over $241,000 in debt, which was only $15,000 less debt than it had carried in the midst of the lockout. Profits had been absorbed in litigation and acquisition of new properties, and reductions in wages and supply costs had been offset by the falling price of silver and declining ore values. The *Journal* warned that unless the price of silver recovered, the company would have to suspend operations and confine itself to exploration and development while awaiting a better price or a new bonanza.[23]

Although 1885 had not been a great year by boomtown standards, it looked better than the tumultuous 1884, and things seemed to be improving. At the beginning of April 1886 the *Tombstone Epitaph* editorialized that "the outlook for our camp is quite encouraging. . . .

Altogether, it looks as though, not perhaps a boom, but good times, [are] near at hand." The *Arizona Daily Star* agreed, saying that "Tombstone is entering upon an era of permanent prosperity. The days of her boom are past forever, but [the] tide of dull times has also past." But within two months of these pronouncements, another crippling disaster struck the Tombstone District.[24]

Just before eleven o'clock on the evening of 26 May, the men in the engine room of the Grand Central's hoisting works were astonished to see flames shoot into the room from the adjacent boiler house. Blasts on the mine's whistle instantly alerted the whole town, but the Grand Central's hoisting works were engulfed before assistance could arrive. "It was a grand spectacle," the *Tombstone Epitaph* reported, "the flames rising to an immense height and illuminating the country for quite a distance around." Attempts by Tombstone's volunteer firemen to battle the blaze came to naught when they discovered that their hose couplings did not match those of the mine. No one was killed in the inferno, but the Grand Central's hoisting works and pumping equipment were totally destroyed and its main shaft badly damaged for some distance into the earth. The next day the *Epitaph* put the loss at two hundred thousand dollars, with insurance covering only sixty-five thousand dollars, but reported that the extent of the damage was difficult to determine, as the fire was still burning fiercely when the paper went to press.[25]

This time neither the press nor the public rose-tinted the situation. A correspondent to the *Arizona Weekly Citizen* informed that newspaper by telegraph the next day that the fire was "regarded as the most disastrous blow that ever happened to this ill-omened city," and that it had "caused a feeling of deep depression throughout the city." George Parsons noted in his diary the day after the fire that no pumping or other work was going on on the hill, and concluded glumly that it "looks as though the Devil surely is foreclosing his mortgage."[26]

So it seemed. The Grand Central quickly announced plans to rebuild, but in mid-July Parsons reported "everything remarkably quiet." One reporter observed at the end of 1886 that the district's three major mines had nearly ceased production. If they found new deposits to match the old ones, everything would be fine. "Otherwise," he concluded ominously, "the place has nothing to hope for."[27]

There were intermittent rumors over the five years after the Grand Central fire that the three big companies would construct a consolidated shaft with a massive pumping apparatus and, thus united, slay the water dragon at last. In the summer of 1888 and again at the beginning of 1890, "dame rumor" danced through Tombstone, but nothing ever came of the portends. Officials of the three companies apparently disagreed over which company should bear what share in the proposed consolidation. While the discussions went on, the Contention's pumps stood still and water rose in the shafts.[28]

In the meantime, the companies discovered no new ore bodies of consequence, and as the 1880s dragged on Tombstone continued to waste away. By the end of 1886 the Grand Central's mill was the only one operating on the San Pedro River, and most of Tombstone's mines were closed, leased, or doing only development work. The *Tombstone Prospector* reported the Grand Central's original holdings just about worked out above the water level in June 1888. By April 1888 the Tombstone Mill and Mining Company had $246,000 in unpaid debt and was laboring under a second mortgage. In 1889 the company made only enough money to pay expenses.[29]

By then even the district's perennial optimists had grown worn and cynical about Tombstone's chances. "The same old story," the *Prospector* snapped in May 1889, "plenty of ore in sight. Mines never looked better. Thousands of tons of ore underground that will not be raised to the surface til something happens which we are not supposed to know til it happens." When Contention managers arrived in town at the end of the year, one man who spoke with them could only conclude despondently that "they have come to go away again."[30]

In 1890 the price of silver began to climb again, thanks largely to the Sherman Silver Purchase Act, and spirits in the silver queen of the Southwest began to improve. Nevertheless, circumstances did not change much in Tombstone. The Contention remained idle, the Grand Central staggered along, and the Tombstone Mill and Mining Company reported employing only seventy men on four properties and shipping but ten tons of ore a day. Then silver prices began to slide again, and whatever chance Tombstone had to recover slid with them.[31]

As if to punctuate the end of Tombstone's bonanza decade, the death blow for the glory days fell in the early morning of 27 December 1891, when the Contention Mine's surface plant caught fire. The fire companies again mobilized and dragged their equipment a mile up the hill, to be frustrated this time by a frozen pipe. Within forty-five minutes the whole hoist house was a mass of flame. The loss, which totaled more than $300,000 and was again insured for less than half its value, included the Contention's $250,000 pumps, which had lain idle since the Grand Central fire.[32]

Even after the Contention's fire, the Tombstone Mill and Mining Company continued to produce, but in 1892 the Contention's directors—again unable to reach an agreement to consolidate with the Grand Central—conceded defeat and dismantled and sold their mill on the San Pedro. The press and the public continued to console themselves with the idea that "there are millions of treasure yet untouched within a radius of a mile of the courthouse. All that is required to bring it [t]o the surface is union of action amongst the mine owners."[33]

But what the press and public had to admit was that there were two issues upon which a revival of mining in Tombstone depended. Over one, the extent of the "treasure yet untouched" beneath the town, the mining companies had some jurisdiction, but over the other, the price of their product, they did not. By 1892 the price of silver, which had been above $1.10 an ounce during Tombstone's bonanza years before 1883, had slumped to eighty-five cents, forcing the *Tombstone Prospector* to admit that "the outlook is anything but flattering."[34]

CHAPTER FOUR

Employment by Mines Practically Nil

The stock market's plunge in October 1929 initially affected few residents of the Verde District. United Verde general manager Robert E. Tally apparently lost a not-so-small fortune, but to most residents the crash did not seem that important at first. As late as mid-February 1930, Tally assured the *Verde Copper News* that although United Verde had reduced production to 60 percent of capacity, wages and prices should hold, and there was no possibility of a shutdown. Tally believed that conditions would improve in the copper market in the late spring.[1]

True, there had been some bad news. In January a group of copper producers, including United Verde and uvx, announced a curtailment of production and a layoff of employees. Stocks of refined copper at refineries had more than doubled by the end of 1929, compared to the previous year. But United Verde went ahead with plans to sink a new shaft and build more company housing in Jerome and Clarkdale. The *Verde Copper News*, after reporting on unemployment and unrest all over the country at the end of March 1930, editorialized that it was "a distinct relief to live where such conditions do not exist." In the same issue the paper predicted that 1930 "should be a great mining year ... [with] a tremendous demand for the base metals." The editor also contended that the industry was healthy because it had been rationalized in recent years.[2]

"Rationalized" was a polite euphemism for a copper cartel called Copper Exporters, Inc., which attempted to fix the price of exported copper and, at least indirectly, domestic copper as well. The system had its flaws. The price-fixing by large copper companies excited competition from small mines and custom smelters, and though the cartel could set any price it desired, once the depression hit, buyers disappeared.[3]

The bubble burst in April 1930, when Copper Exporters, Inc., lowered the market price of copper—held at eighteen cents a pound for over five months after the stock market crash—to fourteen cents. In spite of a general production curtailment, the stockpile of refined copper held by the industry had grown alarmingly. Experts felt that the

Fig. 4.1. Boilermakers at Jerome's United Verde Mine, ca. 1930. Before the bottom fell out of the copper market, these skilled workers were among the highest-paid wage laborers in Arizona's history. (Arizona Historical Society, Tucson, #74074)

price reduction would "result in the rapid moving of unwieldy stocks into consumption," but that did not happen. By October 1930 national stocks of copper on hand, which had been 53,000 tons in April 1929, had risen to 365,000 tons. There was simply no alternative to further curtailment, and on 26 June 1930 United Verde laid off 825 men in Jerome and Clarkdale.[4]

Both Charles Clark, chairman of the board of United Verde, and James Douglas, president of uvx, assured Jerome residents in the second half of 1930 that there would be no further curtailment of production. But they could not speak for Jerome's third large mine, the Verde Central. That company became the first of the three to suspend operations completely. The high prices in the first quarter of the year had availed the Verde Central nothing, as there had been no sales. After April there were lower prices and no sales. The company stopped breaking new ore at the end of July, and on 30 September 1930 Robert H. Dickson, the Verde Central's general manager, announced that the company would suspend operations completely until the price of copper warranted resumption. That decision added another 100 men to Jerome's unemployed. One of these was John Muretic, who had been driving concentrates from the Verde Central to the railhead in Jerome. When Muretic sought employment at United Verde, he found himself standing in line with "hundreds of men who were trying to get back on the job."[5]

By the end of 1930 the situation in Jerome was becoming grim. In December city manager R. E. Moore reported to the *Verde Copper News* that there were only 250 men in Jerome absolutely unemployed, but the paper admitted that many men had only part-time work, had seen their wages reduced, or had "accepted work which is not in their line." While year-end reports showed that the United Verde was still the state's leading copper producer, the product of the state's five largest mines had lost more than half of its value from the preceding year, and output was at its lowest level since 1922. Although the United Verde

Extension continued to pay reduced dividends, the company, which had earned a net profit of almost $3 million in 1929, suffered a net loss of over $1.5 million in 1930.[6]

By October 1930 copper was bringing its lowest price of the twentieth century. uvx stockpiled 85 percent of its product in the second quarter of 1931—despite production curtailments—and mine employment in Jerome, which had been 1,983 in the second half of 1929, sank to 881 in the first half of 1931. By the first week in May 1931 the price of copper had dropped to nine cents, its lowest level since 1894, and there was no demand for the metal. National production had been drastically curtailed, but stocks continued to grow. Faced with these facts, United Verde gave up.[7]

On 8 May 1931 United Verde mine manager Val DeCamp announced that the company would completely discontinue copper production and would close its smelter in Clarkdale. In July the company announced the layoff of another 400 men and a 10–25 percent reduction in pay for salaried officials. The company would resume copper production when the market price of the metal rebounded, and it did not appear as if that was going to happen anytime soon.[8]

As if the United Verde's suspension were not bad enough, a month later the United Verde Extension announced its own plan to suspend operations for the third quarter of 1931. Due to the low price of copper, the company determined to lay off all but a few of its 600 employees for ninety days. There were those who felt that these suspensions were a necessary thing. The *Prescott Courier* editorialized in June that given that the world's copper stockpile had passed one billion pounds, "the mines have no alternative but to close down, just as they did in 1921, until this colossal stock of unused copper is absorbed. . . . The shut down will have a serious effect on Arizona, but in the long run it is possibly the best thing for everyone concerned." Such talk of the long-term good gave little solace to Jerome's unemployed miners.[9]

Several fortunate circumstances saved Jerome from a total shut-down of its mines. One of these was a rock slide into the United Verde's open pit. The company's engineers had been watching an unstable wall of the pit since a previous cave-in in June 1929. On 23 March 1931 it finally gave way, "enveloping the town in dust and making a noise which could be heard for some distance." The slide dumped around one-half million cubic yards of overburden into the pit. Although it caused a curtailment of smelter production in Clarkdale, removing six million cubic yards of overburden from the pit from this and subsequent slides kept a large number of men employed in Jerome for several years. The other good news for Jerome's miners was that the United Verde Extension could not afford to stop mining for long periods, due to a preexisting copper contract with the French government and to the deteriorating physical condition of its mine.[10]

Fig. 4.2. The excavation of United Verde's open-pit mine at Jerome, ca. 1930. Removing subsidence debris from the pit kept a number of miners employed during the worst part of the Great Depression. (Lewis McDonald Collection, Cline Library, Northern Arizona University, #435.7)

These circumstances kept about 650 men at work, but things were bad enough. The summer of 1931 saw the Morenci, Arizona, mines working a reduced schedule; the Old Dominion Mine in Globe, Arizona, shut down indefinitely; and the Magma Mine in Superior, Arizona, like the uvx, with almost its entire force of 700 men temporarily laid off. The *Verde Copper News* reported at the beginning of September 1931 that "a large surplus of skilled and unskilled miners and smeltermen [was] apparent" in the Verde District. Later that month, along with the other major mining companies in the state, United Verde posted a 9 percent wage reduction. Companies idled their smelters all or part of the year in Superior, Humbolt, Douglas, Clarkdale, and Clemenceau. In 1931 the average market price of copper was 8.3 cents, the lowest since 1850—and the price just kept falling.[11]

While it did not seem as if things could get much worse, what followed were the three leanest years Jerome had ever experienced. By the end of 1932 copper brought five cents per pound on the domestic market, and even at that price there was no demand. Large producers operated at a quarter of capacity by then, and because it cost them around seven cents per pound to produce the metal, there was no prospect for a revival of the industry.[12]

These circumstances fully exposed the economic vulnerability of Jerome and many of the state's other copper communities. Arizona's governor, George W. P. Hunt, wrote to the chairman of the Reconstruction Finance Corporation in the summer of 1932 that "many thousands" of copper miners were unemployed in his state, and that "these men cannot produce food to any extent in the mining districts . . . [and] there is no place where they can go to find other employment." The Magma Mine in Superior shut down again that summer, and reopened only for maintenance work in the fall. The large mines at Morenci and Ajo, Arizona, suspended at the same time, laying off 1,000 men altogether. In Jerome, the United Verde Extension reported that "some copper is being sold, but most of our production is being piled up at Clemenceau." The company nearly tripled its stockpile, and reported a net loss of more than $830,000 in 1932. U.S. copper production, almost 890,000 tons in 1929, had fallen every year since and was a mere 200,000 tons in 1932.[13]

The year 1933 brought a new administration to Washington, but the situation in Arizona mining towns did not improve much. The price of copper finally hit bottom at 4.75 cents per pound in January. Production, at last, began to fall into line with consumption, but 684,000 tons of stockpiled U.S. copper remained in search of a market. In August United Verde general manager Robert Tally announced that his company would hire more men and raise wages, but he pointed out that the company did so to combat unemployment in cooperation with the National Recovery Administration (NRA), not because there was any market for copper. He predicted that it would take at least a year to reduce the copper stockpile to a reasonable level at the rates of production and consumption then existing. Hard times remained for mining in Jerome and around Arizona throughout the year. Governor B. B. Moeur cabled NRA chief Hugh Johnson in December 1933: "employment by mines practically nil." Only 3,000 miners had jobs statewide, compared to the 14,500 miners employed in the latter half of 1929.[14]

In 1934 things finally began to turn around. Mines at Miami and Ajo rehired substantial numbers of men in January, but the big news came in April when the copper producers finally agreed upon an NRA copper code. The code did not attempt rigid production control or price-fixing, but the nine-cent price of Blue Eagle copper certainly did aid the industry. The situation was still serious, however. The NRA's director, Hugh Johnson, reported to President Roosevelt at the beginning of May that the industry could shut down completely and that stockpiled copper would meet all domestic needs for eighteen months. In October, United Verde general manager Robert Tally estimated that the industry was producing at only 23 percent of capacity. Nevertheless, miners started to return to work, and Jerome received some especially good news at the end of the year. In the first week of December United Verde announced that its smelter—idle for over three and one-half years—would reopen. Although this event promised only limited production and no immediate addition to the company's payroll of 700, the blowing in of one furnace on New Year's Eve must have seemed to Jerome's residents to portend bigger things.[15]

It did, and in a surprising direction. When William Andrews Clark died in 1925, the future of United Verde had devolved upon his son, William Jr., and grandsons, Charles and William III. Over the course of the next decade, all three of these men died, and the remaining heirs, unable to see a dawn at the end of the depression, decided to sell the property. Rumors circulated at the beginning of February 1935 that the Phelps Dodge Corporation, one of Arizona's largest copper companies, was gaining control of United Verde.[16]

Rumors became reality on 13 February when Lewis Cates, president of Phelps Dodge, assumed the presidency of United Verde. Phelps Dodge acquired all of United Verde's holdings. These included the Verde Central, absorbed by United Verde in 1931, and ninety million pounds of refined copper bars, covering three acres of ground at the smelter in Clarkdale. Upon taking over the property—christened the Phelps Dodge United Verde Branch—company officials announced that there would be no great changes, and operations continued much as before.[17]

Phelps Dodge chose a good time to reopen the United Verde. Mining activity began to pick up throughout the state. Arizona mineral production rose one-third in value in 1935, compared to 1934, with both the United Verde Branch and the uvx among the state's leading producers. Sadly, this revival did not come from a quickening domestic market for copper. It was caused by the Ethiopian crisis, the first of the war scares to wrack Europe in the second half of the 1930s.[18]

Conditions kept improving in 1936. Producers enjoyed a "brisk" demand for copper due to the European situation. By midyear the

United Verde Branch produced more than one hundred thousand tons of ore a month and employed over 800 men at improved wages. By the end of 1936, copper reached its highest price since 1930, at twelve cents, with production way up. The United Verde Branch was right in the thick of it, ranking second in the state in gold, silver, and copper production. Much the same pattern continued for the first few months of 1937, with higher prices, better wages, and, most encouragingly of all, reasonable stockpiles of the red metal.[19]

But just as the resurgent United Verde Branch began to restore vitality to the district, the United Verde Extension began to fade. Warnings of this newest crisis had been coming for some time. In 1930 UVX concluded an exploration drilling program that had yielded no values. In August 1934 a company representative reported that the company's

Fig. 4.3. The United Verde Extension's surface plant in the 1920s. The headframes of the mine's two shafts appear in the foreground, with the white Douglas mansion in the background to the southeast. (Sharlot Hall Museum Photo, Prescott, Arizona, #PB117, F3, album "Photographs, Jerome, Ariz.")

ore bodies would be worked out in nine months, and three months later Jerome's newspaper announced that uvx was having trouble producing a smelting mix from its diminished ore reserves. In 1935 the company began distributing assets to shareholders and bonuses to employees. In January 1937 the uvx smelter at Clemenceau shut down for the last time, laying off 125 men, and a month later it was offered for sale. The company suspended all operations in June 1938, after a long and illustrious career, and paid off almost all of its workers that summer. With uvx gone, the number of miners employed in Jerome dropped below 500 for the first time in memory at the end of 1938.[20]

The recession of 1938, coupled with the demise of uvx, put the Verde District back on the skids. That recession began for copper in the first half of 1937, after the price of the metal peaked at the end of March, and then began another decline. Consumers stocked up in the last half of 1936, and then domestic demand disappeared. The by-then familiar pattern of pay cuts, layoffs, and production curtailments began again at the end of May 1937. Joe Larson, an independent mining engineer, reported that November that "everything in Jerome is very quiet, more so than it has been for a long time." In December 1937 Larson ran into the Clarkdale smelter superintendent and asked if there was any truth to the rumor that the smelter would shut down again. The superintendent, retelling an old story, replied, "We haven't sold a pound of copper for months, and we can't keep on going forever without selling copper, and unless copper picks up we will have to shut down."[21]

By May 1938 the price of copper had plunged all the way back to nine cents, affecting all of Arizona mining. Although the Clarkdale smelter limped along with one furnace in blow, Larson concluded that "if this keeps up I guess we will all be on relief. The rumors are that the P.D. smelter may close too. If that happens this will be a ghost town."[22]

The year 1939 brought more of the same, the old copper camp barely showing signs of life. At the end of April, Larson predicted that things would not improve "unless Hitler and Mussolini kick over the traces." The dictators obliged four months later, stimulating record copper sales in the first week of September, but demand soon settled back down. A price of twelve and one-half cents at the end of the year hardly bespoke a buyers' panic. The district reached another mile-stone in its decline in 1939 when uvx closed out its affairs with a final distribution of assets and the demolition of the Clemenceau smelter. The Clemenceau Mining Company, capitalized at $250,000, replaced uvx, charged with settling that company's remaining liabilities.[23]

As the 1940s opened, Jerome's depression-bound residents could look forward only to another world war. While some might have self-ishly hoped for another war boom, nobody could be certain what lay ahead, though it seemed clear that the future could not bring travails much worse than those of the preceding decade.

CHAPTER FIVE

The Drawbacks Incident to Mining

In his 1883 report to Tombstone Mill and Mining Company stock-holders, company president George Burnham wrote that "in no year in the history of our Company has the management been required to deal with so many of the drawbacks incident to mining for the pre-cious metals." Whether beset by a flood of subterranean water or a flood of surplus copper, the major mining companies in Tombstone and Jerome could no longer turn a profit. That simple fact had many ramifications for the towns of Tombstone and Jerome and for their mining companies. The bust of these mineral districts produced two paradoxical trends in mine ownership and operation: the consolida-tion of ownership into fewer and larger holdings and the return to prominence of small operators who leased parts of larger properties or worked their own small claims.[1]

One has to be careful not to commit the post hoc fallacy when dis-cussing cause and effect in busttowns. One would be at some hazard, for example, arguing that bust caused trees, though there were cer-tainly more of them in Tombstone after 1882 than before. As any mining town matured it underwent recognizable patterns of change. After a serious fire or two, residents stopped building in canvas and wood and started building in brick or adobe. The percentage of women and children in the population increased as a mining town matured. Consolidation of ownership was another normal occurrence in most mining camps.

Consolidation occurred naturally as a mining district aged because deep mining required the technical expertise, manpower, and capital investment that a corporation could best provide. Consolidation occurred during bust because smaller operations, including smaller corporations, lacked the financial resources to weather low prices or other difficulties. This was true of other industries besides mining. Hard times wrought such amalgamations in agriculture and lumbering as surely as they did in mining.[2]

Companies unified through mergers or purchases or as the only alternative to pyrrhic litigation. Another less known but often practiced technique was to acquire an impoverished property at a tax sale for a fraction of its value. Mining men in both Tombstone and Jerome used this method to acquire properties from which they thought they could profit.[3]

There were occasions, like the Bank of California syndicate on the Comstock Lode in the late 1860s, when combination during bust was part of some grand design. But interest in such mergers waned in Tombstone as the bust wore on. Although the newspapers noted trends toward consolidation in 1883 and 1886 and the *Engineering and Mining Journal* reported the Grand Central buying up smaller claims as late as 1888, bickering over who should pay for pumping, rather than bargaining toward a great conglomeration, dominated the last half of the 1880s in Tombstone.[4]

Past a certain point in the decline of a camp, the slight chance of developing a new bonanza no longer justified the cost of consolidation. Sales of mining claims in Tombstone, which had numbered more than 800 in the boom year of 1881, fell to fewer than 150 in 1883, and from there began a slow decline. There were 73 sales in the district as late as 1889, but after the price of silver collapsed in the 1890s, with little possibility of recovery, interest in Tombstone's mineral properties practically disappeared. Only 2 claim sales were registered in the depression year of 1894 and never more than a handful in any other year in the 1890s. In 1901, after the price of silver had recovered somewhat, a group called the Development Corporation of America (DCA) consolidated the principal companies in Tombstone, centralized pumping, and resumed deep mining. But the DCA did not purchase mining claims in a busted camp during a depression in the hope of being able to work them someday. The company's investors made their play during a recovery, when the immediate possibility of working their properties existed.[5]

This is in contrast to what happened in Jerome, where the depression brought opportunities to consolidate. The Verde Central, which suspended operations in 1930, received the first such attention. Although United Verde itself would suspend operations shortly thereafter, its managers saw the opportunity to extend their holdings by purchasing the Verde Central. They entered into negotiations for the property at the beginning of 1931. Verde Central management informed stockholders in March that they could not develop the mine profitably, and that United Verde had offered $250,000 for the company's assets. To the stockholders with $90,000 in cash but also $200,000 in debt and no prospects, United Verde's bid offered a clean way out. They approved the deal, and the Verde Central deeded all of its holdings to United Verde in May 1931.[6]

United Verde's turn came less than four years later. When the Clark family decided to sell they had a number of suitors—including the American Smelting and Refining Company, known as ASARCO, and James Douglas of the United Verde Extension—but Phelps Dodge gained control for $20.8 million. Some people in the industry regarded that price as excessive given the state of the copper industry in 1935, but over the next eighteen years Phelps Dodge proved the wisdom of its purchase. The company repaid the $5 million loan it had needed to buy the property in a little over two years. By the time the ores played out in 1953, Phelps Dodge had doubled its investment in United Verde, clearing a profit of $40 million.[7]

Other companies formed combinations in Arizona during the Great Depression. Kennecott Copper bought the Ray mines shortly after they shut down in 1933 and reopened them to a long and productive life in 1937. But no company was as successful in consolidating holdings as Phelps Dodge, which had done the same thing during the depression of 1921. Under the aggressive leadership of Lewis S. Cates, who became president in 1930, Phelps Dodge purchased the National Electric Products Corp.—makers of wire, cable, and electrical fixtures—in 1930, and the mineral holdings of the Calumet and Arizona in Bisbee and the New Cornelia in Ajo in 1931. United Verde was just one of a series of acquisitions that made Phelps Dodge "the most powerful corporation in Arizona and the third largest copper producer in the United States" by 1940.[8]

Guided by long-range corporate goals and with capable leadership, Phelps Dodge used the Depression to purchase United Verde and other holdings and achieved spectacular success. The Bank of California syndicate on the Comstock Lode is another example of a consolidation during bust that produced a legendary fortune. But Charles Shinn, chronicler of the Comstock, understood the fine line between glory and disaster. "Although borrasca had put them into possession," Shinn wrote of that syndicate's investors, "a few more years of borrasca would utterly smash their fortunes." Consolidation of a district in bust might be considered the ultimate mining speculation. Though the rewards could be stupendous, such ventures ended much more frequently in failure than in fortune. The Development Corporation of America spent almost a decade making a heroic effort to move the water out of Tombstone's mines, but in the end achieved only its own bankruptcy.[9]

The closure or curtailment of the large operations at least partially returned mining to small operators. The growing scale of mining in bonanza districts generally forced out small operators or turned them into employees, but as a district faded, something like a reversal took place. As the bust wore on after the strike of 1884, Tombstone's news-papers made frequent reference to chloriding operations worked by small groups of men who leased the underground holdings or reworked the dumps or tailings of larger companies.[10]

Although one mining historian describes *chlorider* as "not exactly a complementary term"—analogous to *scavenger*—in busted Tomb-stone these men had higher status. The Tombstone Mill and Mining Company resorted to a tributing system as early as 1883, whereby con-tract miners paid a one-third royalty to work selected pieces of the company's property. Although the Tombstone Company's superin-tendent saw little merit in the system then, by 1886 the *Arizona Daily Star* reported that "chloriding has become a permanent industry of the camp," and the *Daily Tombstone* was singing the praises of the "hardy and persevering chlorider." "Our hard times," that paper reported, "have created these chloriders and helped them along wonderfully." The *Daily Star* reported in March 1886 that "the hills are swarming with chloriders, and they are all making money."[11]

One need not believe that last statement to realize how important chloriders became to the Tombstone District. By the 1890s, chloriders and other independent operators were the life of the camp. One wit-ness reported more than three hundred men working in the mines of Tombstone in April 1891, "principally chloriding." The following year the Grand Central leased claims to chloriders for a 20 percent royalty. By the end of 1892, chloriding constituted most of the activity mentioned in accounts of mining in Tombstone.[12]

Gold placering made a comeback in Yavapai County and through-out the West during the Great Depression, as miners and many others without jobs tried to scratch out a living working the creeks. The *Prescott*

Courier predicted that the spring of 1931 would see "a real gold excitement, for almost daily reports come from properties which are being
inspected and old properties which are reviving and producing." Stores
reported increased sales of gold-panning equipment, and banks and
businesses reported greater numbers of people exchanging gold. "A
tip to the man hard up right now," the *Courier* advised its readers in
April 1931, ". . . is to get to the hills."[13]

The trend continued throughout the depression. The *Prescott Courier*
noted in 1932 that "the gold panners are not getting rich but they are
making their bacon and beans, which are something these days." The
paper reported whole families engaged in placer mining. U.S. Forest
Service officials in Colorado, Wyoming, California, and Arizona found
themselves inundated by gold seekers in search of subsistence. The
Verde Copper News reported in January 1935 that miners in Arizona had
produced more than 145,000 ounces of placer gold, valued at over
five million dollars in 1934, which made the state one of the leading
gold producers in the nation. The paper could only guess that "thousands" of people were working Arizona's placer sites. In 1937 Yavapai
County contained over two hundred active placer operations. As late as
October 1938, one correspondent to the *Courier* claimed that more than
one thousand people were placer mining in Yavapai County, and that
placer mining was "the most directly useful industry in the county."[14]

Small lode mines made a similar comeback. In 1929, 39 gold mines
operated in the state. By the end of 1934 that number had risen to
150, with renewed prosperity among silver mines as well. At the beginning of 1938, 275 lode mines operated in Arizona, most of them "individual leases employing from two to ten men and producing only a
comparatively few tons of ore, principally gold and silver."[15]

By February 1938 small lode mines had made enough of a comeback
that their operators saw fit to organize the Arizona Small Mine Operators' Association. That organization sought state and federal legisla-
tion favorable to small operators, as did similar organizations in other

western states. Within a year the Arizona association had 2,500 members, and councils in forty-eight towns, including Jerome. Interested
parties organized a Yavapai County council in April 1938.[16]

State departments of mines assisted the return of the little guy
during the Great Depression by offering classes on gold panning and
the associated arts. The Arizona Bureau of Mines offered a series of
placering classes in Prescott and Phoenix and issued the publication
Arizona Gold Placers and Placering in 1932, "in order that residents
who would rather seek gold than remain idle may be equipped with the
latest information." The federal government contributed to the gold
and silver fever by raising the prices of those metals.[17]

Independent miners targeted other mined commodities besides
gold and silver during bust. Contract mining returned in eastern coal
communities during the Great Depression, as did the leasing or selling
of coal properties to small operators. Another practice called bootleg
mining developed, in which a few unemployed miners dug out exposed
small and shallow seams that companies could not mine profitably.
At first miners used this coal for fuel or barter; later they loaded it
onto trucks, took it into the nearest city, and sold it door to door. One
authority estimates that by 1933 bootleg coal mining employed twenty
thousand men in a thirty-five million–dollar illegal industry. Coal
companies did their best to stop this piracy of their deposits, usually
without much success.[18]

Copper mining remained a playground for the big boys, even during bust. Most of the industrial metals simply required too much
capital investment and technological sophistication to be worked by
small operators. In 1935 the six largest copper mines in Arizona,
which included the United Verde and the uvx, produced 96 percent of
the state's copper. Figures from other years in the decade are roughly
the same. Jerome residents who wished to start a small operation
during the Depression had to go elsewhere in the county. The record
of small operations in Yavapai County kept by the secretary of the

Yavapai County Chamber of Commerce contains no listings for the Verde District.[19]

When a large copper property finally played out, it might revert to small operators, however. At the end of 1939 Joe Larson recorded that lessees were reworking the UVX dump under royalty, moving three carloads of ore a week to the Clarkdale smelter. After Phelps Dodge closed the United Verde in 1953, the company leased the open pit to the Big Hole Mining Company, which removed small pockets of ore for the next twenty years, never employing as many as ten men.[20]

Whether the return of the small operator was a good thing depended upon whether one preferred independence or regular pay. Although newspapers in both eras boosted the small operations with enthusiasm, these were clearly not the preferred modes of subsistence during the boom days, and certainly both towns prospered less with the big concerns shut down. In a moment of unusual candor the *Tombstone Prospector* admitted that demonetization of silver would find "leasers working on chances as their only way of working at all." The Works Progress Administration (WPA) guide for Arizona, published in 1940, reported that "lode gold mining is still a minor industry in Arizona, and placer mining is no longer considered even mildly profitable by most miners."[21]

In 1935 Arizona's governor, B. B. Moeur, warned an Illinois man writing to ask about the possibility of panning gold in Arizona to stay home. The governor replied that while many people made a living panning for gold, many others had been forced to return to relief because of a lack of equipment or food. Placering in Depression-era Arizona produced something like one dollar a day—better than nothing, but hardly comparable to the five-dollar day enjoyed by the state's miners during the boom days. Prescott resident Sharlot Hall concluded that "sometimes they really do pan out a few cents—or once in a while get a dollar or more—but the old diggings are very lean of gold—having been worked over and over all these years."[22]

In each of these cases special circumstances nurtured the small operations that sprouted during bust. In frontier Tombstone they were rooted in the extravagance of the boom days. Chloriders and contract miners worked the lower-grade ores dumped or left untapped in the race to process high-grade ores during the boom period. In the Great Depression massive deflation and crippling unemployment certainly made gold placering attractive, but small operators of that era also benefited greatly from artificial price supports by the federal government. A proof of this lies in the number of mines in Yavapai County during various Depression years. In 1929 the county had 81 operating mines, and though that figure more than doubled to 188 mines in 1932, the proliferation of mines really began after 1933. In that year the county had 191 mines, but the figure rose to 502 in 1934, and 885 in 1935. In the years following the panic of 1907 and in the postwar depression of the early 1920s, the number of operating mines in the county decreased. The difference between the eras was the federal government's price supports for gold and silver during the New Deal, which subsidized the greatly increased number of operating mines in the county after 1934.[23]

Several mining historians argue that the return of the small owner or lessee was simply part of the life cycle of a typical camp, that the final years of decline saw the small operators scavenging the remains. Duane Smith notes in his study of Caribou, Colorado, that small operations were often the death knell of a camp because with weak financial backing, they had neither the resources nor the inclination to do the sort of exploration or development work that might produce a new bonanza. The diminishing number of locations in the Verde District during the Great Depression suggests that exploration and location are the products of at least moderate prosperity. The *Arizona Weekly Citizen* reported in 1890 that a recent plunge in the price of silver "seems to have knocked the enthuse out of our local prospectors and claim jumpers." This study confirms that the return of the small operator, though lauded by newspapers with nothing better to discuss, often marked the beginning of the end for a mining camp.[24]

Bust and its attendant layoffs and wage reductions did not necessarily produce labor agitation or labor-management conflict. The Tombstone standoff—Arizona's first important labor action—was an unusual event in nineteenth-century western mining. Miners in those days usually expressed their dissatisfaction with wages or working conditions by moving on.[25]

Tombstone's union initially received general support from the town's residents. When four hundred miners rallied on 5 May 1884, the proceedings included a brass band and sympathetic speeches by several leading citizens. Local merchants enlivened the event by contributing three barrels of beer and one thousand cigars. Some citizens busied themselves raising subscriptions for the union. Newspaper editorials in Tombstone and Tucson generally sympathized with the miners' cause or at least stayed neutral. The *Tombstone Epitaph* managed to hold on to its neutrality for at least three weeks into the strike.[26]

But as local business conditions worsened, public support for the miners began to fade. The companies could withstand the shutdown; the community could not. The *Arizona Daily Star*, though unsympathetic to the union, painted an objective picture of the situation in Tombstone. "The miners at first had the undivided sympathy of the community," the paper's correspondent reported, "but as the lock-out continues and impending financial ruin begins to cast its shadow over our businessmen, a change of sentiment is noticeable; and should the mine owners import $3 men with the intention of resuming work, self-interest would doubtless induce the business element of the place to support and countenance the movement, regardless of what the Miners' Union might wish or of what action it might take in the matter."[27]

Although Tombstone's union received assistance from unions as far away as Virginia City, Nevada, and Bodie, California, as well as nearby Bisbee, that aid counted for less and less as its local support evaporated. At the end of June a Tombstone correspondent to the *Daily Star* "found the 'Miners Union' standing alone without either the respect, financial or moral support of the good people of Tombstone." At this same time the *Epitaph* climbed down off the fence with an editorial calling for a rational resignation to the realities of Tombstone's situation. "The Tombstone merchants are growing weary, the miners are getting tired of fasting, and the mine-owners are enjoying themselves at the summer watering places," the *Epitaph*'s editor wrote. He concluded that "there is no denying the fact that there is not one in ten of us who can last a month longer," and urged miners to return to work in the hope that renewed prosperity would restore the four-dollar day.[28]

Driven to desperation by their situation, union miners resorted to intimidation and eventually violence against managers, replacement miners, and property—techniques that backfired utterly. As the *Tombstone Epitaph* stated, the attack on the Grand Central "removed every particle of sympathy which this community heretofore held for the Miners' Union," and the newspaper, originally neutral, tore into the union with fang and claw. The companies took advantage of the union's lawlessness to issue the ultimatums that destroyed that organization and scattered its membership. A union miner named Sullivan was right when he complained at the beginning of August that most residents of Tombstone regarded the union men as outlaws rather than citizens.[29]

The progression of Tombstone's labor struggle bears considerable similarity to other nineteenth-century western mine labor disputes, like those in the White Pine District of Nevada; in Silver Reef, Utah; on the Comstock Lode; and in Goldfield, Nevada, after the turn of the century. The labor dispute in Idaho's Wood River District in 1885 progressed almost exactly as Tombstone's had the year before. In all of these cases mines that had run upon hard times tried to lower costs by reducing wages; in each case miners organized to combat the reduction; and in each case the miners' union eventually lost public support when members turned to violence or was starved out of existence when the mining companies would not or could not reopen.[30]

In all of these cases the unions fought a hopeless battle from the start. Companies then had no legal obligation to recognize or treat with unions, and miners found it difficult to bring public pressure to bear against management because of popular attitudes toward the rightful place of unions. Tombstone's union men heard again and again that they had the right to organize a union and to refuse to work for three dollars a day, but that they did not have the right to interfere with those willing to work for that wage. A miner who signed himself "Chlorider" in a letter to the *Tombstone Epitaph* on 6 July 1884 expressed sentiments that the union faced in many of Tombstone's residents. "A union is, no doubt, a good thing, when gotten up for certain purposes and conducted properly," Chlorider wrote, "but when two or three hundred men join together for the purpose of stopping the wheels of industry and ruining the best camp on the coast, which this union has done, then it is no union, but in plain English, a mob."[31]

Unions, in the prevailing opinions of the nineteenth century, should be social, fraternal, and humanitarian in nature; they should not attempt to tell companies how to run their businesses or tell citizens whom they should support. Newspapers and citizens alike, though sympathetic to the miners' plight, were primarily interested in the welfare of their communities. Normally, that welfare was best served by high wages, but as a strike or lockout wore on, it was best served by any wages at all. Other things besides corporate power and attitudes about labor organizations worked against the strength of unionism in a busted camp. The return of small operations gave many a chance to work their own claims, other boom camps beckoned, and miners in a depressed or depleted mining community had very little with which to bargain.[32]

In contrast to Tombstone, Jerome experienced very little labor activity, and no known labor disputes during the Great Depression before Phelps Dodge took over United Verde in 1935. This absence is particularly surprising in light of the tumultuous, often violent labor conflicts around the country in the early years of the Depression, including those in other copper towns such as Butte, Montana, and Bisbee, Arizona. Over one hundred thousand workers struck nation-wide in the summer of 1934, and there were strike-related deaths in Alabama, Texas, South Carolina, Georgia, and West Virginia. That same summer witnessed the formation of Jerome Miners' Union no. 37 of the International Union of Mine, Mill, and Smelter Workers of the American Federation of Labor (AFL), but that local undertook no labor action in the Verde District. Perhaps this is mute testimony to the suffocating paternalism of Jerome's copper companies, but other communities, including Bisbee, had the same sort of corporate control and still had labor-management confrontations.[33]

Several factors probably contributed to the lack of labor activity in Jerome. The breadth and depth of the Great Depression left labor little with which to bargain. Unlike the frontier days, no new bonanzas drew off surplus laborers, and with the market price of copper below its cost of production those who remained employed knew they were lucky to be working. Any miner inclined to raise a fuss realized that a long line of men waited for his job. Further, this disaster was self-evidently not the companies' doing, which reduced frustration with the companies, if not the situation. Union organizers divided among themselves over ideology and tactics, and neither craft nor industrial unionism really suited a mining complex like Jerome. The craftsmen represented by the Verde Valley Federation of Labor were but a small minority amid an army of miners, muckers, and laborers. That AFL organization was simply not willing or able to represent unskilled laborers, and the craftsmen saw little advantage to joining semiskilled and unskilled laborers in an industrial union.[34]

Jerome also had a language problem. Most of the craftsmen were white, but by the 1930s most of the miners and laborers were Latino.

None of the officials elected to run the Verde Valley Federation in 1933 were Latino. The *Verde Copper News* reported in the spring of 1934 that "at first several Mexicans were members of the Verde Valley Federation, but everything had to be interpreted for them and it made the meetings so slow that they discontinued to come." It is not surprising, in view of all of these problems, that organized labor lacked significant strength in the Verde District during the Great Depression.[35]

The companies' previous good treatment of their workers might have been the most important factor in Jerome's labor peace. Both of Jerome's principal mining companies were owned by individuals who exercised personal authority over their operations. United Verde, and to a lesser extent uvx, had spent the 1920s paying competitive wages; aggressively building recreational facilities, hospitals, and residences for their workers; and engaging in other benevolent activities in the community. Those family connections to the business and the workforce continued during the Depression, as did good relations between labor and management.

Both companies undertook development projects to keep their workers employed during the Depression, and United Verde put some workers on staggered part-time employment in order to rehire others. When uvx reopened after its shutdown in the third quarter of 1931, the company did so partly "for the purpose of relieving unemployment among our employees during the winter months." James Douglas warned stockholders in the company's quarterly report that "it is not our intention, however, to sell copper at the present price and our cash reserves will consequently shrink." Robert Tally reminded Jerome residents in August 1933 that although United Verde had not sold any copper in over two years, it had spent twelve million dollars in the district in the previous three years.[36]

Certainly, the companies exploited a certain propaganda value from their actions, but their workers understood the situation, too. Herbert Young, a former company official, recalled later that the Clarks had a reputation within the industry for being generous to a fault with their workers. John Muretic, who helped to excavate the United Verde's new shaft, thought that the company undertook the job more to keep key personnel working than because it was necessary. When mine and smelter committees from United Verde petitioned President Roosevelt in quest of an increased copper tariff in September 1933, they acknowledged that since May 1931, the company had employed an average of 650 men, but "has not produced a pound of copper since the above date and much of the work we have been given to do has been for the mere purpose of keeping us and our families from want."[37]

Circumstances changed, however, after the Clark family sold United Verde to Phelps Dodge in 1935. Tensions between management and labor increased, and union organizers began to work more actively in the district. Again, this was partly a national trend. Labor unions had been organizing and expanding since the right of employees to bargain collectively was sanctified in Section 7a of the National Industrial Recovery Act of 1933 and reaffirmed in the Wagner Act of 1935. As the Depression dragged on, workers' dissatisfaction with conditions manifested itself in the growth of the Congress of Industrial Organizations (CIO), which advocated industrial unionism and broke away from the more moderate American Federation of Labor after 1935. CIO organizing became increasingly aggressive, and a wave of labor actions took place all over the country in the second half of the 1930s. In Arizona these included contentious strikes at Bisbee and Oatman in 1935 and CIO efforts to organize the Miami District in the spring of 1937. By 1940 the CIO was vigorously organizing in most of Arizona's mining and smelting towns. These external conditions were bound to influence labor relations in Jerome.[38]

But local conditions did enter the equation. After the Clarks sold out, the United Verde became a small part of a large enterprise, and many in the Verde District sensed that management had become more remote and impersonal after the Phelps Dodge takeover. With UVX gone, the United Verde Branch became Jerome's only major employer, and by 1938 it found itself confronted by union organizers and the newly created National Labor Relations Board (NLRB). The NLRB ordered an election among Phelps Dodge employees in Jerome and Clarkdale that spring to determine who would represent the company's machinists, boilermakers, electricians, and carpenters. The board excluded two company unions from the ballot, and the AFL won the right to represent the aforementioned trades. Within four months the AFL filed a complaint with the board, charging Phelps Dodge with violating the Wagner Act by failing to bargain in good faith with the new Jerome and Clarkdale unions.[39]

Mining entrepreneur Joe Larson believed that Phelps Dodge management would do whatever it could to retard the labor organization. He wrote to a friend in April 1939 that "it is the opinion of [a] good many people that the P.D. is just using these hard times for an excuse to get rid of their so-called undesirable employees, that is, those that uphold organized labor. The P.D. has no use for organized labor." Larson was convinced that Phelps Dodge had undertaken a series of tactics, including curtailments, to keep the CIO from organizing the district.[40]

Phelps Dodge began a public relations counteroffensive at the end of 1938 with an open house and barbecue for one thousand civic and business leaders. At the gathering company officials reminded their guests that the company was the largest employer in Yavapai County, with a monthly payroll of more than $160,000. But the company's campaign achieved limited results because Phelps Dodge and its United Verde Branch were only part of a national trend. By 1940 labor unions had organized in mining towns all over the state, though one can argue that the increased labor activity resulted from returning prosperity, or was at least independent of the bust.[41]

Phelps Dodge officials did have a point. Even in places like Tombstone, which were not company towns, mining companies contributed to local welfare by the simple act of putting money in circulation. Much has been written about all of the money that departed from these towns in the pockets of the bonanza kings, but many mining towns, kept alive for years by outside capital, never repaid their investors. "Funds from London" permitted one Tombstone mine to reopen after a three-month shutdown in 1888. By 1889 G. C. Willis had sunk around $56,000 into his Tombstone mine and had yet "to ascertain whether the mine has any merit or not." It is quite probable that neither the London investors nor Willis ever saw a return on their investment, but it was their money that gave life to Tombstone.[42]

A quasi company town like Jerome owed the company in more ways than its residents might care to admit. When mining companies closed down, these communities lost not only payrolls, but also leadership, organization, and expertise. What might be called positive paternalism disappeared along with the negative. When the United Verde Branch finally shut down in 1953, so did the town's hospital, and the twenty-six thousand volumes in Jerome's library, funded by the company, were boxed and stored. The municipal government no longer had the management capabilities of company officials elected to municipal offices. United Verde and UVX helped churches and community organizations, and even aided farmers by loaning them tools or giving them shop assistance. The companies also loaned equipment to New Deal projects. Once mining companies ceased to operate, all of that assistance disappeared, and leadership in these busted towns fell to members of the business community.[43]

CHAPTER SIX

Owing to the Dullness of Trade

Ten days after Tombstone's mines shut down, the town's economy had a heart attack. George Parsons slept in on 10 May 1884 because he had stood guard duty into the early morning hours. Upon going uptown later that morning he discovered "intense excitement." The Hudson and Company Bank had suspended operations, and its cashier, his friend Milton Clapp, had departed for Tucson. Parsons, whose own small savings had been swallowed by the bank, spent the next few days trying to salvage the reputation of Clapp—whom most people felt had absconded with the money—and defending Clapp's wife and property against ugly abuse and threats. Rumors circulated that either the bank or Clapp's residence would be put to the torch, and both were placed under guard. Clapp's supporters soon decided that it would be prudent for Mrs. Clapp to leave town. On 13 May, Parsons drove her to meet the train at Fairbank.[1]

The Safford and Hudson Bank, with its headquarters in Tucson, had a branch office in Tombstone by 1880. By 1884 the bank had been reorganized as Hudson and Company, a partnership of Charles Hudson and Col. J. H. Toole. Upon arriving in Tucson, Clapp told the press that in the previous month miners leaving Tombstone because of the labor dispute had withdrawn $100,000 in cash from the Tombstone branch, causing the company's collapse. Clapp and his employers blamed the collapse on insufficient warning of the impending shutdown from the mining companies.[2]

Those not inclined to trust the bank's officials blamed the suspension on bad speculations by the parent bank in Tucson, which owed $71,000 to the Tombstone branch. A committee from Tombstone sent a representative to Tucson to perform a postmortem on the company's books. He discovered an excess of liabilities over assets of more than $140,000 and a host of overdrafts and irregularities. One leading resident of Tombstone expressed the opinion of many when he said he "believed the failure was a d—d swindle and that the depositors had simply been robbed." Hudson and Company was probably a financial house of cards—like many nineteenth-century banks—obliterated in the maelstrom of Tombstone's labor dispute. Whatever the truth, the bank was gone, and the district had suffered a serious blow. Seventy-seven depositors lost over $500 each, and by the end of the decade the company's creditors had recovered less than 30 percent of their investment.[3]

The mines' lockout produced immediate and drastic effects on Tombstone's economic community, with reports of depressed business even before the bank failure. A few days after the bank suspended, a witness claimed that business in Tombstone had fallen "to the ground paralyzed." Within the month another reported that "complete stagnation prevails in all branches of business, and if the present situation continues . . . it will inevitably end in general bankruptcy." Economic conditions did stabilize somewhat with the return of mining after the labor dispute, but the economy never regained its prelockout vitality. Although Tombstone saw periodic minor revivals, its economy continued to slide; by the 1890s the town was described as "woefully depressed."[4]

Jerome's economy experienced something more like a debilitating disease throughout the 1930s. Although one of the town's banks—the short-lived Jerome branch of the Bank of Clemenceau—closed at the end of 1933, it did so without loss to its customers. As in the mining industry, news of the stock market crash caused no panic in the ranks of Jerome's economic community. Even at the end of March 1930 the manager of the Jerome branch of the Bank of Arizona reported that business conditions "might be classified as good . . . [and] I anticipate a very satisfactory summer." This was not just brave talk. As noted before, the market price of copper held steady into April 1930. But after that came the succession of disasters to Arizona's copper industry previously recounted. Although the town's economy rallied briefly in 1933, 1935, and again in 1937, the 1930s largely witnessed decline, unemployment, and business failure in Jerome.[5]

The signal feature of a boomtown economy is an abundance of cash, credit, and easy confidence—money for investment, speculation, and consumption. "You would doubtless be amazed to see the extravagance and display of this frontier town," a Tombstone woman wrote to friends back home at the height of the boom. "A person will not look as long at ten dollars here before spending it as he would at one dollar in the East." Mining engineer John Rockfellow recalled of 1881 Tombstone that "real estate changes hands at fancy prices. Money is plentiful and everything goes."[6]

The reverse holds true in a busttown economy. Capital becomes a scarce commodity, which produces serious deflation. George Parsons, collecting rents in Tombstone in 1886, found it "like pulling hen's teeth to get any money out of people [in] these times." Attorney Perry Ling made the same discovery forty-five years later in Jerome. When he tried to sell a friend's property, he found that nobody wanted to put up any cash, though many were interested in the property "if they could pay next year or some indefinite time in the future when they were sure where they stood."[7]

Price and wage declines, as mine managers liked to remind their employees, occurred naturally as a mining district matured. Wages in the Mother Lode country of California, $20 a day in 1848, dropped to $3 a day by 1859. As transportation improved and laborers flooded into an area, prices began to fall into line with those of established communities. But when bust deflation hit, prices could plunge fantastically. Charles Shinn reported one mill on the busted Comstock Lode, which originally cost $200,000, offered in vain for $5,000, and another, which cost $60,000, eventually sold for $3,000. Mine wages dropped 25 percent in Tombstone and even further in Jerome. Room and board in Tombstone fell by the same 25 percent, as did the price of theater tickets, which visiting dramatic companies reduced in the 1890s in an effort to stir up some business. Sometimes no demand existed at all. After Harry Parker had gone nineteen days without a lodger at his Jerome Hotel in 1938, he gave up, closed the hotel, and sold off the furniture.[8]

Particularly severe deflation struck real estate prices, which all but collapsed in both communities. The tents pressed into service in Tombstone as late as 1881 disappeared, as did the fancy prices for buildings and lots. Town lots had brought between $150 and $400 in 1879, and the following year small two-room houses rented for $25 to $40 a month. An 1881 visitor to Tombstone reported that a desirable commercial lot on Allen Street brought $6,000, and "a shanty that cost $50 to build rented for $15 a month."[9]

How sharply conditions changed. Only two months into the Tombstone labor dispute, a witness reported that real estate had little value and that landlords had voluntarily reduced rents by as much as one-third. In July 1889 a lot containing a furnished five-room house and two cabins could be purchased for $200, and a three-room house with a kitchen could be rented for $8 per month. When Mrs. Vogel tried to sell her two houses and lots—complete with fruit trees and gardens—in 1892, she did not even bother to name a price, saying only that no reasonable offer would be refused.[10]

Jerome experienced much the same thing. By 1934 the town's Home

Owners' Loan Corporation administrator reported many vacancies in apartments and detached dwellings and noted that properties were renting for half of their fair value. Joe Larson wrote to his friend Homer Nihell in California in 1936 that "it looks like all real estate in Jerome is for sale, that includes my house too." He went on to tell Nihell, for whom he acted as agent and rent collector, that "renting [out] properties here in Jerome is almost impossible at the present time." Nihell owned two properties in Jerome: a house that originally cost him $1,900, and a store on Hull Avenue that cost him $2,250 and that he had rented out during the war boom for $75 a month. In April 1937 the store rented for $15 a month, with the rent unpaid since the previous December. Larson had been unable to sell the store, and the best offer he had received for the house was $400. Even the commercial properties that sold did not bring anything near their pre-bust value. Nihell eventually instructed Larson to rent his business property for whatever he could. Something was better than nothing.[11]

Deflation robbed property of its value and destroyed all confidence and credit. Some merchants in busted towns tried to keep customers on credit, but soon abandoned the practice when it became apparent that this was suicidal behavior. In September 1885 Summerfield Brothers of Tombstone—dealers in clothing and general and fancy goods—announced a cash-only policy "owing to the dull times." Credit would be "positively refused in all cases, there being no distinction." Other Tombstone merchants soon followed Summerfield's lead. In the 1930s receipts of the Kovacovich Mercantile Company of Jerome bore across the bottom the legend: "We Sell for Cash Only."[12]

In the most extreme cases, a busted district returned almost wholly to a barter economy. In the spring of 1931 the *Prescott Courier* informed its readers of "larger and larger quantities of flake and nugget gold being brought to town in midget bottles for converting into grub or cash," and reported county storekeepers bartering merchandise for nuggets. A local banker said that his establishment had been buying small amounts of gold just to keep money in circulation. The following year Perry Ling, unable to find a buyer for his friend's property, did manage to find someone to move into it for a month in exchange for work done on the place. Ling also had a cabinetmaker improve his own property in lieu of several months' rent. Harry Amster of the New State Motor Company in Jerome wrote that the firm lost considerable business to workers doing their own auto repairs. Presumably, those skilled in the mechanical arts could swap their talents for other goods or services. Bartering was not confined to busted mining camps; merchants and professionals in depressed farming communities traded merchandise or services for farm products.[13]

Deflation worked to the advantage of some individuals. Buyers of almost everything suddenly found the market most agreeable. The *Tombstone Prospector* reported much real estate changing hands in 1888, with former renters buying homes at low prices. Even in a busttown speculators could not resist temptation. The paper predicted in 1891 that "those who are quietly picking up real estate around Tombstone will reap a rich harvest before many months." Joe Larson purchased the Hayes Apartments and the Haskins house at the end of 1941 because he thought that Jerome still had some life left in it and that housing would be in demand again.[14]

Those locked into their leases could not see any justice in continuing to make their former payments in the face of these new real estate values. Harry Amster petitioned his landlady in 1935 for a reduction in rent on the New State Motor Company's garage building from $100 down to $65 a month. Amster argued that local conditions justified the request and added that he and his partner paid three to four times the rent of any other garage in town. Joe Larson wrote to Homer Nihell at the end of 1938 that "all the business men in town are . . . kicking about paying too much rent."[15]

The absence of cash wended its way through the economic system, inflicting casualties as it went. Merchants could neither collect debts, cover expenses, nor gain the credit necessary to carry themselves through the lean days. Perry Ling, one of Jerome's two lawyers, spent 1930 trying to collect unpaid bills owed the Liberty Garage and the Popular Store. He spent the following year defending those two firms in bankruptcy proceedings. A solid enterprise slowly eroded away, until only a shell remained. Gottlieb Fischer had been in business in Jerome since 1903. After he died in 1939, Joe Larson recorded that "everybody thought he had money; but instead he was broke and had not paid his taxes."[16]

Some types of businesses were more vulnerable to hard times than others. The Dynky Lynx miniature golf course—built on Main Street in Jerome in the summer of 1930—never had a chance; it vanished within a year. Lumber had to be guarded to keep it from wandering off during Tombstone's boom days, but as new construction dwindled so did business at L. W. Blinn's lumberyard. People allowed various types of insurance coverages to lapse during bust days. The Jerome branch of the Bank of Arizona billed 129 life insurance premiums in 1929, but only 16 in 1932; 36 Latinos were billed in 1929, but only 1 in 1932. Restaurants suffered as people economized by eating at home. John Krznarich enjoyed good Mexican food from Jerome's restaurants, but remembered that his family "very seldom [ate] in a restaurant because we didn't have the money. It was a treat when my father used to stop at the old Copper Star Café . . . and bring us a plate of Mexican food . . . on payday." King's Café, which Larson called "the leading eating place here," lasted until the end of the decade, but closed in 1939. One sociological study found that other items on the list of expendable luxuries included alcohol, telephone service, gifts and contributions, clothing, travel, magazine and newspaper subscriptions, and some food items.[17]

Automobiles were another luxury that could be sacrificed in a small town like Jerome. Krznarich, who grew up during the Depression, recalled that he walked everywhere, he did not drive until he was twenty-six years old, and his father never drove. Nevertheless, Jerome, like the rest of the country, had taken a great fancy to the automobile in the 1920s. With a population of around 10,000, Jerome had 1,348 registered automobiles in 1929. The Depression reduced automobile registrations in Jerome to only 681 for 1932. Those who depended on automobile sales or service for their livelihood began to suffer at once.[18]

The town had five registered automobile dealers in 1928. The Tipton Motor Company, also known as the Liberty Garage, sold Fords. Jerome Garage sold Nash cars, and the H. E. Dicus Motor Company sold Dodges. Fred Whitaker and Maurice Goodman sold Studebakers and Chevrolets. Their New State Motor Company went bankrupt early in that year, and E. K. Reese and Harry Amster assumed the company's name and property. They dropped the Studebaker line, preferring to sell Chryslers and Chevrolets. Perry Ling estimated in the spring of 1928 that there were also twelve gas stations and at least fifty storage garages in the Jerome area. By 1932 only New State and the Dicus Motor Company remained as registered dealers.[19]

Andrew R. Tipton and Ersel Garrison formed Liberty Garages, Inc., in September 1928, out of a combination of the Tipton Motor Company of Jerome and the Liberty Garage of Cottonwood. But an auspicious beginning soon came to grief. As the economic crisis deepened in Jerome, people stopped making car payments, and Liberty soon found itself awash in repossessions. Tipton had to sue Frank Daniel and William Back to try to recover the payments they had failed to make on the Model A Fords they purchased in 1929. In 1931 the company held a "special sale [of] repossessed cars."[20]

Money that could not be collected from customers could not be used to settle the company's own debts, and in short order Liberty itself was in trouble. Garrison and Tipton reorganized the company at the end of 1930, with themselves as the sole owners, but they did not leave their troubles behind. They still had almost twelve thousand dollars in debt outstanding in a note from the Bank of Arizona and had entered a five-year lease on a property a few months earlier. In a swift chain reaction Liberty's creditors began clamoring for satisfaction: Jerome Plumbing and Sheet Metal Works, the *Verde Copper News*, the Motor Supply Company, and Rio Grande Oil all wanted the money owed to them. By the middle of 1931 Liberty Garages, Inc., had forty-nine creditors.[21]

Liberty's bankruptcy came at the beginning of June 1931 when the Bank of Arizona foreclosed on its note. Liberty owed its creditors more than eleven thousand dollars by then, and the process of recovery was long and acrimonious. Disagreement at a creditors' meeting in Phoenix reached the point of name-calling, and a Phoenix attorney advised Perry Ling that it was just as well that Tipton and Garrison were not present, "because I am fairly well satisfied that somebody's features would be in a sling if they were." Repayment of Liberty's debts dragged on for years.[22]

Although the New State Motor Company, capitalized at one hundred thousand dollars at the end of 1929, outlived the worst of the Depression years, partners E. K. Reese and Harry Amster had to scratch to survive. In his 1935 letter to his landlady complaining of his "excessive and unbearable" rent, Amster reported a bleak market for the business. Workers could not afford to own or operate automobiles, and there was no demand for storage, even at one-third of the rate charged during better times. New State's force of eight to ten mechanics had dwindled to one, and the company had operated at a loss for the previous five years, staying afloat "by Mr. Reese and the writer working all hours of the day and night and Sundays." Amster requested a 35 percent reduction in his rent to keep his company going and was not the first to make such pleas. H. E. Dicus had asked for a similar reduction of his rent at the end of 1931. By the end of 1937 Amster had given up waiting for local business conditions to improve and had relocated his family to Phoenix.[23]

Some types of businesses were fragile even in the best of times.

Mining town newspapers were notorious for their rapid rise and demise. Tombstone saw eight newspapers founded prior to 1890; they ranged from partisan political sheets to a stock trade paper. The *Tombstone Prospector* had six different owner-editor groups from its founding in 1887 to its demise in 1924, four of them in the paper's first two years. It is easy to understand why. In 1888 the *Prospector* filled space with advertisements for itself, its editor lacking sufficient news and ad copy to fill a four-page, six-column paper.[24]

Newspaper editors expanded and contracted their number of pages and days of publication like accordions with every change in the local business climate. But when the really bad times hit, they pulled in completely in their efforts to survive. When a man named Brook went to work for the *Tombstone Epitaph* in 1886, he served as editorial writer, telegraph editor, reporter, and bookkeeper. Brook claimed it took one-third of his time to do his job and two-thirds of his time to collect

Fig. 6.1. A modern view of the former New State Motor Company building on Main Street in Jerome.

his salary, and that when he quit after nine months the paper owed him seventy-five dollars.[25]

The *Verde Copper News* suffered a similar pattern of decline. That paper had been a biweekly of up to ten pages in 1928, but had been reduced to four pages by the beginning of 1931. At the end of 1932 management announced that they would issue the paper only once a week. By the middle of the following year they had economized further by abandoning subscription editorials in favor of a local society page. At the beginning of 1935, with the Phelps Dodge takeover and end of support by the United Verde Copper Company, it became "necessary to temporarily suspend publication until such time as business conditions warrant printing a newspaper in this district." Such time never returned, and the *Verde Copper News* disappeared after almost forty years of publication. Two later attempts by Jerome's business community to sponsor a local paper did not last long. Jerome's radio station, KCRJ, fared somewhat better, though it too changed hands frequently. Charles Robinson, who got his independent station on the air in June 1930, leased it to another operator only a year later, and sold it to a broadcasting company at the end of the decade.[26]

Other businesses vending commodities with seemingly elastic demand proved surprisingly hearty. Although gambling, drinking, and prostitution would appear to be activities heavily dependent upon good times and ready cash, bust did not eliminate those activities in either Tombstone or Jerome. In both places the number of people engaged in those callings certainly diminished, but only in proportion to the reduction of the population as a whole. The number of men in the Tombstone voting register listing saloon keeping as their occupation declined from thirty-six in 1882 to nine in 1892—a reduction in exact proportion to the decline in the number of registered voters. Maps of Tombstone indicate that the town had thirty-four saloons at the height of its boom in 1882. In 1889, when it had lost roughly half of its boom population, the camp still had thirteen saloons, about 40 percent of its boom-days total. Twenty-two wholesale and retail liquor dealers in Tombstone signed a protest against territorial liquor licence increases in 1887. John Muretic remembered bootlegging to be a very important business in the early Depression days in Jerome, one very few people opposed and about which local law enforcement did little. Clearly, bust did not arrest drinking among residents of either town.[27]

Although more apparent in the flush times of a mining town, gambling and prostitution did not end with bust, either. Tombstone's town council found it necessary to discuss the regulation of prostitution and gambling in 1887 and 1888. The *Tombstone Prospector* noted "three or four faro games [and] a keno game" in town in the former year. The same newspaper reported on a Tombstone faro game that ran all night in 1889, and the federal census of the town in 1900 contains several entries for both "gambler" and "sporting woman." When Jerome's police officers went on one of their periodic vice raids in April 1937, they cited twenty-one different establishments, including a drugstore, for illegal gambling. John Krznarich remembered that the Post Office Cigar Store, on Main Street in Jerome, had a back room that contained slot machines and poker tables in the 1930s and 1940s, and Joe Larson witnessed gambling in 1939. Jerome police continued their monthly arrests of females for "vagrancy" throughout the 1930s. The town's leading madam operated into the 1940s and kept her telephone service throughout the Depression. While Tombstone's decline soon ran the two elite prostitution syndicates out of town, China Mary, who allegedly ran the rackets in Tombstone's Chinese community, died there in 1906.[28]

Theaters, often centers of the illicit trades, did suffer. Like the elite saloons and bordellos, their high operating costs made them vulnerable. Hard times quickly thinned Tombstone's six theaters. The original owners of the Bird Cage Theater closed that establishment and sold out in the summer of 1883 because of the lack of business. The Bird Cage reopened only fitfully until 1886, when it was purchased by a new set of owners, who operated it as the Elite until they gave up in 1892. Tombstone theater managers not only had trouble meeting expenses, but also had problems attracting any traveling companies to play in the district. Professional theater companies bypassed Tombstone for Bisbee by the end of the 1880s.[29]

Drinking, gambling, and prostitution did decrease in busted mining towns, but as long as miners remained they sought such entertainments. Prostitutes worked in Silverton, Colorado, for example, into the 1950s, though the glory days of that settlement had long passed. Saloon keeping, gambling, and prostitution cannot be considered essential services, but their survival in depression days indicates that these activities represented something more than a frivolous expense to miners.[30]

As bust wore on, it eroded the vitality of even the core businesses of a community. The Cochise Hardware and Trading Company of Tombstone—seller of hardware, mining, assaying, and blacksmithing supplies, as well as clothing, furnishings, and other general merchandise—was the largest mercantile establishment in Cochise County in the late 1880s. But it was slipping into trouble. Frank Moore, a stockholder and the manager of the company, wrote to business partner and fellow shareholder Frank MacNeil in April 1888 that their "dividend for this month was knocked in the head by heavy purchases and by failure of [the Grand Central Mining Company] to pay in full. They are $1,700 behind on wood." That summer he reminded MacNeil of interest due on a note, and added that until their debtors paid their balances, "you will understand we must pay this interest from our private reserves." Moore reported business as very slow in Tombstone that summer, and L. W. Blinn, the company's managing director and principal stockholder, further complicated matters by deciding to sell out.[31]

Blinn had arrived in Tombstone and established a lumberyard there in May 1880. He quickly became one of the leaders of Tombstone's business community. Described as "a close calculator, a man of excellent habits, who attends with great promptness to every branch of his business," Blinn had spent years in the lumber trade in California. The opportunities of boomtown Tombstone had beckoned the competent and ambitious from California and elsewhere, but after the town began to decline the reverse held true; Blinn wanted out of the Cochise Hardware and Trading Company so he could use his money to better advantage elsewhere.[32]

Blinn informed his partners of his decision in September 1888, which immediately threw them into a crisis. Business was not good, had little prospect of getting better, and now Moore and MacNeil—who each had roughly half as much invested in the company as Blinn—would have to come up with the money necessary to buy him out. Blinn was their business associate, but Moore and MacNeil were friends as well as business partners. Moore confided to MacNeil that he was upset by the prospect of being forced to sell his share if the partners

Fig. 6.2. The Cochise Hardware and Trading Company, Tombstone, ca. 1890. (Arizona Historical Society, Tucson, #83126)

could not raise enough money to buy out Blinn, saying, "I hate this feeling of being crowded out."[33]

But there seemed to be little else that the junior partners could do. Business remained terrible, and at the end of 1888 the railroad reached Bisbee. Moore reminded MacNeil that this meant that their Bisbee competitors "will land goods at Bisbee as cheaply as we can at Fairbank, and with further haul to Tombstone against us, our margin is decreasing daily. Such must be, it cannot be avoided." In April 1889 Moore reported that "business this month is worse than I have ever seen it. I believe stores and customers [are] generally so far behind that it is necessary to draw in and not attempt to push trade." Two months later Moore told MacNeil that Tombstone was "about all gone up, at least for [a] long time to come," and concluded with resignation that "a man cannot force business."[34]

Moore and MacNeil continued to try to fight their way out of the morass of economic depression. All three partners entertained the idea of relocating the business to Tempe, Arizona, in mid-1889, but nothing came of it. By autumn of that year Blinn had apparently relocated to Los Angeles, despite being unable to sell his share of the business, and Moore informed MacNeil that he believed Blinn to be in much worse financial shape than either of them. Moore hoped that Blinn could be bought out, either by more stable investors or by themselves. Moore's review of the situation in a letter to MacNeil in January 1890 showed the partners playing a hand that contained nothing but bad cards. "I would prefer to sell," he confided to his friend, "for this reason only, and that is the prospects are bad here ... but this I assure [you] we cannot do, and under these circumstances [I] will willingly stand in on a purchase [of the Blinn stock]. . . . Should we acquire the stock, we would as the only means to make a success, be required to largely reduce the recurring expenses."[35]

To that end the partners reduced the stock of merchandise they carried, closed out their clothing and furnishing goods department in

September 1891, and relocated their business farther from the center of town in the spring of the following year. How much any of these measures helped is unknown, but the Cochise Hardware and Trading Company, the largest business house in the county at the beginning of the 1890s, apparently did not survive the decade. The same thing happened to the T. F. Miller Company, founded in 1890, incorporated in 1897, and unquestionably the flagship of Jerome's business district. First came the notice that the company would centralize its offices at its Clarkdale warehouse, then in 1943 the announcement that stockholders had agreed to dissolve the company and liquidate its assets.[36]

Companies tried a number of tactics to forestall dissolution. First they engaged in mortal combat with competitors. The Fairbank ice works fought such a battle with the Tombstone ice works. While both companies lost money in the encounter, the Fairbank company endured two mortgages, almost six thousand dollars in liabilities, and had been driven nearly bankrupt by August 1891. At the end of 1930 Jerome's service stations engaged in a price war, selling gasoline at one-half cent above wholesale. Sometimes these competitions thinned the herd. The Fox Drug Company, which opened for business in Jerome in February 1930, lasted only fourteen months against its better-entrenched competitors, the Post Office Pharmacy and the Service Drug Company, which purchased its remains. Service Drug quit the contest in turn in 1938.[37]

After all sides assessed the damage these battles inflicted, they occasionally decided that cooperation would be a better idea. A group of Tombstone merchants ran an advertisement in November 1892 stating that they would not offer gifts to customers that holiday season "owing to the dullness of trade." At a meeting in May 1932, Jerome's garagemen agreed that only two of their garages would remain open, in rotation, on Sunday afternoons. Consolidation, the ultimate cooperation, occurred occasionally, too. T. F. Miller Company's gas station and the Dicus Garage combined their operations in 1933. Such consolidations generally left their survivors with more strength to face conditions in a busted community.[38]

Reorganization was akin to consolidation, in that it put a new face on an old business and fortified it enough to return to the struggle, while individuals came and went as their finances and business sense dictated. Emil Sydow and William Kieke took over a Tombstone general merchandise house called the Bonanza Cash Store from the partnership of McClure and Herrera in the spring of 1887. The new partnership of Sydow and Kieke lasted until January 1890, when they dissolved it by mutual consent. Sydow apparently retained the Bonanza Cash Store, while Kieke opened or took over the New York Store. Kieke made a go of that enterprise until the end of the year, at which time he filed for

Fig. 6.3. Lower Main Street, Jerome, photographed in the early 1930s from almost the same place as figure 2.4, but facing the opposite direction. The Popular Store appears in the middle of the block on the left. Compare this photo to the modern view on page 203. (Lewis McDonald Collection, Cline Library, Northern Arizona University, #435.6)

bankruptcy. Kieke's insolvency had the usual cause: "the impossibility to make collections." The *Tombstone Prospector* noted that "Mr. Kieke is a man of great energy [and] no bad habits," and hoped for his eventual recovery "from the fatal effects of hard times." Both Kieke and Sydow tried; each remained in the district into 1892. That summer Kieke finally gave up and relocated his business and person to Albuquerque.[39]

The Popular Stores, Inc., with branches in Jerome, Prescott, and Flagstaff, also tried to reorganize its way out of trouble. Unable to collect money owed to it, the company eventually collapsed under the weight of seventy-six creditors. By October 1932 creditors were fighting over the carcass of the Popular Store, which the district court in Phoenix declared bankrupt that December. By then the company retained perhaps 10 percent of its former value, but that was not the last incarnation of the firm. Although five individuals owned the Popular Stores, Max Krause and I. L. Bacharach were the prime movers. These two men purchased the merchandise and fixtures of the Jerome Popular Store at its bankruptcy sale in April 1933, rented a different and presumably less expensive property in Jerome in which to do business, and opened the Popular Shoppe, which Krause managed for several more years.[40]

If a company had establishments in more than one community, its managers could reduce their involvement in a busted town without

Table 6.1 Summary of Garnishments, Attachments, and Suits to Recover, Jerome Justice Precinct, 1929–1932

Year	Garn.	Attach.	Suits	Total
1929	191	19	7	217
1930	137	31	8	176
1931	43	25	8	76
1932	20	16	2	38

Source: Yavapai County, Justice of the Peace Monthly Records, Sharlot Hall Museum and Archives, Prescott

suffering too much injury. That is precisely what L. W. Blinn had in mind when he announced to his partners that he was selling his interest in the Cochise Hardware and Trading Company and sinking that money into his own lumber enterprise. Blinn founded the L. W. Blinn Lumber Company in Tombstone, but by 1885 he had outlets in eight Arizona and New Mexico cities, and he continued to expand the business. By the time he relocated to Los Angeles in 1889 he had eight establishments in California, five in Arizona, and one in New Mexico. By 1903 his empire "extend[ed] over a score of towns and cities of California, Arizona and New Mexico." With this degree of dispersion, Blinn could shift his resources around as the health of local economies dictated. Some people discovered this idea only after things turned sour in their locale. Tombstone photographer C. S. Fly opened a second photography studio in Phoenix in 1892 and another in Bisbee in 1897, but without success in either case. In July 1931 Daniel's Men's Shop relocated to Flagstaff due to "recent adverse business conditions in Jerome."[41]

Although litigation was more common in flush times with abundant money for such activities, legal action could be used to recover monies owed during bust. Perry Ling spent the 1930s rattling his saber at people on behalf of a number of Jerome businesses, trying to collect what he could. But litigation had its limits. Garnishment proceedings were of little use against people who did not have jobs and were themselves in financial trouble. Consequently, the number of civil actions in the Jerome Justice Precinct—the lower court with jurisdiction over Jerome and its environs—fell dramatically after 1929. The year 1932 saw only one-tenth as many garnishments as 1929, while the number of suits to recover and attachments of property also fell off during those years. The total number of civil actions in the Jerome Justice Precinct declined sharply every year, from 217 in 1929 to 38 in 1932. Legal action could also be self-defeating because of the social component accompanying business transactions in a small town. Tombstone restaurateur Quong Kee Gee put the problem succinctly: "No go to court. When they have money they pay. In court, lawyer make 'em pay, then lawyer take money. Quong have no money and no friends."[42]

So even lawyers, who might have been expected to do well collecting debts and litigating bankruptcies, found money in short supply. They often recovered only a fraction of the value of their services. Perry Ling wrote to a friend in September 1931: "I have been so darned busy the last couple of weeks that I haven't had time to turn around, and probably have collected fifty bucks in real money." After Ling petitioned the bankruptcy referee in November 1937, claiming a fee of $150 for his part in the six-year-old Liberty Garage case, he was awarded $85.35—his prorated share of the estate administrative expenses. Ling received less than 75 percent of his fee in the Popular Store case, and this time, too, several years elapsed before he saw the money. Norman Wykoff, Jerome's other lawyer, called his career "a meager living," and neither he nor Ling stayed in town much beyond World War II. Tombstone, being a county seat, provided better pickings for lawyers. The town, which may have had forty lawyers in the boom days, still had eight in the late 1880s, and at least three after 1900.[43]

Other professionals stayed on after the glory days, though they too had trouble collecting their fees. They bartered their services occasionally, and eventually moved on. Some, like Tombstone doctor George Goodfellow, continued to keep an office in town and make regular visits even after relocating their main effort to another town. But these professionals, with their highly portable trades, also showed surprising persistence in some cases. A Tombstone business directory listed two dentists in the town in 1889. Though the number of doctors in Jerome fell by half between 1929 and 1932, from six to three, it increased again in the mid-1930s. The town was well supplied with doctors thanks to its large hospital, which served the whole Verde Valley. Not until after World War II did doctors disappear from Jerome, often relocating their practices in the healthier towns of the Verde Valley. Several studies of nineteenth-century mining camps have shown that professionals persisted in proportion to their boomtown numbers or even increased slightly in percentage during bust, the latter being the case in Tombstone in the ten years after 1882.[44]

Artisans and laborers also suffered the effects of bust, and often more quickly than merchants or professionals. The collapse of Hudson and Company in Tombstone took the savings of many of them with it, and even in a town experiencing a more gradual decline conditions could be severe. With the high-paying mining jobs gone, many were unemployed, others were underemployed, and those who continued to work often did so with a substantial reduction in pay. John Krznarich remembered the people in his neighborhood living "from payday to payday. They had a rough time." Survival strategies for these people included mobility, an idea that worked much better in the frontier West of the 1880s than in the Depression-era West of the 1930s. Workers also used the tactic of taking several part-time jobs rather than one full-time job, and, as discussed previously, bartered their skills for other goods or services or for room and board.[45]

An interesting feature of these economic communities is their proportional declines. Tombstone experienced a modest relative decline in the importance of mining. Miners dropped from 51 percent of those listed in the county voting register in 1882 to 40 percent at the depth of the bust in the late 1880s, with a slight rebound to 44 percent of those registered in 1892. The town saw a moderate increase in agricultural and professional workers, due to the increased importance of agriculture and county government to Tombstone's subsistence. All other occupational groupings remained in relative proportion, staying within a point or two, and often less, of their boomtown percentages. One thing that this suggests, for Tombstone at least, is that while merchants suffered hardship, they did not suffer disproportionately because they lacked the mobility of capital or labor. Businesspeople seemed almost as willing and able to abandon Tombstone as members of other occupational groups.[46]

Something similar could be said of Jerome. An examination of the town's telephone directories and fire insurance maps for the 1930s shows that bust reduced the number of businesses, but did not much alter the composition of the business community in terms of the proportion of different types of businesses. These sources indicate significant reductions in the number of grocery stores, meat markets, hotels, and restaurants, and the elimination of some specialty stores such as a confectionery, bottling works, taxi service, and barber. Vendors of commodities with elastic demand and operators of small establishments probably suffered the worst. If specialization of businesses marked the growth of a district, bust produced a reverse process, resulting in fewer and more generalized establishments. When the final shutdown came in 1953, what remained of Jerome's business district almost completely disappeared.[47]

All of the above evidence suggests that, even in bust days, both of these communities depended heavily upon mining and proved unable to alter their economic orientations. Tombstone quickly learned that bitter lesson. "The profound silence which now pervades our almost deserted streets," wrote one resident,

> affords a striking and disagreeable contrast to the scenes of bustle and activity that in days past prevailed in this then thriving camp. Six weeks have elapsed since the closing down of the mines, and the absolute dependence of Tombstone upon the mines has been fully demonstrated. Heretofore it was a frequent boast of Tombstone . . . business men that our commercial industry was dependent only in a small degree upon the mines; that the trade from Sonora and from the surrounding country would always insure for Tombstone a prosperous business future. But as before stated the short period of six weeks has been amply sufficient to show the baseless foundation of this hopeful theory.[48]

Mining had no substitute in the short term.

In a sense, then, the merchants, professionals, and artisans who made up the heart of these economic communities speculated on mining to the same degree as the mining men. Frank Moore, of the Cochise Hardware and Trading Company, expressed his ambivalence about present risk versus potential reward when he and Donald MacNeil were considering assuming a large loan to buy out L. W. Blinn. "I have great and abiding faith in the future of this camp," Moore wrote to MacNeil in the fall of 1888. "In time, perhaps in two or three years from now, this town will boom, and a rich harvest will await those who remain, but we neither of us can afford to go too far on the proposition. . . . I believe the future will be all right . . . still there is a doubt as to the outcome of Tombstone, and a failure would mean ruin to us."[49] Such was the gamble facing the merchants, professionals, and artisans of a busttown in the 1880s or the 1930s.

Frank Moore took some comfort from the lessons he had learned. He wrote to MacNeil that "our experience here has been valuable, our losses doubly repaid by experience gained. You are aware that we are far better qualified today to hold our own in the business world than six or seven years ago." The business world of the nineteenth century could be a hard school, and the economic community of a mining camp doubly so. While studies indicate that a long-term economic decline might permit adaptation, most busted mining towns died too quickly to offer that chance. When they gambled and won in a successful mining town, merchants often came out at or near the top of the hierarchy in their communities and enjoyed a comfortable station, at least until the mines played out. William Andrews Clark, for one, built his initial fortune in freighting, retailing, and banking, not in mining. When they lost, most merchants, professionals, and artisans joined that polyglot of peoples on its way to a new start in another town—but who remained behind?[50]

CHAPTER SEVEN

Two Chinese, an Irishman, a Frenchman, and a Negro

Like any other mining town, Jerome was a place where cultures, as well as individuals, could collide. Y. C. Shen, a photographer by trade, had a dispute over some property with another businessman, jeweler Enrique Guerrero, at the beginning of 1934. But rather than resort to the American justice system, Shen turned to a tactic more traditional to his Chinese culture, that of shaming his adversary. Shen hung a large sign in front of his store in Spanish, accusing "Enrique Watchmaker bandit" of stealing his watch and refusing to give him a ring that he had paid for. Shen denounced Guerrero on his billboard as a "shameless man," but Guerrero demonstrated a better grasp of the American system. He had Shen arrested for criminal libel. Shen received a one-year suspended sentence with a promise to the judge that he would behave himself in the future.[1]

It is certainly a threadbare observation that mining towns were cosmopolitan places. But it is not unreasonable to ask whether and how the demography of boomtown days changed during bust. The first problem confronting someone studying the demography of a mining camp is to establish a total population for the place. Newspaper editors and other town boosters often complained that the census taker underestimated the total population figure that their town deserved. There is some justice to their claims; trying to take a census amid the chaos of the boom days was neither an easy nor a wholly accurate task. Still, the past and present claims of town boosters are just as suspect. It was in their interest, for a variety of reasons, to claim as large a population for their communities as possible. Estimates of a peak population as high as 15,000 have been offered for both Tombstone and Jerome—estimates accurate in neither case.[2]

In both Tombstone and Jerome we are fortune to have a census conducted during the period of peak population. Cochise County conducted a census in 1882, right before the water troubles began to take hold in Tombstone. Federal census takers enumerated Jerome in April 1930, before the Depression began to have serious manifestations in the community. In the case of Tombstone we also have biannual voting registers and census tabulations from 1890 to provide at least some demographic information throughout the bust. The federal census of 1940 provides an end mark to the Depression in Jerome. It occurred before the events of World War II began to influence the town's population.[3]

Using only contemporary official and unofficial estimates, one discovers general agreement about the peak populations of Tombstone and Jerome. The Cochise County census of 1882 put Tombstone's population at 5,300, and while several sources thought that estimate too low, their own figures were not much greater. The editor of *The*

Table 7.1 Population Estimates for Tombstone in the Bust Decade

Year	1882	1884	1889	1890	1891
Population	6,000†	4,000*	3,000*	2,000†	1,800*

Sources: See nn. 4 and 6 of this chapter.
*Contemporary estimate. † Author's estimate.

Tucson and Tombstone General and Business Directory, published in 1883, thought that Tombstone had a population of "at least 6,000 people," and his own estimate of "fully 1,000 more than the census returns of last year" would put the town's population at 6,300. An Arizona business directory seconded the Tucson and Tombstone directory, putting Tombstone's population at 6,000, and a further confirmation comes from a citizen who later estimated the peak population during his residence from 1881 to 1884 as "six or seven thousand." Even the highest population estimate of the time, that of an insurance company in 1882, granted the town a population of less than 8,000. Given these estimates, a peak population for Tombstone of around 6,000 seems in order.[4]

Jerome's greatest population—though larger than Tombstone's—certainly never reached the 15,000 that has been claimed for it. The official 1930 census figure was 4,932 for the town and 7,755 for the Jerome Justice Precinct, which included the unincorporated areas surrounding the town. The editor of the *Verde Copper News*, while reporting the official census figure of around 5,000, claimed that "fully half of the inhabitants live outside the incorporated city limits." If true, that would produce a total population of 10,000 for the town. United Verde officials, in a 1930 article for the *Mining Congress Journal*, also put Jerome's population at "about 10,000," which seems a reasonable peak population for Jerome and its immediate vicinity. No total

Table 7.2 Population Estimates for Jerome in the Bust Decade, Incorporated Area and Total

Year	1930	1934	1936	1940
Incorp. Area	4,932*	3,000*	4,185*	2,295*
Total Population	10,000*			4,000†

Sources: See nn. 5 and 7 of this chapter.
* Contemporary estimate. †Author's estimate.

population figures are available in this period for the unincorporated company towns of Clarkdale and Clemenceau.[5]

What happened to the total population of either town after the bubble burst is a more ticklish matter. One source claimed that Tombstone's population declined from 7,000 to 2,000 in six weeks during the strike. While that might be an exaggeration, both towns quickly suffered substantial declines. One would assume that the population would decline most sharply soon after the mines shut down, and would then continue to diminish gradually. A business directory from 1884 reported the population of Tombstone "considerably diminished" at 4,000. Two sources put the town's population at 3,000 in 1889. The census of 1890 listed 1,875 inhabitants, which the *Tombstone Prospector*'s editor thought was too low. He maintained that Tombstone's winter population "can safely be figured at 2,200." It seems reasonable to grant the town an average population of 2,000 in 1890, about one-third of its 1882 peak. The *Prospector* estimated a town population of 1,800 in the second half of 1891.[6]

Perry Ling thought Jerome had only about 5,000 residents at the beginning of 1932. A report to the Home Owners' Loan Corporation put the town's population at around 3,000 in April 1934, doubtlessly a figure for the incorporated town. A WPA estimate of two years later gave a higher figure of 4,185, incorporated. The 1940 census recorded 2,295 inhabitants, with an estimated metropolitan population of 3,700. One of Jerome's school officials cited chamber-of-commerce figures the following year of 2,193 for Jerome and 4,578 for the district. Even using the highest figures for these later years, Jerome experienced greater than a 50 percent decline in population during the 1930s.[7]

An important point to keep in mind when considering busttown demography is the terrific general mobility in the United States in the nineteenth and twentieth centuries. Mobility was normal in mining towns especially, but also in urban areas generally and on the agrarian frontier. Various studies have indicated that persistence over a decade in the nineteenth century ranged from one-quarter to one-half for the general population of a community, regardless of the type or health of the local economy. Specific groups could experience even lower rates of persistence. Scholarship has also revealed that the total turnover in a population could be at least several times the net gain or loss. Partly because of the evanescence of many mining towns, persistence in those places seems to have been lower than in other types of communities, even in the best times. Just over one-quarter of a sample of persons drawn from Tombstone's 1880 federal census can be positively identified in the county census of 1882. If the probables are added, that proportion increases only to one-third. Several historians have argued that what occurs during bust is less an accelerated out-migration than a normal out-migration exaggerated by a complete cessation of in-migration. It is also important to remember that as any mining community matured, its demographic composition changed regardless of its economic fortunes. The increased presence of women and children is only the most obvious example of this. Still, bust itself might produce demographic changes.[8]

One might expect bust to produce ethnic tensions, that majorities or elites might attempt to hold their position by forcing ethnic minorities out of the community. Certainly, nativist sentiments can surface during hard times. One person wrote to the *Tombstone Epitaph* that the leadership of the Tombstone strike consisted of "a handful of men, most of whom are foreigners who cannot rule their own country, much less ours." The *Arizona Daily Star* claimed that four-fifths of the strikers were foreign-born, and that it was time for Americans to resist "this invasion of foreign ignorance." Fifty years later, a state representative from Pima County blamed "insufficiently restricted immigration" for "serious economic, political and social problems in the State of Arizona and throughout the Southwest," and a citizen of Superior, Arizona, who wrote to the governor in 1933 argued that the only solution to the Great Depression in Arizona was "strict citizen employment in all public and private industry and works and whole-sale deportations of all indigent aliens."[9]

Tombstone experienced a fit of nativism with an anti-Chinese movement in the spring of 1886. Anti-Chinese feelings surfaced in mid-February of that year, with calls by the *Daily Tombstone* for white citizens to force the Chinese out of the city. The newspaper soon set about helping to raise five thousand dollars for a white-owned and -operated steam laundry. A few days later came a call for the organization of an anti-Chinese society, which occurred at a mass meeting on 16 February.[10]

Those at the meeting decided to try to expel the Chinese from Tombstone through an economic blockade. The attendees, who included a large number of leading citizens and businessmen, agreed to a pledge to discourage "by all honorable and lawful means, the employment of Chinese, in any capacity in Cochise County." The Tombstone Anti-Chinese League gave thirty days' notice to all citizens to cease employing or patronizing the Chinese or face a blacklist, and urged its members to inform on those in violation. Stanley Bagg, one of the leaders of the movement, also attacked the Chinese through a city ordinance.[11]

Bagg headed a group that placed a petition before the town council in June 1886, asking it to move all laundries outside the city limits. The town's attorney reported that the council had no power to do this, but said that the council could regulate sanitary conditions in those establishments. Although Bagg thereupon pronounced himself satisfied with sanitary conditions in the laundries, that November he got himself elected to the council and reopened the issue in February 1887. Bagg sponsored an ordinance, which passed the council four votes to none, declaring laundries to be health hazards—and therefore a public nuisance—and making it a misdemeanor to maintain a laundry within the city limits of Tombstone. Chinese residents contested the ordinance on the grounds that the city had no right to drive out that

Fig. 7.1. The Jerome Chapter of the Woodmen of the World's float for the Fiestas Patrias (Mexican Independence Day) on Main Street in front of the Miller Building, 16 Sept. 1928. The dual allegiance of Jerome's Latino population is indicated by the symbols adorning the float and banner. (Arizona Historical Society, Tucson, #64489)

particular class of business. Chinese leaders also protested that the Chinese were peaceable, taxpaying residents who held jobs that nobody else wanted and had as much right to live in Tombstone as anyone else. They also feared that they would not receive just compensation for their property if forced to leave.[12]

The idea for a white-operated steam laundry died within a month, but several white women started laundries or advertised their willingness to do so, and the economic blockade seems to have forced some Chinese residents to leave the city. Charlie Lee Kong, owner of the leading Chinese mercantile house in Tombstone, said that so many Chinese were out of work that he could not collect the money owed to him and would relocate his business to Tucson.[13]

The Tombstone Anti-Chinese League eventually claimed more than 800 members, which, if true, would have been around one-quarter of the population, but the movement failed to maintain its early momentum. Prominent citizens withheld their support, disagreements arose within the organization, and the anti-Chinese newspaper turned its attention to other issues. By mid-August that newspaper reported that 60 to 75 Chinese still lived in town. A few days later the league reorganized itself into the Anti-Chinese Secret Society of Cochise County, ending the town's overt anti-Chinese movement.[14]

Jerome, too, saw the exodus of large numbers of an important ethnic group during its bust. As the mines closed in Jerome and throughout Arizona, many of the Mexicans who were the backbone of the

Fig. 7.2. Although bust adversely affected the town's social life, residents still commemorated the important occasions. The queen and her consorts pose for Jerome's Fiestas Patrias celebration of 16 Sept. 1935. (Arizona Historical Society, Tucson, #64454)

mining labor force lost their jobs and returned to Mexico. They had the active assistance of American charitable organizations and the Mexican government in this migration. Almost 300 of Jerome's Mexican citizens returned to Mexico in mid-1930 and another 250 or so in the second half of 1931. The American Red Cross and Jerome's Associated Charities paid to transport indigent Mexicans to the border, usually at Nogales, and the Mexican government paid the cost of getting them the rest of the way home. Along with the voluntary relocations came the deportations of a number of indigent illegal aliens, though these probably did not amount to more than 200 Jerome residents. Almost all of these repatriations took place in the first three years of the Depression.[15]

While the anti-Chinese activities in Tombstone and the Mexican repatriations from Jerome might seem unequivocal examples of bust-induced nativism, the circumstances surrounding these events need to be kept in mind. Tombstone had seen anti-Chinese activity in its boom days, too. In July 1880 some residents held anti-Chinese demonstrations and threatened to riot if the Chinese did not leave. The Chinese held their ground, however, and the anti-Chinese agitation went into remission. The anti-Chinese movement of 1886 coincided with an upsurge of anti-Chinese activity all over the Pacific slope, including in the Pacific Northwest, California, and Arizona. On 2 September 1885 rioters in Rock Springs, Wyoming, killed 28 Chinese, injured another 15, drove away hundreds more, and destroyed almost $150,000 of their property. Five Chinese residents were hanged in an anti-Chinese action at Pierce City, Idaho, in 1885. Anti-Chinese rioting got so bad in Seattle in February 1886 that the governor of Washington Territory had to declare martial law in the city. Prescott, Arizona Territory, had an experience very similar to Tombstone's at about the same time, as did Idaho's Wood River District. Tombstone's nonviolent anti-Chinese movement may even have been a reaction to the regional agitation. The *Arizona Weekly Citizen* reported that

Tombstone's movement "originated contemporaneously with similar ones on the Pacific Coast," and the meeting that formed the league was called ostensibly to deal with "the influx of Chinese" to Tombstone.[16]

Deportation and repatriation of Mexican nationals were unique neither to Jerome nor to the Great Depression. Companies often guaranteed repatriation to their Mexican laborers in the event of unemployment, and as the Depression deepened, Mexican nationals headed south en masse. The Mexican Consulate in Phoenix estimated in June 1931 that 14,000 Mexican citizens had returned home from Arizona since the first of the year, and Arizona provided but a small part of the total. Around 125,000 Mexicans repatriated from the United States in 1931, mostly through Texas, and another 80,000 the following year.[17]

A second point to consider is that nativist actions did not command universal support—or even respect—among native whites. George Parsons wrote in his journal in April 1886: "Attended Anti-Chinese meeting tonight at Schieffelin Hall and heard some fellows make asses of themselves. Bagg at their head. . . . Boycotting don't [*sic*] go with me and there is no reason for an anti-Chinese movement here." Although the town council enacted Stan Bagg's antiwashhouse ordinance unanimously, when Bagg offered another resolution a month later, instructing the health officer to actually enforce the new law, it died for want of a second.[18]

As various individuals pointed out, one complication with any simplistic scheme to deport all foreigners and close the borders during the Great Depression was that although many of the adults in question were foreigners, many of their children were American-born. A *Prescott Citizen* editorial doubted whether such exclusion would do much good anyway. It maintained that mining companies would simply substitute Eastern European labor at the same wages and opined that Mexican laborers spent their earnings locally and "form[ed] no dangerous element in the citizenry of the country." The editorial concluded that "Arizona and the rest of the country should let well enough alone."[19]

Finally, if these were serious efforts to drive these ethnic groups out of Tombstone and Jerome, they were distinctly unsuccessful. The Chinese constituted 4 percent of Tombstone's population in 1882, and more than 3 percent in 1900. The Chinese community flourished through the second boom period, which ended about 1910, and Tombstone had a few Chinese inhabitants into the 1930s. Other Chinese took up farming or ranching in the surrounding area. There is also no evidence of a pogrom against either of Tombstone's two other substantial minority groups, the Irish and the Mexicans. While the Irish suffered a modest decline as a percentage of the population, most of them were miners and suffered the decline because of their occupation rather than their ethnicity.[20]

The Latino percentage of Jerome's population probably increased during the 1930s. It certainly did not decline from nearly 60 percent in 1930. Several Latino fraternal organizations remained active at the end of the decade, and "Mexican" enrollment in Jerome's school system remained constant throughout the decade. Spanish speakers remained such an important element of the community that officials announced election returns to the public in both English and Spanish in the latter half of the 1930s. After World War II one witness found Jerome's children "imbued with a racial tolerance remarkable in a community that is potentially a hotbed of conflict." Clearly, white residents did not universally react with nativism in a busted town, and intolerance was not the inevitable result of an economy in decline.[21]

That said, it is also true that economic self-interest often lay at the bottom of popular indignation against ethnic groups. The most brazen example of this was Tombstone business leader R. W. Wood's sponsorship of the white steam laundry. He was the heaviest investor in and leading proponent of that scheme, though he was later blacklisted for employing Chinese labor in his other businesses. Tombstone's newspapers took to insulting each other over the anti-Chinese issue, and one concludes that eventually the issue became merely a stick with which to beat the rival paper. Editor James J. Nash of the *Daily Tombstone* accused the *Tombstone Epitaph*, which he preferred to call the "morning Chinese Joss," of "prostituting its columns for mercenary gain" by opposing the anti-Chinese movement, and he urged his readers to boycott the *Epitaph* and its advertisers. Chinese leaders, in turn, accused the *Daily Tombstone* of fanning the anti-Chinese fire. Each paper sought political and competitive advantage by abusing the other over the Chinese issue.[22]

Politicians as well as newspapers sought to capitalize on the issue. Tombstone's mayor, C. N. Thomas, served as one of the presidents of the Anti-Chinese League, and Stan Bagg, one of the movement's leaders, won a seat on the council, though it is unclear whether he won election because of his anti-Chinese views or if his new position merely gave his views wider expression. A Tucson newspaper went so far as to say that Tombstone's anti-Chinese movement was manipulated for "political purposes merely, and is being converted into a stepping stone to official preferment by adroit schemers."[23]

Although the anti-Chinese agitation derived from a variety of motives, getting rid of the Chinese might eliminate some competition and create a few jobs, so perhaps Tombstone's anti-Chinese agitation was bust-driven. Proponents of Chinese exclusion usually argued its advantages in economic, rather than racial, terms. In one of his messages to the league, President Thomas called for action against the Chinese "still in our midst taking the bread that rightly belongs to our wives and children from their mouths" during Tombstone's hard times. Jerome experienced the opposite situation; Mexicans left town because they had no jobs, not because other members of the community tried to force them out of employment for economic gain. The Mexican Consulate in Phoenix estimated that unemployment was the reason that over 99 percent of Arizona's Mexican citizens returned to Mexico. In both cases, the relationship of bust to overt anti-ethnic activity is ambiguous.[24]

Of course, other possibilities existed besides physical expulsion from the community. Different groups could be excluded economically, marginalized to such an extent that no overt action against them would be necessary. It is therefore worth examining the economic fates of those members of different ethnic groups who chose to remain in these districts to see if they suffered patterns of economic discrimination as the result of the busts.

Due to serious limitations in sources, it is difficult to determine how bust affected minority business opportunities. Ethnic businesses tended to be smaller concerns, less likely to receive attention in newspapers or to appear in government records than such large white-owned enterprises as the T. F. Miller Company or the Cochise Hardware and Trading Company. It is only through the accidents of historical record keeping that we know about Pasqual Nigro's saloon and mining interests and Sol Israel's bookstore in Tombstone or Guadalupe Amayo's grocery store and Francisco Madrid's Palace Lodging House and mercantile establishment in Jerome—or Scott and Gibson's grocery of Jerome, for that matter.[25]

Such evidence as does exist, however, shows no obvious pattern of decline among minority businesses as compared to those owned by whites. The only ethnic group that shows up in the same proportion in Tombstone in both the censuses and the voting registers is the Irish. The 1882 voting register listed 15 of 301 Irish as merchants or businessmen. The 1892 register classed 2 of 41 Irish in those categories. In both cases they represented just under 5 percent of their ethnic group. Tombstone had a substantial Jewish community that included many of the town's merchants. Although many of these people moved on when bust came, Tombstone still had enough Jewish merchants to close down the business district for Rosh Hashanah in 1889.[26]

Jerome—like most other Arizona mining and smelting towns of its day—had a Latino mercantile community, largely quartered in the Latino section of town, which served its Spanish-speaking clientele.

Jerome also had a "Chinese colony" into the 1930s, many of whose inhabitants were members of the business community. Yee Hong Song operated the English Kitchen from the 1920s into the 1970s. He faced competition for part of that time from Yee Chee, who opened the American Café at the end of 1933. As with those of their white colleagues, some minority businesses foundered almost immediately when struck by hard times, while others showed remarkable persistence.[27]

Ethnic loyalty through patronage probably helped some of these ethnic businesses to survive. When John Muretic tried to get a cash advance before payday from the Bank of Arizona, Assistant Manager Phil Luna gave it to him, saying that if he did not, Muretic would just get the money from Joe Pecharich, one of Jerome's leading merchants and a member of the Croatian community. Pecharich did this for just about every Croatian, and they probably paid him back by taking their business to him whenever possible. As long as a large-enough ethnic community survived, these businesses could as well.[28]

The careers of individual ethnic laborers are even harder to trace than those of merchants. In 1882 four-fifths and in 1892 three-quarters of Tombstone's Irish who registered to vote listed themselves as miners. The same proportions probably held true for Tombstone's Mexicans, who were predominately teamsters and laborers. An *Arizona Weekly Citizen* correspondent reported "many Mexicans" in Charleston in mid-1879. Latinos probably concentrated in the smelter towns, as they tended to work in milling and transportation rather than mining. But Tombstone was the heart of the district, and it had a considerable Latino presence. It would appear that bust neither denied nor created special opportunities for the Irish and Latino workers of Tombstone.[29]

The record of uvx hirings from before and after its shutdown in the second quarter of 1931 indicates the same lack of bust-related employment discrimination against Jerome's Latinos. An analysis of the three-year boom period before the suspension and the five-year period between the suspension and final shutdown in 1937 shows no change in hiring practices by the company—save one in favor of men with dependents. While the percentage of men listing Mexican nationality dropped slightly, the percentage of Mexican miners, the best-paying and most prestigious job, rose to the same degree. The percentage of Mexican laborers, the worst-paying and least-prestigious job, fell significantly. The percentage of all foreign-born and Spanish-surnamed employees fell slightly, while the percentage of foreign-born and Spanish-surnamed miners remained practically the same, and that of laborers fell. Here, too, economic opportunities for ethnic workers do not seem to have been significantly altered by bust, though Arizona's mining companies did sometimes discriminate in favor of U.S. citizens during the Depression.[30]

Women, like the Irish or German Jews, constitute a somewhat special case. All of these groups, whatever the discriminations against them, had the advantage of being white and generally did not face the language and cultural barriers that obstructed the Mexicans and Chinese. Nevertheless, women were a minority in Tombstone, and their working life shows a complex character.

Business directories and advertisements from the camp's early days identify a number of working women. Many of them, like boardinghouse keeper Mary Tack, had jobs that might be considered extensions of domestic activities. Others, such as Mrs. James Younge, seamstress, and Molly Fly, boardinghouse keeper and photographer, worked to supplement their families' income or assisted in their husbands' enterprises. But some women had less traditional employments. Nellie Cashman operated the Nevada Boot and Shoe Store and the Russ House Hotel; Bertha Leventhall managed the general Miners' and Merchants' Store; Mrs. G. W. Stewart ran a millinery shop; Lizzie Newell was a livestock dealer; Mary Anderson was a saloon keeper, with three other women listed in business directories as liquor dealers; and Mrs. Marion Webb, M.D., practiced medicine. Roughly one in five of the women

Table 7.3 Comparison of United Verde Extension Hirings Before and After Its 1931 Shutdown

Hiree Background	Before		After	
Total Entries	1,237		621	
Number Listing Nationality	1,235		619	
Mexican Nationals	709	(57.4)*	320	(51.7)
Other Foreign-born and				
Spanish-surnamed	244	(19.8)	117	(18.9)
All Foreign-born and				
Spanish-surnamed	953	(77.2)	437	(70.6)
Total Miners	455		156	
Mexican Miners	197	(43.3)	78	(50.0)
Other Foreign-born and				
Spanish-surnamed Miners	152	(33.4)	43	(27.6)
All Foreign-born and				
Spanish-surnamed Miners	349	(76.7)	121	(77.6)
Total Laborers	178		72	
Mexican Laborers	122	(68.5)	36	(50.0)
Other Foreign-born and				
Spanish-surnamed Laborers	17	(9.6)	12	(16.7)
All Foreign-born and				
Spanish-surnamed Laborers	139	(78.1)	48	(66.7)
Number Listing Marital Status	1,234		617	
Single	552	(44.7)	194	(31.4)
Men Listing Dependents				
Beyond a Wife	520	(42.1)	294	(47.6)
Mean Age/Median Age	32.0/32		35.5/35	

Source: United Verde Extension, Employee Records, DB271-2, Sharlot Hall Museum Archives, Prescott.

* Percentages in parentheses.

listed in business directories from 1880 to 1884 had a nontraditional occupation.[31]

Some of these women were veterans of the mining frontier. Nellie Cashman, a native of Ireland, had much more than her business interests to occupy her time. Cashman was known for her charitable work; she was treasurer of the Irish Land League and a leading organizer of Tombstone's Catholic church. She also gained notoriety for her community leadership and her interest in a personal bonanza, "she being no exception to the large majority who speculate in mining stocks. She likes the frontier, and is . . . well known by the prospector and miner, and . . . universally liked and respected . . . for none of the 'old boys,' as Nellie calls them, ever came to her for a meal or a grub stake and met with a refusal."[32]

Women like Cashman and her business partner, Kate O'Hara, may have liked the bonanza camps because they offered more economic opportunities than those available to them elsewhere. Some women, Cashman included, possessed the same desire to see the elephant as their male counterparts, and single women could turn Tombstone's demography to their advantage. One woman reported to her friends in the East that "the masculine sex decidedly preponderates, and marriageable girls or women are appreciated as they deserve."[33]

One might expect bust to destroy business opportunities for these women, but the meager data that exist contradict that assumption. The census of 1880 showed almost 8 percent of Tombstone's women employed, nearly 3 percent in nontraditional jobs. The census of 1900 recorded that 24 percent of women held jobs outside the home, 9 percent of them in nontraditional jobs. This rise in female employment could be related to the bust, with women entering the workforce to supplement the reduced incomes of husbands or fathers, or it could be independent of local economics, caused instead by national trends in female employment. One suspects the former to be true because

Tombstone lacked the typing pools, switchboards, and factories at which that era's females found employment in the East.[34]

An examination of Jerome's decline confirms that the first hypothesis is more probable. In addition to a prior assortment of rooming-house keepers, clothiers, and small proprietors, a number of women—Latino and otherwise—started businesses in the depths of the Great Depression. Some of these establishments were beauty shops, occasionally located in the woman's residence, but many other women located their businesses along Main Street in direct competition with the rest of the business community. The number of married women who engaged in these enterprises suggests, again—given that culture's attitudes about married women's employment—that these women worked to supplement reduced family incomes. Perhaps women gained greater employment opportunities during bust, but only in exchange for a decline in living standards for themselves and their families.[35]

So what happened to the demography of the busttown? Economic opportunities for different ethnic groups were neither especially hindered nor enhanced, so one might expect these communities to retain much the same demographic profile. But changes occurred in both communities. Tombstone experienced demographic changes simply as an aging frontier community. Jerome, a long-settled twentieth-century town, can serve as a control community to help separate those trends resulting from maturation from those resulting from bust.

The three principal ethnic groups in Tombstone, the Irish, Mexicans, and Chinese, held their proportions in the community or increased slightly in percentage during bust. In the 1882 county census, Irish, Mexicans, and Chinese constituted 10, 8, and 4 percent of the total population, respectively. By 1900 Mexicans constituted 11 percent, Chinese 3 percent, and Irish 3 percent of the total population. But the sharp decrease in Tombstone's Irish residents is misleading because the decline in Irish nativity was a national trend. Almost fifty years had elapsed since the great wave of Irish immigration. If one counts as Irish those people who listed both parents as natives of Ireland in the 1900 census, the total returns to almost 10 percent of Tombstone's population.[36]

Unfortunately, the manuscript censuses for Jerome for 1930 and beyond are presently unavailable, and to further complicate matters, published census tabulations listed Latinos as "other" in 1930, and as either foreign-born or native white in all other years. It is therefore impossible to tell precisely what happened to Jerome's Latino population, either native-born or foreign. Evidence that exists suggests that Latinos increased as a percentage of the town's population. Latinos constituted just under 60 percent of the population in 1930 and retained their majority through the 1940s, A writer visiting in the early 1950s thought that they constituted three-quarters of the town's population.[37]

In the 1930s the decision to stay or leave might be made for a foreign-born miner because mining companies apparently retained and rehired U.S. citizens whenever possible. In both Tombstone and Jerome the percentage of foreign-born residents decreased. But it is difficult to separate what happened in Tombstone and Jerome from what happened throughout the United States. The declines of the foreign-born in both of these towns coincided with declines in immigration generally, during the depression of the 1890s and after World War I.[38]

Clearly, bust did not diminish the cosmopolitan nature of either town, except, perhaps, at the bitter end. In 1936 the *Jerome Chamber of Commerce News Bulletin* ran an article listing by name residents native to thirty-eight states and twenty-five foreign countries. A reporter visiting Tombstone in 1905 claimed to have witnessed a card game involving "a Swede . . . a Pennsylvania Dutchman . . . two Chinese . . . an Irishman . . . a Frenchman and a Negro." Members of particular ethnic groups came to town for the same reasons as their white counterparts. Like whites, their decisions to remain or leave were dictated more by perceived economic opportunities than by ties to the ethnic communities of particular places.[39]

In both Tombstone and Jerome, the proportion of females climbed throughout the periods under consideration. Only one resident in seven was female in Tombstone in 1880, and even in Jerome in 1900 females numbered fewer than one in four. Two generations later, females constituted slightly less than half of the population of both communities. While this is typical for mining towns, one important clue lies hidden within the figures for Tombstone. By 1900 mining had all but ceased in Tombstone, but a revival of activity in the years between 1900 and 1910 more than doubled the town's population. In that decade the increase in the percentage of females in the population effectively stopped. Mining was a male activity; a revival in mining attracted mostly males, while a decline chased more males away sooner. In Jerome the mining companies helped increase the percentage of females during the decline through their practice—general throughout the state—of laying off unmarried employees first. In the prevailing opinion of the day, single men could fend for themselves and thereby keep married men with dependents employed as long as possible. Laid-off married men might have increased the female preponderance in Jerome still further by leaving their families in town while they went elsewhere in search of work.[40]

Table 7.4 Percentage of Female Inhabitants, Tombstone and Jerome, 1880–1950

Year	1880	1890	1900	1910	1920	1930	1940	1950
Tombstone	14.6	n.a.	40.1	41.0	47.1	—	—	—
Jerome	—	—	22.4	29.8	37.3	42.5	46.1	49.9

Source: Published federal census data for the years indicated.

Youth also predominated in mining, and when the decline came the young were among the most mobile members of the population. As they departed the adult portion of the population aged. The mean age in the Tombstone voting register, thirty-six in 1882, rose steadily to forty-one in 1892. The mean age of the total populations of these places probably stayed about the same or even declined during bust because of a significant increase in the percentage of children during the bust periods in both towns. But a breakdown of Jerome's population reveals that percentages shifted among particular age groups

Table 7.5   Percentage of Tombstone's Population Under Age 16, 1880–1920

Year	1880	1882	1900	1910	1920
Percentage	10.3	12.7	32.5	30.7	34.9

Source: Federal manuscript censuses and territorial census of 1882.

Table 7.6   Percentage of Jerome's Population Under Age 21, 1920–1950

Year	1920	1930	1940	1950
Percentage	32.5	42.4	43.6	45.3

Source: Published federal census data for the years indicated.

Table 7.7   Age Distribution of Jerome's Population (in Percentages), 1930–1950

Year	<5	5–14	15–24	25–34	35–44	45–54	55–64	65<
1930	12.6	20.6	18.2	21.1	15.8	7.5	3.0	1.1
1940	11.1	20.8	18.9	18.3	12.3	9.0	6.1	3.5
1950	12.6	23.9	13.9	15.4	10.8	8.1	7.9	7.4

Source: Published federal census data for the years indicated.

within the total population. Between 1930 and 1950 the percentage of those under the age of fifteen remained practically the same, while the percentage of those from the ages of fifteen through forty-four declined significantly, and that of those over fifty-five rose sharply. The same thing happened in miniature in the town's school system. While total enrollment declined almost 43 percent between 1930 and 1940, enrollment of high school students remained roughly the same. It was younger families with elementary school–age children who left the town in search of brighter prospects. These towns became the residence of an ever older population.[41]

There may have been changes beyond those simply demographic. Unemployment, business failure, and general economic dislocation may have done more than change the character of the population. They could have changed the nature of social interaction and the character of the society itself.

CHAPTER EIGHT

The Painful Necessity of Closing the Church

Just before midnight on the evening of 11 May 1890 Tombstone's police chief, Dick Gage, entered McDonough and Noble's saloon and pool hall on Allen Street, walked to the back of the building, drew his pistol, and blew his own brains out. "The causes that led to this rash act are several," the *Tombstone Prospector* informed its readers the next day, "although the immediate cause was his inability to meet a note of $100 due today." Gage, age fifty-nine, had informed a creditor earlier in the evening that he had eighty dollars of the amount, and would get the rest. The creditor told him to leave what he had at the billiard parlor and to forget the rest, but Gage promised to get it, proceeded to lose the eighty dollars at a faro table, and then killed himself. Gage had discussed his sagging fortunes with the *Prospector*'s editor a few days before and had indicated that his concern was not for himself, but for the well-being of his wife and five children.[1]

Certainly, suicide is the most dramatic display of personal distress imaginable, and others besides Chief Gage escaped from bust in that manner. Both the silent partner and the cashier of the failed Hudson and Company Bank ultimately committed suicide. Sometimes the emotional stress, economic dislocation, and uncertainties of bust could badly traumatize individuals, but it is important to look beyond these sensational cases to examine general indicators of social and psychological duress in busted communities. Such indicators might be less dramatic than individual episodes, but they may permit us to discover trends in the social relations of busttowns.[2]

The changes bust brought to these societies are frequently ambiguous. Several scholars have commented upon the caustic effects bust inflicts upon the social health of communities through increased economic competition, malicious gossip, and damaged town pride. Sociologists have carried out most of the examinations of the influence of boom and bust on the social health and structure of communities. Their studies of modern boomtowns have concluded that boom brings increased social tensions, manifested by greater crime and violence, alcohol and drug abuse, number of divorces, and alienation.[3]

Sociologists have not devoted much attention to bust, per se, but their studies of mass unemployment have produced less consistent results than their observations about booms. They are generally agreed that the mass unemployment that accompanies the collapse of a town's economy ought to produce insecurity, stress, and lowered self-esteem among those involved, leading to increased alienation, alcoholism, number of divorces, violence, and suicide rates. One informant, visiting the copper towns of southeast Arizona during a bust period in the 1980s, found "a pervasive tension that I had never witnessed in a community before. I was particularly struck by the pain in the eyes of middle-aged men as I caught them at home in the middle of the day watching game shows on television rather than at work."[4]

But when they try to quantify the damage, these same researchers have often failed to discover obvious correlations. One team that studied plant closures and mental health believed unemployment to be "one of the major environmental sources of personal stress," but also concluded "that none of [their research] groups was severely affected by the closing, one or two years later. Indeed, psychopathological behavior was no more prevalent among individuals in our samples than one might expect among the general public." Several researchers have noted an almost universal disregard of available mental health services by unemployed workers and have concluded that the stresses caused by unemployment can usually be managed without professional help. Although most of these sociological observations are drawn from the era of the modern welfare state, the historical evidence from Tombstone and Jerome shows similar ambiguities.[5]

Statistical records of social behaviors are neither abundant nor comprehensive for the bust periods of either Tombstone or Jerome, but existing evidence betrays no significant discontinuities one could associate with bust. Suicide and homicide statistics would seem to be good barometers of stresses caused by bust, but these give mixed indications when viewed over an extended period rather than episodically. Several authors felt Tombstone's legendary boomtown violence to be overestimated, but even if so, the number of murders seems to have dropped sharply after the bust. One author calculated a total of twenty-five murders in the town during the years 1879 through 1884, but only seven during the years 1885 through 1890. No statistical evidence remains—if any ever existed—concerning the number of suicides in Tombstone during the frontier period.[6]

Coroner's inquests for Jerome from the years 1932 through 1947 contain five rulings of suicide and five of homicide. The town had as

Table 8.1   Aggregate Statistics for Homicide and Suicide, Yavapai County, for Equal Periods Before and After 1929

Years	Homicides	Suicides
1925–1928	37	21
1930–1933	19	31

Source: Arizona State Board of Health, *Arizona Public Health News* for the years indicated.

many fatalities from mining accidents during the period as from suicides and homicides combined. Only one of the suicides could be considered a product of economic distress, that of a woman who could not afford a medical operation. Four of the five suicides covered by the inquest records occurred after 1940, when Jerome had already entered the modest recovery brought about by World War II. In none of the five cases judged murders by inquest juries could economic motives or stresses be positively established. Total figures covering all of Yavapai County for the four-year periods before and after 1929 show a sharp decline in the number of homicides, but do show a significant increase in the number of suicides. Modern sociological research has established a crude correlation between national rates of unemployment and suicide.[7]

Bust seems to have had less effect upon mortality and public health generally than on the suicide rate. Again, there is some evidence of a rise in mortality during the Great Depression. The national mortality rate among members of the Alianza Hispano Americana— the Latino death-benefits society—rose between the years 1929 and 1933, then fell back slightly by 1936. One study of mass unemployment shows no significant variation between local and national death rates, however, and evidence from Yavapai County records fails to show any increase in mortality or decline in public health during the Depression.[8]

Table 8.2   Aggregate Statistics for Births, Deaths, and Infant Mortality, Yavapai County, for Equal Periods Before and After 1929

Years	Births	Deaths	Infant Mortality
1926–1928	1,672	1,275	164
1930–1932	1,651	1,299	159

Source: Arizona State Board of Health, *Arizona Public Health News* for the years indicated.

State public health figures for Yavapai County reveal a slight increase in deaths after 1929. Although United Verde Hospital officials reported the baby business "extremely poor" at the beginning of 1931, the number of births in Yavapai County stayed essentially the same in the three-year periods before and after 1929. The number of infant deaths declined slightly in the three years after 1929, compared to the three years that preceded it. Certainly, this slight decline in infant mortality reflected the slight decline in the county's total population during the same period, and it may also be attributed to advancements in public health. At the end of 1933 the state's board of health reported diphtheria, smallpox, and typhoid almost absent from Yavapai County. That good news can partly be attributed to a joint immunization effort undertaken by Jerome's town government and the United Verde Hospital. The county's rate of infant mortality declined slightly in the first few years of the Depression when compared to the years before 1929. Clearly, bust did not necessarily manifest itself in public health problems or increased mortality.[9]

Bust did seem to have a salutary influence on the amount of crime occurring in these two towns, though crime also probably diminished as the towns matured. A business directory reported of Tombstone as early as 1883 that "the disorganized days, attending the bringing

Table 8.3   Total Criminal Cases and Total Felony Cases Registered in Jerome Justice Precinct, 1929–1932

Year	Cases	Felonies
1929	116	23
1930	63	22
1931	44	9
1932	29	11

Source: Yavapai County Justice of the Peace, Monthly Reports for the years indicated.

together of a heterogeneous mass of people, have now passed. Law and order are today as much respected, and have as much control, as anywhere." Although comprehensive crime statistics do not exist for that period, one suspects that the one arrest made by Tombstone's police in the month of May 1894 ranks far below any monthly arrest total from the boom days.[10]

The rate of crimes, violent and otherwise, decreased in Jerome as well. John Krznarich, who grew up in Jerome during the bust period, later remembered "very little juvenile delinquency . . . [and] no drug problem." Records from Jerome's judicial apparatus suggest a correlation between bust times and decreased crime. The total number of criminal proceedings entered in the record of the Jerome Justice Precinct fell by three-quarters from 1929 to 1932, and the number of felony cases declined by half. The surviving record of the Jerome Police Court—which handled lesser crimes committed within city limits—shows a significant resurgence of criminal cases after 1933, when New Deal programs and a revival of mining brought a measure of prosperity back to the district. As in Tombstone, once mining effectively ended in Jerome, criminal activity all but disappeared. From 1954 to 1959 the town's police force made only five arrests—all for drunken and disorderly conduct. The rump mining town still had a tavern.[11]

The general decrease in crime notwithstanding, several types of crime might be classed as bust-related. Perhaps domestic violence fits this category, but because that crime was not classified as such in those days, it is impossible to tell. The only time that problem drew editorial comment in Jerome's newspaper was during the relatively prosperous year of 1925. Both Tombstone, in January 1888, and Jerome, in December 1934, suffered a rash of burglaries, some of which involved entry into vacant buildings. The problem got bad enough in Tombstone to elicit a mayoral proclamation offering a one hundred–dollar reward for the arrest and conviction of the offending parties. But again, these episodes do not obscure an overall decline in criminal activity during the bust periods of Tombstone and Jerome.[12]

Significant alterations in the rates of marriage or divorce would be another obvious statistical indicator of social discontinuity in a busted community. Once again, however, the surviving historical record is inconclusive. Marriage and divorce records for Tombstone are clouded not only by the transience of the frontier, but also because Tombstone was the Cochise County seat, and thus hosted all of the county's divorce proceedings and many marriage parties from outside the city. In neither surviving marriage or divorce records are there substantial changes in rates after the onset of decline, though marriage and divorce may have been more easily obtained on the frontier than in the East. Many of Tombstone's divorce cases list abandonment as a charge, suggesting a poor man's solution to an unhappy marriage that may have been more frequently practiced during bust.[13]

Marriage and divorce records do not exist for Jerome itself, but those of Yavapai County suggest a rough correlation between economic conditions and marriage and divorce. The peak year for marriages in the years 1928 to 1935 was 1929, which also saw the most divorces of any year until 1935. The lowest totals for both in the period occurred in 1932. The county clerk reported that the twelve marriage licenses issued in December 1931 "was the lowest figure registered in this county for several years," and the *Jerome Chamber of Commerce News Bulletin* regarded increased matrimony among the town's schoolteachers as

Table 8.4   Yavapai County, Marriages and Divorces, 1928–1935

Year	1928	1929	1930	1931	1932	1933	1934	1935
Marriages	268	394	298	304	240	262	310	279
Divorces	87	96	94	91	70	82	93	101

Source: Prescott Courier, 2 Jan. 1936.

a "prosperity sign" in 1936. On the other side, several studies have indicated that unemployment does not cause the dissolution of stable marriages. What bust does, if anything, is finish off unstable marriages. One author has suggested that people may even cling to their marriages as their last island of stability during bust.[14]

There has been some disagreement among historians as to whether mining towns were inhabited by "hardened individualists who paid little attention to community affairs unless their personal interests were threatened" or by people who sought to replicate the communities and social systems that they had left behind. The evidence gathered for this study supports the latter view. Residents of both Tombstone and Jerome put much of their energy into building fraternal and community organizations, less formal socializing, and a host of recreational pastimes. These activities were also vulnerable to bust.[15]

Although the social effects of bust upon individuals may or may not have been severe, the effects of bust upon a town's social institutions could be quite serious. In both Tombstone and Jerome, religious organizations found themselves struggling for survival almost immediately. Tombstone had four churches by 1883, and the secretary of the Ladies' Aid Society for the Methodist church reported that church out of debt in 1882. Yet trouble followed directly in the wake of bust. When the boom days disappeared, so did money for the collection basket. Three weeks into the strike George Parsons reported that his church suffered "poor attendance and collections." Things did not get better; two years later he mentioned "more church collecting today, but it goes very, very slowly."[16]

No money in the collection basket turned inevitably into no minister in the pulpit. In January 1885 one of the Tucson newspapers reported that neither Tombstone's Episcopal nor Methodist churches had pastors. A guide to the area recorded that Tombstone still had four churches in 1889, but indicated that "the chief trouble is experienced in obtaining pastors ... and much of the time there are vacancies in the pulpits." By the mid-1890s Tombstone had no resident minister of any denomination.[17]

This problem had several solutions. Itinerant ministers occasionally reopened a church for a Sunday or two. Circuit-riding ministers made more regular visits to town. In 1886 George Parsons reported a good service performed by a Mr. Haskins, a missionary "with extended jurisdiction reaching to Phoenix and Tucson if not further." Occasionally, a higher church official might pay a visit and give a sermon. The Reverend Bishop Kendrick did this at St. Paul's Church in Tombstone several times. A church might also share or borrow the services of a minister from a nearby town. Both the Catholic and Methodist churches of Tombstone shared a minister with Benson, twenty-five miles distant, and the rector of the Episcopal church in Prescott officiated at a funeral at the Episcopal church in Jerome. Finally, churchgoers left with no other choice could travel to those nearby towns. In 1890 the *Tombstone Prospector* began printing the schedule of services at the Catholic church in Bisbee.[18]

Lay preachers also filled the void. George Parsons found himself drafted into this duty in 1886. On 10 October he noted "S[unday] School as usual, but church unusual as reading the service and carrying it through devolved upon me." Parsons also recorded when other people pinch-hit for the pastor. But as the lay preachers departed with the rest, even that option disappeared. Loss of lay leadership hurt a church just as seriously as the lack of a minister. The Ladies' Aid Society of St. Paul's Episcopal Church of Tombstone replaced both its president and its vice president within a few months in 1885. By the spring of 1888 the monthly meetings of that organization occurred less and less frequently, due to difficulty achieving a quorum. The same things happened in Jerome half a century later.[19]

Eventually, these conditions could prove fatal to a church. After his congregation dwindled from sixty-three down to seven, most of them his relations, lay pastor Sabino Gonzales closed the Mexican

Fig. 8.1. George Parsons photographed in front of Tombstone's Episcopal church on 7 April 1929. The caption penciled on the reverse of this photograph reads: "I was confirmed there by Bishop Dunlop on Feby 24th 1884 & was [the] first & only one at that time. Great Pilgrimage & celebration on above date & photo was made that morning early. Geo. W. Parsons." Perhaps Parsons confused this April date with that of the first Helldorado Days celebration that October. (Arizona Historical Society, Tucson, #5188)

Methodist church he had built in Jerome. It is now a private residence. The town's Haven Methodist Church saw its congregation drop from three hundred to three before the local economy began to revive. Jerome's Episcopal church did not survive. After United Verde shut down in 1953, it was closed and deconsecrated, and its fixtures removed to other churches or vandalized. Tombstone's Episcopal church closed for ten years after 1892, to be revived by a resurgence in mining. Tombstone's Methodist church eventually closed, had its charter removed to Bisbee, and its building was ultimately demolished. In March 1895 a Presbyterian newspaper announced word from Tombstone's minister of "the painful necessity of closing the church . . . on account of the hopeless financial condition of the town, its population having shrunk from 2,000 to 500 during his stay there." In 1900 the synodical mission ordered that church building removed to Clifton, Arizona, where it still stands.[20]

In the face of busttown conditions churches occasionally consolidated in order to keep some religious activity going. In Tombstone several denominations combined their Sunday schools under the town's postmistress, Laura Crable, using the Episcopal church building. When the town revived somewhat after several denominations had departed, a group of residents formed a Congregational church in 1902. Although none of them were Congregationalists, the members of several different denominations coexisted as such. In 1934 Armistice Day and Thanksgiving Day services in Jerome were held in the Methodist church and performed by the Methodist minister, but included the combined Methodist and Episcopal choirs. After mining ended, Jerome's Episcopal church merged with the one in Clarkdale.[21]

Some churches survived. The Catholic churches in both towns had enough support to keep them afloat even during the worst times. Jerome's Haven Methodist Church survived the suspension of mining, and a visiting church official reported in 1934 that the church was "one of the few churches in the country that would be able to close its fiscal year with no debts." When prosperity returned, church organizations could as well. After 1900, during Tombstone's second boom period, local Presbyterians reorganized and held services in city hall and on vacant lots until they built a new chapel in 1904.[22]

Fraternal orders could be even more tenacious than religious denominations. A number of Tombstone's early fraternal orders, including the Ancient Order of United Workmen, the Improved Order of Red Men, the Knights of Pythias, and the Independent Order of Odd Fellows, continued to function a generation after the boom. Numerous other organizations, like the Grand Army of the Republic and the Masons, remained active at the end of the 1880s, after six years of hard times. The same hardiness was evident in the Moose, Masons, Elks, and Knights of Columbus in Jerome. The American Legion post in Jerome not only survived the Great Depression, but played an important role in relief and recovery efforts as well.[23]

The maintenance of these orders was not a luxury affordable only to the elite. Jerome's local of the Croatian Fraternal Union and its Cristobal Colon Encampment no. 59, Woodmen of the World, both survived into the second half of the 1930s, and the Simon Bolivar Logia no. 13, Alianza Hispano Americana, remained a pillar of Jerome's Latino community through the end of World War II. Tombstone's Hebrew Benevolent Society remained active in 1886. Some orders even seemed to gain strength in the face of hard times. In April 1885 the *Tombstone Record* reported the local chapter of the Ancient Order of United Workmen, a fraternal insurance society, "growing at an unprecedented rate." In 1934 Jerome's lodge of the Woodmen of the World also reported quite a few new members.[24]

Perhaps even more astonishing than the mere survival of so many fraternal orders and social organizations was the creation of new ones. In the depths of depression, with mining companies, businesses, and churches boarding up and governments scaling back, the residents of these towns went on forming numerous social and fraternal organizations. The Patriotic Order of Sons of America instituted a lodge in Tombstone in 1887, with an inaugural membership of about forty. Residents even formed a lacrosse club in 1890. Depression-era Jerome was a veritable beehive of social organizing. This fever for organizing produced everything from a revived chamber of commerce to the Mexican Mothers' Club, a softball league, an athletic club, a Spanish club, a symphony orchestra and a military band, the 4-H Club, the PTA, a junior philatelic club, a tennis club, a shooting club, and a young persons' dance club called the Top Hat. Many of these organizations originated in the second half of the decade, under slightly—but only slightly—better economic conditions.[25]

These organizations survived for several reasons. One was their national, rather than local, character. Most of these organizations drew their strength as much from their nationwide or even international membership as they did from the locale in question. They were thus somewhat immune from local conditions, though that was less true during the Great Depression. They were also a bargain; for the nominal cost of a few dollars a year, a member got not only good fellowship, but also entertainment, religious affiliation, or even life insurance. One historian has suggested that these voluntary associations may even have been strengthened by the passing of the boom, as the transience of the population and the availability of the professional entertainments of the flush times disappeared. Formal social organizations also played an undeniably important social role in their communities. Another historian argues that, at least in the nineteenth century, religious, fraternal, and social organizations were the community, providing a sense of belonging to members of a very mobile society, no matter where they might find themselves residing at the time.[26]

Even with their importance, fraternal organizations were not immune to bust. The Alianza Hispano Americana suffered more than a one-third decline in its national membership between 1929 and 1933. Between 1925 and 1940 the Improved Order of Red Men lost 300,000 members, the Knights of Pythias over 500,000, and the Freemasons and Odd Fellows about 1 million between them. A number of smaller orders vanished during the Great Depression, and thousands of individual lodges went under or merged with other chapters. This was not only because of the hard times. In the first part of the twentieth century these orders had seen their welfare functions (injury, health, and death benefits) superseded by governmental and commercial services, while the importance of their other functions (entertainment, fraternity, and assimilation) had waned due to social and cultural changes in the United States.[27]

Still, the local situation mattered, and again money became the issue. Jerome's chapter of the Alianza Hispano Americana saw its cash

Fig. 8.2. Volunteer firemen pose before the firehouse of Tombstone's Engine Company no. 1. Even with its volunteer labor, the company suffered from lost membership dues within a few years of the beginning of the town's decline. (Arizona Historical Society, Tucson, #27673)

reserve dwindle by half between the end of 1929 and the end of March 1931. The AHA economized in June 1931 by discontinuing its telephone service and its rental of the meeting hall in the Miller Building. The secretary of the Highway 79 Association of Yavapai County, a good-roads booster club, noted that the organization had 109 members in 1938, "but few [had] paid their dues." Jerome's American Legion Auxiliary voted to drastically reduce dues for 1933, and Tombstone's Engine Company no. 1 just decided to start over. Its members voted "to rescind all dues up to March 1st 1888, and take a fresh start." As with churches, the worst case would be the dissolution or relocation of an organization, as when Jerome's Odd Fellows and Rebekahs surrendered their charters and transferred their members to the society's Prescott chapters.[28]

People kept up their amusements in the bad times, too, though on a reduced scale. Bust probably had the least influence on these towns' informal social activities. It did not curb the enthusiasm of Jerome's bridge-playing set, and Tombstone's nine continued to defend their town's honor against all comers. Sporting activities continued to fascinate residents of both communities, and both continued to do well despite their general declines. Tombstone's baseball team routinely defeated teams from nearby towns and military posts, and could occasionally afford to venture considerable distances in quest of other prey. The team traveled to Tucson and Phoenix with some regularity and even journeyed to Albuquerque, New Mexico, in June 1888 for a five-game series in which Tombstone's lads were "pretty severely handled." Perhaps that treatment was assuaged by the campaign of 1890, in which the team gained a series of decisive victories, and the *Tombstone Prospector* bragged that they could "lay claim to being the champion club of the territory." Nor were Tombstone's sporting events confined to baseball. In October 1885 a witness reported residents "having horse and foot racing by moonlight."[29]

Jerome's sportsmen had similar success during the Great Depression. The high school teams, the Jerome Muckers, won the northern

Arizona football championship in 1932 and again the following year and captured the basketball championship in 1937. Boxers at the Jerome Athletic Club built a record of twenty-three wins, three losses, and five ties against regional competition by the spring of 1934. The town's baseball team, the Jerome Miners, reorganized in 1934, won the Grand Canyon League the following year, and competed in the Western Regional Semi-Pro Baseball Tournament in Flagstaff against Arizona and California teams in 1938. John Krznarich later remembered that the town always seemed to have good football, basketball, and baseball teams.[30]

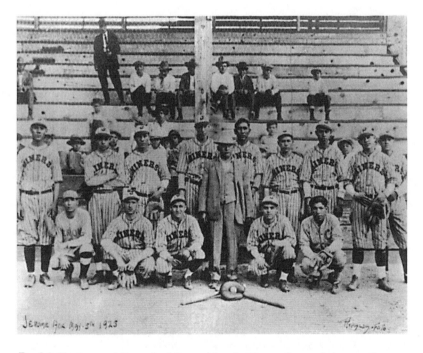

Fig. 8.3. The Jerome Miners (or Mineros) baseball team of 1925. As long as times were good, the mining company could afford to pay for those fancy pinstripes. (Sharlot Hall Museum Photo, Prescott, Arizona, #SP-R-101P)

Social activities and celebrations continued as well. Residents of Jerome continued to raise a Christmas tree in front of Miller Hall each year, attend the occasional society wedding, and picnic on Mingus Mountain. In August 1931 the volunteer fire department staged such a picnic with a variety of activities—including a ladies' baseball game and a men's bathing beauty contest. In 1886 George Parsons described Tombstone's amateur theatrical talent, which included himself, as "quite remarkable," and as late as the end of 1892 the *Prospector* paid the same compliment. Patronage of Jerome's library actually increased in the first half of the 1930s, despite a substantial reduction in the town's population. Reports from the town's librarian indicate that the number of patrons checking out books remained roughly the same, while the number of books they checked out continued to rise slightly every year from 1929 through 1933. At the end of 1933 the library had 667 active members.[31]

The party went on while the ship went down for several reasons. The loss of discretionary income prevented people from paying for professional entertainments. Things like picnics and potlucks, dances and baseball games, cost very little. One historian points out that the bridge clubs so fashionable in the 1930s "required only four ladies and a pack of cards." None of the sports, games, and other diversions described by historian Russell Elliott in his reminiscence of his youth in McGill, Nevada, in the 1920s required much money. While appealing for donations Jerome's librarian noted "that many people find the magazine racks and reading rooms a real privilege in these times of depression." John Krznarich suggested that sociability might have been improved by bust. He remembered a great closeness among his friends, engendered in part by the fact that all of their families lived from payday to payday. And some people even viewed the celebrations and recreations as a psychological antidote to bust. The American Legion post in Jerome, sponsors of the Independence Day celebrations of 1931, urged everyone "to forget the hard times for one day and get out and celebrate with the kids." Even the departure of friends from town might be looked upon as an occasion for a bridge party.[32]

But there were limits; eventually, even the informal recreations and celebrations began to suffer from a lack of reasons and means by which to celebrate. The largess squandered on social activities by mining companies and leading merchants soon disappeared. At the end of 1932 United Verde closed its employee clubhouse, in which company officials had taken such pride, "until the present economic condition improves in the vicinity." It was not reopened until the beginning of 1935. As bust wore on, company ball fields, pools, parks, and playgrounds deteriorated. When a mining company shut down for good, the towns had to take over maintenance of those facilities, if it was taken over at all.[33]

Jerome's mining companies and merchants rendered what support they could. The town baseball teams were special favorites. The manager of the 1936 team "expressed his sincere appreciation to the mining companies, the business men and the fans for the excellent support given during the past season." The town's bands also received such support, and the J. C. Penney store took upon itself the expense of printing tickets for the high school football team in 1931. But bust often forced social organizations to cease soliciting donations from mining companies and merchants and to rely upon such things as dances and private contributions to meet expenses. The *Verde Copper News* announced that Jerome's baseball team of 1932 would be self-sustaining, with "no solicitations of business houses or professional men."[34]

In cases where expenses could not be met, an activity simply had to be foregone—a more and more frequent occurrence as bust wore on. It was indeed a sign of depression days when the *Tombstone Epitaph* announced in 1886 that "the Fourth of July will be spent by our citizens in dignified retirement and quiet contemplation." George Parsons, who did not share the paper's need to put the best face on things, called that Independence Day "a very quiet one . . . as we are all busted." The *Tombstone Prospector* reported of the 1889 Thanksgiving Day celebration "an entire absence of horse racing, baseball, turkey shooting, etc., which formerly made the day a lively one." But during the modest revival of 1890, the town hosted a complete and elaborate Independence Day program, suggesting that these celebrations would return quickly enough with a slight infusion of money.[35]

The same thing happened in Jerome. The Mexican Independence Day celebration of 1928 had included fourteen arrests, mostly for drunken and disorderly behavior. By 1931 Mexican residents reduced what was normally a two-day affair to a one-day celebration "owing to the present depression." Attorney Perry Ling wrote to a friend that it was "the quietest celebration that the Mexicans have ever put on; haven't seen a single drunk all day."[36] .

The following May, Club Estrella threw itself into the breach upon discovering that no other club or lodge intended to sponsor a Cinco de Mayo celebration. The club did not solicit business houses, according to Jose Lopez, chairman of the arrangements committee, but instead sold tickets to individuals "to defray a few small expenses." The American Independence Day passed even more quietly in Jerome in 1932, marked only by an American Legion dance. People who wanted to enjoy a larger observance had to travel to other towns. There was no Cinco de Mayo celebration in Jerome at all in 1933.[37]

The sporting life suffered a similar fate. Jerome's high school football team scheduled only five games in 1932, this after a meeting of area schools that limited the number of games and size of the squads, and at which many delegates "voiced their approval of using present equipment until unusable"—all of this in an effort to reduce expenses. An attempt to revive American Legion baseball—dropped for economic reasons in 1932—got a boost when a local sports enthusiast agreed to help sponsor the team in 1933. But apparently nothing came of the effort. The same poverty prevailed in the adult game. The 1936 town team took the field "after five dull years" for baseball in the Verde District.[38]

All in all, bust produced mixed effects upon the social lives of individuals and the vitality of social institutions in these towns. Because neither of these communities disappeared their social activities had to continue in some form. Their declines produced a perceptible diminution of social activities and severe emotional injury to some people, like Chief Gage, but the structures of social organization and intercourse remained much as before. The same could not be said for the physical structure of these communities.

CHAPTER NINE

As Tombstone Has
Empty Houses to Burn

One of the more spectacular vistas in Arizona may be observed when standing next to the former New State Garage on Main Street in Jerome, looking northeast. To the left, in the far distance north of Flagstaff, the San Francisco Peaks jut over the horizon about fifty miles away. In the middle distance to the right lies the red rock country of Oak Creek Canyon and Sedona. All across the view in the center stand the colorful rock formations of the north wall of the Verde Valley itself. A Yavapai County Chamber of Commerce pamphlet from the 1930s lauding the scenery on the Prescott-to-Flagstaff highway hit the mark when urging readers to take the drive into the Verde Valley: "Mere words cannot describe it although many writers have attempted it." These extraordinary scenes have remained much the same for many thousands of years.[1]

It is in the near distance where things have changed. Clarkdale still lies directly northeast, four miles away and more than eighteen hundred feet below. Although its large stack is gone, the eviscerated remains of the United Verde smelter and its slag piles are still visible. Closer in, the view has changed even more in the past fifty years. The headframes of the uvx are still visible directly below, as is the white mansion that Jimmy Douglas built nearby, but the Little Daisy Hotel stands a gutted hulk to the left, and the little settlement of Daisytown, which surrounded the mine, is gone. Also missing from the modern view is the "Mexican Colony" that lay in the Bitter Creek Gulch along Rich, Juarez, and Diaz Streets below Hull Avenue, as are the houses and facilities at Hopewell and on Sunshine and Company Hills. A ninety-degree turn to the right reveals that the four-story T. F. Miller Building no longer exists and that many of the lesser buildings in the central business district have also vanished.

As mentioned previously, bust produced a surplus of dwellings, business houses, and real estate in an affected community. That was

Fig. 9.1. The Verde Valley photographed from the west end of Main Street, Jerome. The San Francisco Peaks appear at the far upper left, with the Douglas mansion visible at center-right. The headframes of the United Verde Extension lie just to the left of the Douglas mansion, with the gutted remnants of the Little Daisy Hotel at center-left. Clarkdale and the remains of the United Verde's smelter lie in the middle distance at the center-right.

more than an economic condition for those who remained behind;
it had significant consequences for the physical community as well.
Towns full of tent houses and crowded hotels quickly became collections of largely deserted streets and buildings. After Jerome went
into decline, five-bedroom houses could be purchased for fifty dollars
and rented for practically nothing. A rash of arsons of Tombstone's
deserted buildings in 1900 did not trouble one newspaper, which reported that "the damage was not material, as Tombstone has empty
houses to burn."[2]

Eighteen years before, at the height of its boom, the plat of Tombstone Township consisted of fifteen numbered streets running east to
west, and eight named streets running south to north. The town-site
company's plat was overly optimistic. The town of that era covered at
most eleven streets east to west, and six south to north. The company
laid out the town's grid of streets on a mesa just to the north of Tombstone's hills and mines. First Street lies farthest west. Toughnut Street
is the southernmost east-west street, with Allen, Fremont, Safford,
Bruce, and Fulton Streets laid out to the north. By 1882 the town had
made the standard mining-camp progression from canvas and wood
to more durable construction. The *Arizona Weekly Citizen* reported
Tombstone's residents initially reluctant to build in adobe, "but after
two destructive fires, and four years of high insurance, they have concluded that after all there is virtue in the mud bricks." That was a lesson learned the hard way in many mining towns of the West.[3]

Along with more permanent construction came a rationalization and
segregation of functions and activities as a town aged. Mining always
occurred largely outside of Tombstone's limits on the hills to the
south. By 1882 business centered on Allen Street between Third and
Sixth Streets, although quite a bit of business was also conducted on
Fremont Street between Third and Fifth. The liquor trade dominated
the north side of Allen Street east of Fifth Street. East of Sixth the
cribs and "female boarding" establishments of Tombstone's prostitutes

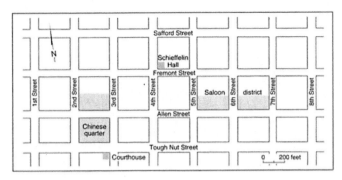

Map 9.1. Tombstone.

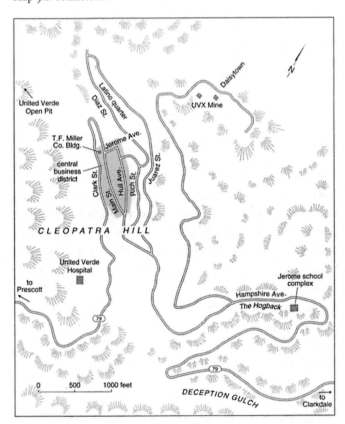

Map 9.2. Jerome.

joined the saloons. Another segregated community, that of Tombstone's Chinese, centered at the intersection of Third and Allen Streets.
But "Hoptown" lacked rigid definition, and many Chinese residents
worked and lived intermingled with non-Chinese neighbors. Enterprises such as livery stables and the lumberyard existed on the periph-
ery of the business district. Residential neighborhoods surrounded
the business district, particularly to the northwest.[4]

Unlike Tombstone, which never filled the boundaries outlined for
it, Jerome overflowed its corporate limits in a patchwork, though not
haphazard, manner. Perry Ling observed in 1933 that "the town limits
cover about one-half of the buildings which go to make up Jerome,
and the line exists only on the map, as the buildings are continu-
ous and extend beyond the old limits for a considerable distance on
three sides," north, east, and west. Its severe topography also shaped
Jerome. It meant that life was easier for workers who lived in proximity to their jobs, a reality that fostered the settlements of Daisytown, Hopewell, and Sunshine Hill. That these three neighborhoods
were company-built marked another difference between Tombstone
and Jerome. United Verde maintained four small settlements beyond
the town limits adjacent to the mine, in addition to its dormitories,
apartments, and houses within Jerome. These last included a collection of dwellings on "Company Hill," above Clark Street behind
the Catholic church, which housed United Verde administrators and
engineers.[5]

Those not in the employ of the mining companies might work in the
business district, which clustered around the T. F. Miller Building, on
the southwest corner of Main Street and Jerome Avenue. Hotels and
small businesses occupied most of the buildings along Main Street and
Jerome Avenue, while those on Hull Avenue were largely devoted to
artisans of the automotive and metal trades. Illicit activities took place
along the northern end of Hull Avenue, west of Jerome Avenue, and
farther down the hill.[6]

Jerome residents lived in a number of neighborhoods other than the company settlements, some of which bore a strongly ethnic complexion. Down the hill behind Hull Avenue on the northwest side of town in Bitter Creek Gulch lay the "Mexican Town." On the east side of town lay the middle-class residential area, and directly below that to the north lay the "hogback ... inhabited chiefly by Slavs." As in Tombstone, these ethnic subdivisions were not absolute. Company officials—not language barriers or ethnic preferences—determined who lived in company housing; however, some integration existed even in the uncontrolled neighborhoods. Deception Gulch, below the hogback along the road to Clarkdale, was a jumble of ethnicities. Vensuello Vermudo, Mrs. Carlota Ruiz, and the Valdez family shared "the gulch" with Mrs. Hawkins, Mr. and Mrs. Owens, and Susie Feindel. The same sort of thing occurred on the hogback above. When a house owned by Angelo Moreno caught fire there, Assistant Fire Chief Joe Pecharich, who lived two doors down, helped put out the flames. Latinos like Daniel Gonzales, who operated a hotel on Clark Street, and Marie Vasquez, the hairstylist, also worked and owned businesses outside of their understood "colony."[7]

Bust had only a limited effect upon the location of these ethnic neighborhoods. Jerome's Latino quarter, below Hull Avenue, extended from the switchback from Main to Clark Streets down to where Main and Hull Streets converged at the Hotel Jerome. It remained in much the same place in the late 1930s as it had in the mid-1920s. Tombstone's Chinese section also remained centered in the same place, though it shrank as the bust wore on. "Hoptown," which occupied two square blocks between Toughnut and Fremont and Second and Third Streets in 1882, shrank by 1904 to the heart of the former area, occupying the southwest corner of Allen and Third and the north side of Allen Street between Second and Third.[8]

Departure of ethnic group members sometimes caused their neighborhoods to shrink, as with Tombstone's Chinese section, but in Jerome the Latino quarter did not at first disappear so much as it dissolved into the rest of the town. Latinos became an ever greater percentage of the population between the onset of the Great Depression and the final mine closure in the 1950s. In 1941 one writer attributed some of the Latino relocation to subsidence damage in their former quarter, but subsidence destroyed only two and one-half blocks of the nine-block "Mexican Colony." What remained could have more than housed the greatly reduced Latino population. But the housing surplus caused by bust meant lower rents for all, which gave some of the remaining residents the chance to improve their accommodations.[9]

"People in Jerome always kept looking for a better house to move into," recalled Rafaila Rodriquez of the 1940s, "trying to get moved into a house where it would have . . . either a big porch or some kind of a yard . . . so we moved several times." This practice was not con-fined to Latinos. In March 1933 Mr. and Mrs. Cowley moved from the settlement at the 500 level of the mine just below central Jerome to a house recently vacated "up on the hill." A month later Mr. and Mrs. Barnes improved their housing in much the same way. As the popula-tion dwindled residents could do their social climbing literally on the hillsides of Jerome.[10]

As with a freezing victim, vitality ebbed from the extremities toward the hearts of these communities. The preliminary census for 1940 esti-mated Jerome's incorporated population as 2,193, and its metropoli-tan population as 3,700, meaning that almost 60 percent of residents lived within the incorporated town. The 1930 census taker estimated that around half of the town's residents lived within the city limits. Charles Shinn had noticed the same trend in Gold Hill on the Com-stock Lode fifty years before, recording that "the lessening population retires year by year from the outskirts, leaving shanty after shanty to rot there, and occupies the better buildings."[11]

At about the same time as Shinn made his observation, Joe Bignon moved several of his houses from Eighth Street to Sixth Street in

Tombstone. The same trend applied to businesses as well as homes. In December 1930 Scott and Gibson announced that they would close their branch store on upper Clark Street in Jerome. They found it "more economical to operate a single establishment," but in so doing they left the people in the upper Clark Street neighborhood without a store for the first time since the postwar revival of 1923.[12]

The problem involved more than simply too many buildings and not enough people. That situation occurred even when an overbuilt boom camp settled into a period of stability. A map of Tombstone in 1882, undoubtedly the camp's best year, shows vacant buildings and empty lots. Another part of the dilemma was that basic upkeep of the physical community depended heavily upon prosperity. A reporter visiting Jerome in 1948 found a strong sign of pessimism among the locals "in the failure of either private householders or businessmen to paint or remodel their buildings." The *Verde Copper News* noticed much the same thing fifteen years before, recording in 1933 that "the depression has scored heavily against any display by shopkeepers." The physical deterioration of a busted community often began long before it became a ghost.[13]

Several historians have noted that boom often produced great municipal undertakings designed to improve a town's infrastructure and that the bust that followed left these towns with overgrown and heavily indebted municipal services. That pattern applies to both of the cases examined here. Both Tombstone and Jerome entered their bust periods with impressive, recently completed civic improvements.[14]

The Huachuca Water Company completed its system—a vast improvement over its predecessors—into Tombstone in 1882. That same year the city gasworks opened. In the flush days of the 1920s, Jerome built a school complex that became the pride of the community. In November 1928 voters approved a thirty thousand–dollar municipal expenditure to improve city hall and pave all of the primary streets. That program was completed by mid-1930, and two years later Jerome's clerk, R. E. Moore, lauded improvements to the town's streets and side-walks, sewers, and fire protection system over the previous six years. "Although a considerable amount of money has been spent in these various improvements," Moore wrote, "the expenditure has been justified and was demanded by the taxpayers of Jerome. The streets were too narrow to permit two cars to pass in safety and the dust and mud intolerable. The fire protection was inadequate, and by reason of extending our lines and installing our equipment we were able to secure a good reduction on fire insurance rates."[15]

At first these projects had been quite necessary and might have seemed an improvement even in the early days of bust, as fewer people placed demands on better services. The *Verde Copper News* reported Jerome's fire losses for 1932 as the lowest in many years. These municipalities also worked hard to keep up vital services despite backsliding economies. The Town of Jerome did its best to keep the streets and sidewalks repaired, purchased a new garbage truck in 1934, and made repeated improvements to its fire protection system. The town had roughly the same firefighting capabilities in 1940 as it had in 1930, with about half of the population. Tombstone's city government looked into building a sewer system in 1888 and maintained three fire companies after 1900, as it had during the flush times.[16]

But eventually these overbuilt infrastructures could not be sustained. At some point the towns had to retrench. In the case of transportation systems, removal rectified the problem. After the United Verde Mine shut down, the Verde Tunnel and Smelter Railroad had no reason to exist, and its tracks were quickly torn up. "Sandy Bob" Crouch's coronation as king of the Tombstone stage operators in 1883 was a crown worth wearing for only the next three years. In June 1886 Crouch announced his intention to ship stock and coaches to the boom at Kingston, New Mexico, and he sold his Tombstone stage lines the following year. After that Tombstone stage operators came and went rapidly.[17]

Systems of infrastructure built into the community that could not be removed soon had to battle insolvency. The operators of that grand and marvelous wonder of the age, the Huachuca Water Company, soon found their enterprise assailed as a rapacious monopoly by the local press and city government because of its high rates. In October 1883 Mayor Carr recommended that the city council purchase the works of the Tombstone Water, Mill, and Lumber Company to "protect the people against the exorbitant prices charged by the Huachuca Water Co." The city purchased the Tombstone's works for around thirteen thousand dollars. The mayor defended his administration against charges of extravagance by arguing that the purchase would shield Tombstone's residents from "the yoke of the Huachuca monopoly." He believed that the purchase would pay for itself by reducing water rates by half. The Huachuca Water Company thereafter faced continual competition and litigation.[18]

To put it simply, the eastern directors of the Huachuca Water Company had gambled and lost. The Indiana company had spent almost one-half million dollars to construct a water system capable of supplying a population of twelve thousand, "which, in the judgement of the investors was to be expected, due to the richness and extent of ore bodies in Tombstone and vicinity." Bust reduced the population to under 20 percent of that anticipated, and the company inevitably foundered. When the company's bonds came due in 1902 it defaulted. Its properties were deeded over to new investors, who reincorporated the works in 1912 with a capitalization of one hundred thousand dollars. In 1942 that company offered the works to the city for only forty thousand dollars, but Tombstone probably acquired them for even less when it took over the bankrupt enterprise in 1947.[19]

Nor did the competition that the Huachuca company excited succeed. In July 1888 R. W. Wood, one of the lessees of the city works, notified the council that he intended to shut them down "on the ground that they were not self-sustaining." City officials spent the next four years engaged in gymnastic maneuvers to keep the works operating or at least keep them enough of a threat to prevent the Huachuca company from raising its rates excessively. The council and taxpayers hoped to accomplish that goal without spending any money to make necessary improvements to the city system, so the town saw four lessees come and go within three years. Boosters and old-timers must have thought it a retrogression when a tank-delivered water service, Banner Water Works, began operating in Tombstone at the end of 1892. With the formation of the Tombstone Consolidated Mining Company after 1900, the city system seems to have been abandoned in favor of water supplied from the mine's pumps by that company's subsidiary, the Tombstone Improvement Company.[20]

It hardly mattered who competed against the Huachuca Water Company; too few patrons remained to sustain the infrastructure, even in cases where a monopoly had been maintained. At the end of April 1888 the *Tombstone Prospector* editorialized that there was "a splendid opening" for business concerns willing to compete with the existing gas and ice companies, both managed by R. W. Wood. The editor complained about the prices charged by the gas and ice works and claimed that town residents were "tired of trusts and monopolies."[21]

Wood—who had taken his share of criticism for his business practices—returned salvo the following day in a letter to the editor. He offered to open his company's books to inspection by an unbiased committee and refund to customers any amount that the committee deemed to be excessive profit. His company sold ice more cheaply than it had been available previously, he pointed out, and maintained that "if the price of wood and water were lower and the population larger we might be able to sell ice cheaper." Wood stated that the reductions the editor demanded "would more than represent our profits" and added that he had "no ambition to pose as a public benefactor when it is in conflict with business principles."[22]

Jerome faced similar problems. The few hundred residents remaining in the town at the end of the 1940s made little use of "one of the finest hospitals in the state, both regarding building and equipment, and professional services rendered," and Phelps Dodge closed the facility in 1950. Residents thereafter used the hospital opened in Cottonwood in 1945, headed by the former chief surgeon of the Phelps Dodge hospital. Jerome's post office faced a similar threat of closure and survived only because former residents made it a point to purchase stamps there for a number of years. Declining school enrollments and growing debt forced the Jerome school superintendent to close the Main Street School in 1939.[23]

School superintendents were not the only ones who made decisions about the fates of structures in busttowns. Mining companies

Fig. 9.2. United Verde built the United Verde Apartments, on Clark Street, to house its workers during the boom years. Although these buildings remain, all of the company housing constructed on Company and Sunshine Hills and at Hopewell has been removed.

determined the disposition of their often numerous structures. In Jerome, Phelps Dodge stripped many of the dwellings on Company Hill, Sunshine Hill, and the 500 level for salvage; removed them whole to other company operations; or sold them for use elsewhere. The town also lost its principal commercial building when the company razed the T. F. Miller Building as an unstable fire hazard. The United Verde Extension's Little Daisy Hotel, abandoned when the company closed in 1938, was gutted for salvage in the 1940s.[24]

Nor were mining companies alone in their systematic destruction or reuse of the buildings of busted camps. A modern observer might be astonished by the degree to which people recycled the physical components of busted mining towns, for reuse locally or at other sites. During its various busts components of Virginia City, Nevada, were scattered to Austin, Reno, Carson City, Tonopah, and Goldfield. Some Comstock structures may even have ended up in Las Vegas. Some authors have attributed this practice to the scarcity of lumber in the arid West. In the frontier Southwest wood could be scarce and expensive, but scarcity did not always motivate this process. Recycled lumber could be transported great distances, even with timber closer at hand. When Goldfield, in the Cripple Creek District of central Colorado, went into decline after 1900, two men began buying up buildings in the town, dismantling them, and removing them to Wyoming oil boomtowns. Further, the woodless West vanished once the nation's rail network permitted lumber to be moved anywhere economically. But people continued to remove buildings and materials from busted towns.[25]

That task was easier when the town in question had a railroad. When the Canadian Pacific removed its shops from Donald, British Columbia, in 1897, the railroad offered to move any building in town to anywhere else on the line without charge; everything went. But even without a railroad, residents often moved structures—sometimes large ones—considerable distances. When Runnymede, Kansas, declined, the three-story Runnymede Arms Hotel was moved at least twenty-five miles south into Oklahoma. In an instance of the surprisingly common practice of sectioning big buildings for relocation, the large opera house and the hotel in Ulysses, Kansas, were both cut into three sections and removed to a new town site three miles away. In 1958 Oliver Tremewan purchased an apartment building in Mountain City, Nevada, which he and his father, Charlie, had built two decades before—in Rio Tinto.[26]

Some relocations occurred more than once. The first permanent buildings in Tonopah, Nevada, were hauled in from the faded camps of Candelaria and Belmont. When Tonopah moved on to erecting more substantial structures, some of these buildings moved on to the new strike at Goldfield. St. Luke's Church, built in Rosita, Colorado, was removed to Silver Cliff and then again to Westcliffe. Many of the houses in Maiden, Montana, were removed upon that town's decline to Kendall. When Kendall busted in turn in the 1920s, some houses moved again, becoming dwellings or outbuildings on nearby ranches. Spring Valley, Wyoming, got many of its houses from Almy in 1899, and when its mines closed in 1905 many of those same buildings reappeared in Hanna. In 1884 Benson, Arizona, received the stamp mill from Galeyville, east of Tombstone. After a few years of operation in Benson, that mill migrated to Mexico.[27]

Thus, buildings could tramp from boomtown to boomtown much as people did. During the Texas oil boom the fifty-room Rooney Hotel moved from Big Lake to Fort Stockton in 1925, removed from there to Pyote in 1928, and thence to Hobbs in 1930. The twenty-two-room Higgins Hotel, built in Wink, Texas, in 1927, moved sixty miles to Mentone in the late 1930s. One historian of the oil booms observes that it was quite possible for a booming oil-field worker to sleep in the same hotel building in several different towns. By the 1920s oil companies built completely mobile houses for their workers, which they relocated from camp to camp on rail flatcars. By the 1930s modern mobile home trailers began to appear in the oil fields.[28]

But even fifty years before, Tombstone had "mobile homes," though these did not move as easily or neatly as the modern versions. After the boom days people tore down perhaps hundreds of Tombstone's houses and relocated them to nearby camps, such as Pearce and Bisbee. One longtime resident of Tombstone claimed those towns "were largely built out of Tombstone houses in the early days." One person advertised in the *Tombstone Prospector* in 1890 for "a frame house of two or three rooms. Want it to take down and move away."[29]

The problem of mobile homes in Tombstone became worrisome enough to some people that in the spring of 1887 Councilman Stan Bagg introduced an ordinance to retard the trend. Approved four votes to none, the ordinance made it unlawful to remove any structure—occupied or not—from within city limits without a permit from the mayor or a councilman. The permit had to be countersigned by the chief of police, describe the structure, and indicate from and to where it was being moved. While the ordinance may have had little appreciable effect on the preservation of the town's physical character, it does indicate that the physical dissolution of Tombstone alarmed some of those who remained in the town.[30]

The same sort of exodus occurred in Jerome, with houses from Company Hill moved by flatbed to Prescott, Oak Creek, and the Verde Valley. After they decided to relocate to Prescott in 1944, Rafaila Rodriquez and her husband tore down their house in Jerome and rebuilt it on a lot they owned in Prescott. While her husband continued to work in the mine, Rafaila and her children dismantled the house. They salvaged everything from windows to floors to plumbing; they even sorted and saved the nails. When off shift her husband would drive everything to Prescott and put it on their lot. Although the Rodriquez family may have gone to the extremes of salvage because of wartime shortages, their case is extraordinary in degree rather than kind.[31]

Many salvaged materials also stayed in town, to be reused by the remaining residents. By 1936 Jerome children daily carried home loads of kindling salvaged from the debris piles of demolished buildings. Joseph Curtis, born a few miles north of Tombstone in 1882, recalled that his father bought old houses in the busted camp for twenty-five to fifty dollars each and used that material to improve their ranch. By 1900 surplus houses in that other great silver queen of the 1880s, Leadville, Colorado, were being dismantled for firewood.[32]

The desire to reduce taxation also prompted people to demolish buildings. Governments tax a vacant lot at a much lower rate than one with a structure on it. With bust wearing on and no prospect of tenancy, many property owners demolished their structures rather than pay the higher taxes. This action apparently took place quite frequently in Jerome. The *Verde Copper News*, railing against excessive taxation, opined that property owners might soon reach the point where "it will be cheaper to tear down buildings, thus creating unimproved properties, or allow the property to be taken over by the state, than to pay taxes."[33]

Then there were the depredations of vandals. After uvx closed the Little Daisy Hotel in 1938, vandals broke most of the windows, "carried off the basins, tubs and toilets, [and] even dug the pipes out of the walls" before the company could sell the building for salvage. These scavengers could be completely brazen. One of Jerome's residents who remained after the closures came across two tourists waddling down Main Street one day carrying a "souvenir" bathtub.[34]

A little publicity could be deadly. Shaniko, Oregon, had been forgotten until newspaper stories about its quaintness appeared; the vandals appeared shortly thereafter. After the *Great Falls Tribune* ran an article about Comet, Montana, in 1970, one resident reported his grandmother's house torn apart, her piano stolen, and all the doors and windows in town removed. He believed that more damage had been done to the town in three weeks than in many years previously. People scoured these historic places, removing bricks from old coke ovens to make fireplaces and tearing out weathered wood to panel their dens. Most of the brick buildings in Aurora, Nevada, remained in good shape for decades after the town's abandonment, but, as one historian records, "when used brick became fashionable for the fireplaces and patios of expensive homes, Aurora was doomed. Its brick buildings were sold to construction firms from southern California, and almost nothing of the town remains today. Yet, in a manner of speaking, Aurora lives on in the homes of Beverly Hills, Pacific Palisades, and Malibu." This lack of historical consciousness might seem outrageous— were it not itself a historical tradition. By 1869 vandals had removed doors, windows, furniture, and even whole houses from Meadow Lake, California.[35]

When contemplating the fates of mining sites, it is important to think in larger terms than individual towns. A better unit of organization, the mining district, encompasses not only the place where ore is removed from the ground, but also the milling, smelting, and transportation centers that support that undertaking. If these supporting towns had a previous existence as farm towns or supply centers, they might survive the mining camp they served. If not, the death of the mining camp usually doomed these sisters as well.[36]

The Tombstone District consisted of the silver camp and also the mill towns of Charleston and Contention City, several isolated mill sites, and the rail town of Fairbank, all of which lined the San Pedro River. The bad news for those towns began with the discovery of water in Tombstone's mines and subsequent plans by mining companies to relocate their water-hungry mills in proximity to their mines. The strike of 1884 further damaged the mill towns. By the end of June 1884 the *Arizona Daily Star* reported the businesses of Contention, Fairbank, and Charleston "prostrated.... Trade is at a standstill." Charleston lost two-thirds of its registered voters between 1882 and 1886. In the latter year its mills were dismantled and removed. Charleston's post office closed in October 1888, a month before the post office at Contention City closed.[37]

With less and less material to transship, Fairbank dwindled away. Contention City lingered until the turn of the century, when the resolution of a land dispute in favor of a cattle company forced the remaining residents to leave. One magazine described Charleston in 1889 as experiencing "a season of prolonged and comparative quiet." Even the town's status as the closest civilian outpost to Fort Huachuca could not save it after the Tombstone Mill and Mining Company's mill shut down. Built primarily of adobe, most of Charleston later washed away in floods, although much of its lumber was salvaged and removed to the gold excitement at Pearce in 1890. Enough of Charleston's ghost survived to serve as "Little Tunisia" for the Ninety-third Infantry Division from Fort Huachuca during World War II—an unusual but

Fig. 9.3. The adobe ruins of Charleston in the Tombstone District, ca. 1890. The site served as "Little Tunisia" for United States Army infantry training during World War II. (Arizona Historical Society, Tucson, #44522)

not unique fate. During the war the army used Ashcroft, Colorado, to help train the Tenth Mountain Division, and Spenceville, California, hosted house-to-house combat training.[38]

The Verde District consisted of Jerome and its two smelter towns, Clarkdale and Clemenceau. Unlike Jerome, Clarkdale was a company town, solely created by the United Verde Mining Company. By 1930 the company had built more than five hundred dwellings in Clarkdale, along with two hotels, a golf course, and a substantial clubhouse. United Verde managed its unincorporated town through the Upper Verde Public Utilities Company, which provided the police force and maintained the streets and sanitation systems. The utilities company and Clarkdale were part of the package when Phelps Dodge bought out United Verde in 1935. Phelps Dodge maintained the system until it shut down the United Verde Branch in 1953. It then owned a company town for which it had no use. Phelps Dodge demolished some of Clarkdale's structures, sold most into private ownership in the 1950s, and the town eventually incorporated and survived. Clemenceau, the UVX smelter town, quickly got out of the mining business after UVX shut down in the late 1930s. The smelter soon disappeared, and nearby Cottonwood, never really part of the mining district, absorbed Clemenceau.[39]

Clarkdale may be counted among a lucky minority; many former smelting, mining, logging, and farming communities completely disappeared. Any company town could be blotted out of existence on the orders of company officials far away. John Spratt, who grew up in the company coal town of Thurber, Texas, wrote that when its time came, "the reduction of Thurber was ruthless. The company ordered everything on the surface to be torn down or dug up." At Rio Tinto, Nevada, the company even uprooted the trees and replanted them somewhere else. Phelps Dodge carried out this treatment in a number of its towns, including the coal town of Dawson, New Mexico, once the largest company town in the West. Dawson's coal washer went to Harlan, Kentucky; its church to the local diocese; and four hundred of its dwellings to two

Albuquerque men, who sold or dismantled them. When the dust settled only a few houses, two smokestacks, and a cemetery remained.[40]

A variety of noncompany ghost towns have suffered numerous interesting fates. A surprising number of these places are now underwater, particularly in California and along the Columbia River, where huge hydroelectric projects are especially prevalent. But even places like Calliham, Texas; Randolph, Kansas; and Bonito City, New Mexico, have been submerged by the U.S. Army Corps of Engineers or the Bureau of Reclamation.[41]

Towns also die from natural causes. Residents of mining towns usually lived in morbid fear of fire, and not without cause. Almost every mining town of consequence in the West suffered at least one serious fire. Such a disaster could be overcome if a town was booming, as happened twice in Tombstone and thrice in Jerome. But if the fire occurred after the decline had begun—as it did at Quijotoa, Arizona; Anaconda, Colorado; and Hamilton, Nevada—it might be the final step on the way to oblivion. When a fire destroyed thirty-seven buildings worth over one-half million dollars in Rawhide, Nevada, in 1908, most of the town's residents decamped as quickly as possible. Much the same thing happened in Goldfield, Nevada, fifteen years later. A number of towns, including Olalla, British Columbia, were already ghosts when razed by fires. Nor did the fire have to reduce the entire town. Residents largely abandoned the lumber town of Witney, Oregon, after a fire destroyed its sawmill in 1918. Kelleyville, Texas, died after a fire wrecked the Kelly Plow Company's plant, which the company rebuilt in another town.[42]

Floods and avalanches also threatened many towns in mountain valleys. Avalanches crushed Alta, Utah; Illecillewaet, British Columbia; and Masontown, Colorado, while Panamint City, California; and Mazuma, Nevada, among others, washed away in floods. Occasionally, tornados destroyed prairie towns, like Sumner, Kansas, or Mobeetie, Texas, and sometimes the fates just seemed to conspire. Burke, Idaho,

was almost obliterated by fire in 1923, having been previously damaged by a snowslide and a forest fire thirteen years before.[43]

In an ironic form of cannibalism, mining often consumes mining
towns. Early placer camps, like French Gulch and Parks Bar, California, had their very foundations undermined in the frenetic sifting of
their gravel deposits. The likelihood of this occurrence increased with
the evolution to hydraulic and dredge mining. Residents of Moore's
Flat, California, disassembled their town and rebuilt it one-quarter of
a mile away, "the caving of the surface of the hydraulic mine [having]
caused the necessity of moving." The La Grange hydraulic mine eventually buried the village of Oregon Gulch, California, under twenty or
more feet of debris. Only the town's church was salvaged and relocated.
Underground mining, the next phase in the typical development of
mineral extraction, usually caused the least disturbance of the surface ground, though it did produce the mine dumps and mill tailings
that ultimately consumed neighborhoods or even whole towns, like
Kimberly, Nevada, and Kokomo, Colorado.[44]

The most recent phase of mining—with its massive open pits, leach
pads, and tailings dumps—most threatens the structures of the past.
The ever expanding Lavender Pit in Bisbee consumed the settlements
of Upper Lowell, Jiggerville, and the Johnson Addition, and left Lowell itself with little reason to survive. Nevada Consolidated Copper
Corporation's Liberty Pit has destroyed several renditions of Ruth,
Nevada. Kennecott Copper Company's Ray Mines Division uprooted
the three settlements of Sonora, Barcelona, and Ray, Arizona, and
transplanted them to the town of Kearny eleven miles away, all in
order to create an open-pit mine at their former sites. Some of these
places were much larger than hamlets. Santa Rita, New Mexico, a town
of about six thousand inhabitants, has been replaced by the huge Santa
Rita open pit. Much the same eulogy could be delivered over Monarch,
Wyoming; Metcalf, Arizona; or Winfield, Colorado.[45]

If a community survived destruction by its own mining, it still had to deal with the environmental residua of its boom and bust. Tombstone experienced some environmental problems due to mining. The city has had to fill some holes caused by mine subsidence, but because most of Tombstone's mining occurred in the hills outside of town, these problems have been slight. The same cannot be said for the subsidence of Jerome.

Mining, with its excavations and blast vibrations, occurred under the town of Jerome, and the town itself was built into steep hillsides. The same Verde Fault from which Jimmy Douglas made his fortune ran through the town and added to its geological instability. Smelter smoke from the previous generation may have reduced local vegetation enough to cause erosion, which increased the subsidence problem. By the late 1920s residents understood that what had seemed at first to be an interesting novelty was actually a serious problem. At least ten buildings in the business district were condemned in 1928. By 1930 cracks in the Shea Building were "widening daily," and plans were afoot to hold the Bank of Clemenceau building together and in place with steel cables. The subsidence problem got worse in central Jerome as the 1930s progressed. Although one United Verde official claimed that the town's subsidence problem was "in no way related" to subsidence in the company's mines, business owners and town officials plagued by it thought otherwise.[46]

Property owners in a large area of the central business district of Jerome watched their investments literally slip away. One engineer wrote in 1936 "that Jerome offers some striking examples of the extent to which buildings may be racked and distorted without falling." By the mid-1930s many of the buildings in the central business district, which lay roughly on Main and Hull Streets between the Clark Street switchback and the junction of Main and Hull, were in serious trouble. Property owners eventually had to demolish dozens of buildings declared unsafe by the town's government. Casualties included the post office, the city jail, the J. C. Penney store, and the Ritz Theater.[47]

The Ritz did not go down without a fight. Its owner, Phil Pecharich, disputed the town's contention that his building was beyond rescue. The town clerk notified Pecharich of the decision reached at the town meeting of 12 November 1935, that he must make a series of repairs to his building or see it closed as unsafe. In March 1936 an architect recommended to the council that the Ritz be torn down. Pecharich first threatened to sue the city if it tried to have his building demolished, then agreed to have another examination by a three-man panel of engineers, the city to bear the cost and both sides to abide by the decision of the majority. The results came in only five days later. All three engineers agreed that the building was unsafe, and Pecharich had to construct another theater to replace his demolished Ritz.[48]

Subsidence continued throughout the decade. Joe Larson reported in March 1937 that the back of the Jerome Meat Company building had

Fig. 9.4. Jerome's central business district in the early 1930s, photographed from Cleopatra Hill. Compare this to the modern town shown in figure 9.5. (Lewis McDonald Collection, Cline Library, Northern Arizona University, #435.4)

Fig. 9.5. A modern view of central Jerome from Cleopatra Hill. Much of the downslope side of eastern Main Street was lost to the subsidence that plagued the district in the 1930s and 1940s.

fallen out. That same month John Perez wrote to his landlord detailing the repairs he had made to the building he rented. Perez claimed that buildings in Jerome needed constant repair at continual expense. He told his landlord that if he had not made the repairs, the building would have been ruined already.[49]

In an effort to forestall litigation and close out his affairs in Jerome, Jimmy Douglas authorized uvx to make extensive purchases of damaged buildings. The company purchased almost all of the buildings mentioned above, as well as many others. uvx paid for Phil Pecharich's new theater. Joe Larson believed that uvx also took advantage of tax sales to acquire damaged properties inexpensively. But even with this liberal policy uvx could not purchase every property or satisfy every property owner, and so lawsuits resulted.[50]

Some property owners settled with the mining companies rather than face the uncertainties of litigation, and hindsight validates that course. Plaintiffs had great difficulty proving their cases against United Verde and uvx, despite all of the evidence suggesting at least their partial responsibility for the subsidence damage. The first property owners to try their hand in court were Jose and Pilar Franquera, who alleged in a 1931 lawsuit that uvx undermining had destroyed their twelve-room house. A federal court jury favored the company after four hours of deliberation.[51]

As the decade wore on, the number of legal actions against the companies increased. Joe Larson thought that all property owners in the city had a case against the mining companies, and he mentioned a discussion among some of them about the possibility of a class-action lawsuit. These property owners held their fire, however, pending the outcome of a suit by John Sullivan, owner of the Sullivan Hotel.[52]

The Sullivan suit was the most important litigation in the subsidence disputes by a private citizen. Sullivan sued uvx in April 1938 for damages to the hotel and his Wigwam Rooming House. He claimed that the mining company had undermined his properties, which were uninhabitable in spite of his best efforts, and he asked a total of $87,200 in damages. uvx countered by blaming Sullivan for lack of upkeep and disclaiming any responsibility for Jerome's subsidence. Although the suit seemed to make the question of blame for Jerome's subsidence the issue, uvx effectively avoided judgment on that matter. The mining company won the case in June 1939 after they produced a letter that Sullivan had signed in 1934 releasing them from damages. After the Sullivan decision, the lawsuit fever broke. Merchants such as Emil Kovacovich preferred to settle rather than risk losing all, and the legal contests diminished in importance.[53]

Jerome's property owners probably felt that the mining companies were trying to extract themselves from their obligations as cheaply as possible. Joe Larson, who attempted to get money out of uvx for one of

Homer Nihell's properties, wrote to Nihell that "they surely don't pay anything unless they are forced to." Nihell replied to his friend: "I really don't think they have any intention of paying a fair price, they figure it is the [Christ]mas season and the dear public has grown whiskers."[54]

For their part, mining officials may have felt that property owners were trying to cash out of a busted real estate market at their company's expense. Larson admitted at various times during the Depression that real estate was not selling, that everyone was short of cash, and that "one could not borrow two bits from the bank to apply on real estate here." His advice to Nihell to take a hard line with UVX because "they can well afford to pay your price for it, and sooner or later they

Fig. 9.6. Subsidence damage to the Kovacovich Merchantile Company building on Main Street, ca. 1935. Emil Kovacovich, among other merchants, sued the mining companies to recover for damages caused by the subsidence, but withdrew his claim after hotel owner John Sullivan lost a similar suit against the companies. (Lewis McDonald Collection, Cline Library, Northern Arizona University, #435.38)

will have to," certainly smacks of an opportunism that company officials may have suspected of many claimants.[55]

Evidence indicates that the subsidence cases divided residents
into pro- and anticompany factions and cost the companies some of
the esteem that they had previously enjoyed in the community. Even
given this divisiveness and loss of prestige, it is hard to imagine that
environmental questions would have received the attention they did—
particularly among those not making subsidence claims—had mining remained the heart of the town's economic order. In part this is
because that era did not emphasize environmental quality as later generations would, but also it is because these were mining communities
whose residents understood both the opportunities and the costs of
that endeavor. The *Verde Copper News* was probably right when it claimed
during United Verde's shutdown that "the sulphur smoke about which
nearly everyone in town griped would be welcomed back with shouts
of joy.... [L]ocal residents would cough their heads off and think it a
great privilege." Mining undeniably transformed the local environment in many ways, but outrage about those transformations is more
the product of a later era.[56]

The clamor that existed over subsidence problems centered more
on who should bear the costs of repair. The Town of Jerome also made
claims against the companies. By 1931 the *Verde Copper News* reported
that "numerous breaks" in the city water mains caused by moving
ground had been repaired within the previous year. But not until August
1935 did a group of city taxpayers—mostly prominent businessmen—
petition city council, asking them to bill UVX for subsidence damage
to town property. In December of that year the council instructed the
town clerk to present UVX with a bill for nearly $3,000 in subsidence-
related repairs. The town continued to fight subsidence damage to its
properties piecemeal for the next two years.[57]

The council renewed its attempt to gain compensation for subsidence damage in the fall of 1937. On 9 November the council ordered

Fig. 9.7. A modern photograph of lower Main Street, Jerome, showing the main area of subsidence. Compare this to the 1930s view in figure 6.3 on page 127.

the preparation of a statement of damages, which was presented to the council on 23 November. The list itemized fifty-eight structural, thirty-four sewer, and seventeen fire system damages and valued them at a total of $91,706.25. At its meeting of 15 March 1938 the council instructed the town clerk to notify Jimmy Douglas and uvx manager George Kingdon that the town intended to bring suit over the listed damages, to submit a bill to uvx for same, and to inquire whether the issue might be arbitrated between uvx and Phelps Dodge and the town to avoid the litigation. On 12 April, Town Clerk R. E. Moore and Town Attorney Perry Ling reported that at a conference, uvx officials had stated "that they were in no degree responsible for any portion of said damage," whereupon the council unanimously instructed Ling and Moore to file suit and to secure the services of an engineer to collect the necessary data. The engineer submitted his itemized report of damages to the town in July, putting the total damages at $134,871.16.[58]

The Town of Jerome filed suit against uvx and Phelps Dodge in U.S. district court in July 1938, claiming to have been wrongfully and negligently undermined, alleging subsidence damage all over town, and asking for the compensation figure determined by the engineer. There the matter stood until 22 February 1939 when town and mining officials met at a special session of the town council. At that meeting town officials agreed to accept $53,500 as full settlement of all claims, past and future, against both uvx and Phelps Dodge and to drop the suit. A month later the council put the $47,150 remaining from the settlement after expenses into a newly established town improvement fund, upon which the council would draw "for improvements to the streets, sidewalks, fire and sewer mains and other like improvements." As part of the agreement, the mining companies continued to deny any liability, but one official pointed out that because, as principal taxpayers, they would pay a large share of the repairs anyway, it was in their interest to end the controversy and have those repairs made quickly.[59]

The settlement did not end the controversy, however—or the subsidence. In March 1941 Joe Larson reported that the state had taken over road repairs beyond the city's capacity and was patching Main Street with truckloads of fill, which did not stay in place for long; Larson mentioned that during recent heavy rains both Hull and Main Streets had settled one foot per day and that a fill that raised Main Street five feet lasted less than a week. As late as 1948 crews rebuilt the highway through town every six months or so. By then uvx was long gone, and Phelps Dodge would discontinue its operations within five years. The burden of continued repairs to subsidence damage had fallen to Jerome's government, which faced a host of other problems brought about by bust.[60]

CHAPTER TEN

Due Economy in Town Administration

Dr. F. W. Boville, a member of the Prescott School Board, had made a simple suggestion, and now he found himself besieged. Boville had proposed a ten-dollar increase in the county school tax levy, but in the depths of the Great Depression in July 1932, he found his plan assailed as a "bunch of foolishness." That critique came at a meeting held in Prescott on 11 July packed with representatives from Jerome, Clarkdale, and other Yavapai County towns. "It was a stormy session from start to finish," the *Verde Copper News* reported the next day. "It was virtually impossible for a speaker to give his version of the proposed increase . . . without being interrupted by another present." The disaffected from Jerome and elsewhere accused Boville of trying to make them support Prescott's schools, and one of Jerome's representatives declared that "what this county needs is sanity in looking at the tax problems. . . . Instead of talking [about a] raise in the per capita tax, we should cut $10 from the present tax." The *Verde Copper News* concluded that the meeting offered a "sure sign by the great majority of those present that they did not favor a raise of any kind in anything that sounded as if it might increase taxes."[1]

Boville retreated from his suggestion two weeks later. In a letter to the county's school superintendent withdrawing the idea, Boville admitted that "when we asked for the additional county per capita levy, I did not know it would cause such a storm of protest from the Jerome and Prescott businessmen." If the storm surprised him, the sentiment behind it should not have. The ferocity of the reaction might be attributable to lean times, but the sentiment had as much to do with that era's general attitudes toward taxation and the role of government—attitudes sharpened under the duress of bust.[2]

Things started out well enough for both Tombstone's and Jerome's town governments. By the end of their boom periods, each town had engaged in municipal improvements, and both had sound financial structures. Mayor John Carr informed Tombstone's residents that revenue from licenses, rents, and fines would be "ample to pay the running expense of the city" for the year 1884. At the end of 1928 the *Verde Copper News* reported that Jerome had "a smaller outstanding indebtedness per capita than any other town in Arizona."[3]

Those situations changed quickly. Bust affected the financing of a town's government almost immediately. Within a few years of the onset of the economic decline, Tombstone's government was in serious trouble. The town council found it increasingly difficult to conduct business due to the departures of various city officials from the district. Delinquent tax payments plagued the town, which had little choice but to settle for the best revenues obtainable. The council manipulated accounts to pay the salaries of city employees, or attempted to pay them with city warrants, which employees and others providing services to the city resisted or refused. In the second half of the 1880s Tombstone's government struggled with a growing debt and increasing difficulty meeting its obligations. Jerome, too, faced ever longer odds.[4]

Local government in those days could hardly be termed an ambitious undertaking; neither Tombstone nor Jerome had a large or pretentious municipal bureaucracy. In the opinion of most residents from the frontier era into the 1920s, municipal government had only limited duties to perform. These included protecting the citizenry from criminals, infectious diseases, and fires. Often residents moved to secure even these protections only after learning repeated hard lessons, as in the case of fire protection. Town government might also be asked to build and maintain road networks, or sewer and garbage disposal systems, but that concluded the list of its duties.[5]

None of this agenda aroused much interest among the populace; for the most part the sport of local politics had few followers. Although occasionally more tumultuous, city elections in Tombstone and Jerome usually drew voters by the dozens. A historian of Nevada's White Pine District calls its miners "politically passive," and believes they were "hardened individualists who paid little attention to community affairs unless their personal interests were threatened . . . [and] were exceptionally casual about paying their property taxes." That ethos, though less true of Jerome in the 1920s, might have seemed familiar in frontier Tombstone, at least in politics.[6]

When moved to express themselves, Tombstone's residents of the 1880s voted predominantly Republican, though both Cochise County and Arizona Territory had Democratic majorities. Residents of Jerome, Yavapai County, and Arizona voted mostly Democratic in the 1920s. The Democrats normally prevailed throughout Arizona from the frontier era until after World War II. While Tombstone usually had newspapers representing both political parties, the *Verde Copper News* was a moderately Republican organ. Its editor had more sense than to waste his energy trying to get people to vote the Republican ticket in county and state elections, preferring to express his partisanship on national issues. Partisan politics had little significance locally because everyone agreed about the most important feature of a town government.[7]

The most desirable town government spent the least money. The smaller and less active it was, the better. Tombstone's elected city establishment at the height of its boom consisted of a mayor and four councilmen, a police chief, tax collector, assessor, recorder, treasurer, and attorney. Jerome, though a larger city in a later era, had a proportionate number of town officials and employees. A historian of California gold camps holds the functions of their town governments to be "promoting business and holding expenses to a minimum." That assessment could also be applied to the governments of Tombstone and Jerome.[8]

Even during the boom days many people believed "economy" and "efficiency" to be the most desirable qualities of government, paradoxically, even, the most important qualities of all. The *Verde Copper News* opined in 1926 that "the very strictest economy should be the chief requirement of those who will seek state office this fall." That same year the paper praised Jerome's "splendidly good administration" of the previous two years and maintained that "the thing that appeals to all of us is due economy in town administration. . . . Assuredly we should vote for the men who are able and willing to continue the present adequate system of economy." The newspaper challenged opposition candidates to tell the electorate how they would bring about "greater economy and more efficient administration." The same urge for low-cost administration extended from local to national government.[9]

Neither newspapers, mining companies, nor business leaders opposed what they considered to be worthy improvements. The *Verde Copper News* supported the proposed improvements to the town hall and Jerome's streets during the bond election of 1928. Good roads improved business, after all. The monster to be battled was excessive taxation, and mining company officials realized who would bear most of the tax burden. When the Arizona State Board of Equalization published its 1929 valuations for Yavapai County, more than half of the total figure covered productive mines. The county's standard-gauge railroads valued another 13 percent of the total. Town lots and improvements, the wealth of merchants and residents, accounted for only 9 percent of the county's assessed valuation.[10]

Once things began to go sour, the normal belief in the evils of taxation and excessive government took on a new urgency, and criticism of the degree of taxation crescendoed. The fact that Jerome did not have as high a tax rate as nearby Prescott did not seem to matter. At the time of the United Verde shutdown, the *Verde Copper News* editorialized that "tax gatherers overlook the simple fact that in times of distress, the more taxes that are taken from the people, the less money there is for productive enterprise. And it is from productive enterprise that all taxes and employment must eventually come." The editor went on to prod government officials, arguing that individuals and industries had readjusted to the hard times and that taxing authorities ought to do the same, "or admit to the people that they are incompetent to adjust government expenditures to government incomes." Soon, heavier hitters than the newspaper found the range.[11]

At the end of 1931 uvx president Jimmy Douglas wondered in a letter published in the *Prescott Courier* why the citizens of Arizona did not "rise up in their wrath against unwise, unnecessary, wild-eyed extravagance on roads, schools, and general administration." Did they not know, he queried, "that the goose that lays the golden eggs is sick?" He asked and answered: "Who is going to pay for the expenditures I have noted? Not the mines." In 1934 the assistant general manager of uvx blamed high taxes for much of the plight of Arizona's mines, and a few months after that the president of United Verde, Robert Tally, sounded off on the subject of taxation. Tally unloaded on government officials in a *Courier* article headlined: "Truth about Mine Taxation Set Forth; Industry Can't Bear Existing Burden and Live; State Must Cut Its Budget."[12]

Tally opened his article by stating his opinion that certain politicians were using the mine taxation issue to stir up class prejudice for their own gain. He then presented figures in support of his case for lower taxes. Tally believed that excessive taxation hurt all property owners, but especially mines, which he held to be overtaxed in the first place. He thought that solving the crisis would require streamlined government and equitable taxation. "Discrimination against the mines in the matter of taxation or otherwise," he concluded, "is detrimental not only to the industry itself but to labor, agriculture, and to the general public."[13]

For their part, government officials found themselves in the maw of an ore crusher, squeezed from all sides without apparent means of escape. As the economy collapsed so did the property valuations upon which tax revenues were based. At the end of 1931 Governor George Hunt wrote to a fellow governor that Arizona's valuations were "rapidly decreasing," and he described the mining industry that had previously paid almost half of the state's taxes as "prostrated." By 1934 valuations statewide had sunk even further due to the shutdown of most of the major mines—the valuation of a producing mine being much greater than that of a dormant one. Certainly, a rough correlation existed during the 1920s and 1930s between the price of domestic copper and assessed valuations in Yavapai County. Because the mining industry paid most of the taxes, local and county governments suffered accordingly during shutdowns. In 1929 the assessed valuation of the county's mines, mills, and smelters totaled over $58 million, this being more than 61 percent of the assessed valuation of all property in the county. By 1935 the total valuation for all property in the county was only $44 million, only one-third of that figure being assessed against mining properties.[14]

A serious decrease in nontax revenues formed the second jaw of the crusher. In 1929 the Town of Jerome's justice system collected $17,735 in fines. In 1933 the same source provided only $585 in revenue. Although income from fines improved somewhat after the mines reopened, the best year of the decade, 1937, brought the town less than $3,000 in revenue from that source. Those unable to pay their fines had to be punished by incarceration, which placed yet another financial burden on the city. The town's magistrate mitigated that problem somewhat in the mid-1930s by sentencing troublemakers to time on the city woodpile, sandpile, or garbage truck or at the city yard. In one case a convict satisfied the needs of justice with the payment of three cords of wood. But this creative solution to crime and punishment during bust did not compensate for the absence of boomtown justice revenues, and bust also greatly diminished business license income. At the beginning of 1882 licenses accounted for over half of Tombstone's income. As the bust dragged on and businesses boarded up, that license revenue declined. Jerome, too, experienced a substantial reduction in license fees and sales taxes by 1932. Neither town was unique in that regard.[15]

Caught in the maw, town officials had to choose between two evils: increasing indebtedness or raising the property tax rate. Tombstone's officers—their maximum tax rate established by law—chose the former course and Jerome's the latter, neither with conspicuous success. Tombstone's town council set the tax rate at 0.9 percent for fiscal 1884 but thereafter into the 1890s set it at 1 percent. That, with a 0.5 percent surtax for bond redemption starting in 1888, produced the legal maximum of 1.5 percent. The town government paid for this low tax rate with an ever expanding indebtedness. Tombstone's $11,547.60 debt of 1883 climbed to $14,173.16 by the end of 1889. By the spring of 1890 Tombstone's debt had surpassed $17,000, and was near or over the legal limit, depending upon whose figures one believes.[16]

In the face of declining valuations and revenues, Jerome more than doubled its tax rate between 1929 and 1933. Jerome's residents paid a basic tax rate of 1.674 percent in fiscal year 1929–1930, with an additional assessment of 0.323 percent for bond redemption and interest payment. By fiscal year 1933–1934, the basic rate had swollen to 3.59 percent, with another 0.97 percent added for bond redemption and interest payment. The basic rate rose as high as 4.3 percent before the decade ended, with the surtax peaking at 1.172 percent, fortunately not in the same year.[17]

The attempt by Jerome's officials to keep town revenues stable by raising assessments ran head-on into the general desire of merchants and citizens to cut their expenses. Even with the increased rates, taxpayers were not paying as much as they had before the bust due to their plunging valuations. Nevertheless, taxpayers in financial distress in both communities viewed the taxes they paid as onerous and excessive, and they equated raising the rate with raising the tax. At first their protests consisted of potshots from disgruntled individuals, but soon enough disgruntled individuals organized committees and formed associations.[18]

As he did with many other issues, George Parsons involved himself in the tax question in Tombstone. He and other taxpayers gathered on several occasions in early 1886 to help defeat a school bond issue that Parsons regarded as a "steal, $8,500 [being] more than twice the am[oun]t necessary." The bond issue failed by more than three to one, and those at the meetings took heart from their victory and organized themselves into a taxpayers' association. Tombstone's chamber of commerce proved to be another useful vehicle for protesting license or tax rates before the town council.[19]

Jerome's taxpayers apparently carried the idea even further. In 1931 thirty-two taxpayers petitioned the town council, "urging them to utmost economy in the budget for the next fiscal year." Several taxpayers sat in on the council's budget deliberations that year "and discussed the tax situation at length." The following year tax protesters decided to do more than offer advice. In April 1932 tax-protest leaders called a meeting aimed at selecting a committee to appear before various government bodies in quest of lower valuations. This time they also strove to form a permanent taxpayers' organization. Committees appeared before local, county, and state tax authorities seeking tax cuts as large as 35 percent, evidently with some success. In May 1932 the group adopted articles of association as the Jerome Taxpayers' Relief Association and began bombarding the town council with recommendations for budget cuts. The following year the Taxpayers' Relief Association began an annual practice of riding shotgun on the town's budget process, examining the proposed budget in detail, then voting their endorsement if they found it acceptable.[20]

Those with the means to do more than protest took matters to court. Beginning in 1929 and continuing throughout the 1930s, various mining companies sued Arizona state, county, and local tax authorities annually, claiming excessive valuations. United Verde, uvx, and Phelps Dodge were all among the numerous mining companies throughout the state that went to court to win lower assessments. The companies usually won. Although one newspaper editor called one of uvx's victories "a stern challenge to 'the spenders' that there is a limit to exactions which a government can levy upon harassed taxpayers," those court decisions meant far more than symbolic victory for the contending parties.[21]

The United Verde Extension's first triumph over the state tax commission at the end of 1931 resulted in an award just short of $200,000, plus costs and interest. Of that amount, the Jerome School District owed almost $30,000, and the Town of Jerome over $10,000. By the following spring uvx had won two more suits worth more than an additional $300,000, and that June a judge ordered the governments involved to pay the amounts they owed within three months or levy special taxes for the purpose. Jerome's budget for 1932–1933 included $20,000—more than one-third of the total budget—for a refund to uvx. That same year the Jerome School District owed over $26,000 to United Verde as the result of another judgment. The mining companies sought and expected prompt repayment. Jerome's attorney, Perry Ling, wrote to George Kingdon, general manager of uvx, insisting that the entire amount of the refund could not be raised in the coming year. Ling asked if the refund, owed for three different years, could be paid back over three years. Kingdon responded that his board of directors had already decided that the money should be repaid within the year, and he was certain that there would be no change in that decision.[22]

The editor of the *Prescott Courier* believed that these tax suits came about because all property in Arizona had been overvalued since the beginning of the Depression. He thought that mining companies took exception in the courts, rather than through petitions and visits to taxing authorities, because they had the money to do so. Holders of small properties could not take that tack because they would see any tax savings they realized disappear into legal fees. The same editor pointed out that governments lost more than the amounts awarded in the tax suits. The litigation itself cost quite a bit of money, and contested tax dollars could not be put to use until their disposition was determined.[23]

Reducing the cost of local government provided the only way for town officials to relieve all of these financial and constituent pressures. In Tombstone the economy measures began almost immediately. During the miners' strike in June 1884, Councilman Schmieding introduced a resolution to reduce the salaries of city officials. Although the council postponed action on that resolution, the issue came before it again six months later. The council ultimately cut not only salaries, but also employees. At the end of 1882 the town employed an auditor, license collector, treasurer, recorder, attorney, clerk, police chief, and four policemen. By the end of 1890 town officialdom had been consolidated to an attorney, treasurer, police chief, and two officers. In June 1882 the town spent $1,086 on salaries; by June 1890 that figure had been reduced to $495.[24]

Tombstone's councilmen did more than cut positions and salaries to reduce the town's debt; they saved money wherever possible. The council had the gas shut off in all city buildings, returning to oil lamps, and canceled the insurance on city hall. For this last move they were chided by the editor of the *Tombstone Prospector*, who believed that canceling the insurance amounted to false economy. The council produced a 50 percent reduction in miscellaneous expenses in Tombstone's budget between 1882 and 1890. The town spent a total of $1,748.66 in June

1882, and only $878.72 in June 1890. A municipal government that cost over $20,000 a year in 1882 cost around $10,000 in 1889.[25]

Jerome's councilmen cut back, too. By mid-1931 they had reduced Jerome's budget almost one-third from the previous year, with help from a provision that all city employees take a thirty-day involuntary furlough. Wage reductions followed within six months. None of these measures completely satisfied those members of the community who met the following spring to organize the taxpayers' association, and the town fathers redoubled their efforts to reduce expenses. The council dropped some town officials from the payroll, reduced the salaries of those who remained on several occasions, and discontinued stipends for the town band and the volunteer fire department's annual banquet. Again, dramatic savings resulted from these economies. The town budget ($87,000 for fiscal year 1929–1930) was steadily reduced until the town operated on less than half of the last boom-year budget (under $38,000 during fiscal year 1934–1935). Although the council increased the budget slightly during the recovery of the mid-1930s, by fiscal year 1939–1940 the budget reached its lowest figure of the decade at only $36,000.[26]

During the Great Depression such financial disasters struck local, county, and state governments all over the United States. Many of the dust-bowl towns suffered similar trauma and engaged in the same ruthless economies. During the frontier days damage was more localized but could be just as severe. When Meadow Lake, California, started to go under in 1866, residents voted more than two to one to abolish paid city officials and the police force, only two months after the legal incorporation of the town. Fire protection was one of the luxuries of town government allowed to lapse after Caribou, Colorado, busted, an economy ultimately fatal to the town—though few people remained to suffer the consequences. Even with relentless cost cutting, municipal solvency could disappear under mounting debt. When the debt became intolerable and the choice came to insolvent government or no government at all, a minority in both Tombstone and Jerome worked for the latter.[27]

A movement to disincorporate Tombstone appeared within five months of the end of the labor dispute. In January 1885 some residents circulated a petition, "very generally signed," asking the territorial legislature to disincorporate the town and reincorporate it in a less expensive form. This attempt came to nothing, as did another in the spring of 1887, and for the next few years the war against government expense was confined to protesting rates of assessment before the city council and voting against efforts to raise money through new taxes.[28]

The disincorporation movement returned again in earnest in the spring of 1890. It surfaced when Mayor C. N. Thomas published a notice stating the impossibility of reducing expenses further under the existing charter, saying that the city debt had nearly reached its legal limit and could not be retired, and suggesting that a meeting be called to seek a way out of the crisis. The same newspaper that carried this announcement reported that businessmen and taxpayers had pressured the mayor to call for the meeting. From that meeting came a drive for disincorporation that culminated in an election, held on 30 April 1890.[29]

According to the *Arizona Weekly Citizen*, this time it looked as if the disincorporationists would win, despite considerable opposition. Neither Mayor Thomas nor the newspaper had any doubt about the cause of the movement, it being, as the paper put it, "dull times and lack of funds with which to maintain a costly city government." A committee organized by the disincorporation faction to investigate the question reported, not surprisingly, in favor of disincorporation, urging local taxpayers "to halt and consider if it is not time that we should endeavor to lift the heavy burden of taxation that is upon us." The committee's report called the existing city government an extravagance, and concluded: "We do not think the present times will permit us to indulge in such extravagance, more especially when we can enjoy the same advantages at one-third of the money."[30]

But when election day arrived, those voting against disincorporation carried the day. They won by 15 votes out of 171, then got up a band and paraded the streets in celebration. The *Tombstone Prospector*, which had favored disincorporation this time around, opined that the movement lost because small taxpayers ganged up on the license-paying businessmen who favored the idea, but the disincorporation movement faced other problems. If everyone who had signed the petitions for disincorporation had also voted for it, the measure would have carried easily. Clearly, disincorporation had not been the burning issue the *Prospector* made it out to be. Further, the election offered three options: to disincorporate completely, to reincorporate under a smaller government, or to retain the existing system. The first two options probably divided disincorporation sentiment, handing the election to those opposed to either plan.[31]

Whatever the reasons, the *Prospector* concluded that the disincorporation idea was dead "and that the only way out of the difficulty is to run affairs as cheaply as possible." This Tombstone did with a vengeance. The monthly expenditure of $1,748.66 in June 1882 had been reduced to less than $900 by the time of the 1890 disincorporation movement. By January 1891 the council had reduced monthly expenditure to a little over $300, and during the depression of 1893 expenses were carved to the bone. In June 1899 the town spent a total of only $92.90. This austerity, combined with the willingness of the territory to bond the city's debt, permitted the town to crawl out from under $15,000 of debt and pull back from the brink of bankruptcy in a little over two years.[32]

Some residents spoke of disincorporating Jerome during the dark days of 1932. These rumblings prompted town clerk R. E. Moore to editorialize against the move in the *Verde Copper News*. "Why take this step backward?" he wrote. Moore believed that the town council could "do all the curtailing necessary ... but furnish sufficient revenue to maintain a good economical and efficient government. The citizens expect and are entitled to proper police protection, adequate fire protection, and a good standard of sanitation must be maintained." He lauded the budget cuts already made, predicted more of the same in the future, and concluded by advising his readers to "cheer up and don't let the 'calamity howler' disturb you. We are faring much better than other communities." Moore's arguments carried the day, for the town has remained incorporated to the present. Although that threat had been beaten back, the town government still had to live with the Jerome Taxpayers' Relief Association, which spent the rest of the decade standing watch over its operations, arguing for tax reductions, and endorsing suitably lean town budgets.[33]

Something like the same battle took place over that other important but separate government function: education. Those tumbling property valuations, tax protests, and lawsuits hit the school districts as hard as they hit the towns. Yavapai County School District no. 9 of Jerome had a valuation of more than $53 million in 1929, but only $16.5 million in 1935, and a little over $15 million in 1939. Even though the district raised its tax rate, the actual revenue generated dwindled from almost $165,000 in 1929 to a little over $62,000 a decade later. To a certain extent decreased enrollments balanced falling revenues. Tombstone's public school lost one-sixth of its peak enrollment by 1889. The Jerome school system, which did not draw from the surrounding area as Tombstone's did, experienced an even greater decline. From its 1929 peak of 1,919 students, district enrollment dropped to 1,102 for the 1939–1940 school year—a decline of more than 40 percent.[34]

Even with lessened attendance the financial crises and budget cutting were severe. At the beginning of 1886 the Tombstone school had a $700 deficit, and the school board asked citizens to vote a tax increase to help keep the school afloat. When that increase was not immediately forthcoming, the school system shut down indefinitely on 1 February. By 15 February the board had temporarily solved its money problem and reopened the school, but the Tombstone school confronted a similar cash crisis in April of the following year.[35]

The Jerome School Board began to retrench in earnest after the school year ended in the spring of 1931. It reduced the 1931–1932 budget for the district to just under $176,000, down from $207,000 the previous year. The following year the board hacked an additional $51,000 out of the budget, another 30 percent reduction. The budget for fiscal year 1933–1934 was just under $100,000, the lowest for the district since the 1919–1920 school year. By April 1936 the district's operating expenses dropped to under $84,000. Again, cost saving became the most important consideration. The *Verde Copper News* congratulated the school board in 1933 on maintaining essential services while making "a magnificent showing in savings to the taxpayers." The Jerome Taxpayers' Relief Association also pronounced itself pleased with the board's efforts.[36]

As with town government, these reduced budgets were produced by draconian cost cutting. Salaries provided the most conspicuous target; they constituted almost $48,000 of the Jerome School District's $51,000 cut for fiscal year 1932–1933. Eliminating eleven and one-half teaching positions, two principals, and seven other employees produced that saving. The district reduced from sixty-seven teachers in the 1930–1931 school year to thirty-nine at the beginning of the 1932–1933 year and stayed at roughly that level for the rest of the decade. In the years before the teacher cuts fell into line with declining enrollment, these reductions meant more classes per teacher and a much higher number of students per class. The Tombstone board relieved the school's financial crisis of 1886–1887 largely by eliminating several teachers' positions and reducing the salaries of those who remained.[37]

School boards cast about for other ways to save money. In Jerome these included omitting or curtailing sports, music programs, and field trips, and eliminating such specialized offerings as preschool and kindergarten classes and adult and vocational education programs. The Jerome School Board also reduced capital expenditures to the minimum and renegotiated the district's bonded debt. Tombstone's board authorized a shortened school year and unscheduled vacations, and paid part of its teachers' salaries with vouchers to meet that district's financial crisis.[38]

In the worst case a district might close schools or consolidate. Although the Tombstone district could not reduce from its one school building or consolidate with any nearby district, the Jerome district ultimately did both. Jerome's school board closed several school buildings during the 1930s. In 1951, as Jerome's population continued to shrink, voters determined to consolidate the Jerome and Clarkdale School Districts. When Cottonwood's schools joined the system at the end of that decade, all of the western Verde Valley had unified into one school district. In none of these measures did the school boards of Tombstone and Jerome act much differently from boards in other places confronted by economic depressions or declining populations.[39]

The economic dislocation caused by bust seems to have had little influence on the nature of local government, politics, and leadership in these two communities. One sociologist notes in her case study that unemployment itself did not alter the political views of those affected. Whether the same is true of Tombstone is unknown. In Jerome politics leaned slightly to the left. About 3.5 percent of the electorate voted socialist in one general election during the Depression, and the town had one communist voter. Jerome had always been overwhelmingly Democratic, however, and so it remained; furthermore, this political shift manifested itself in state and national offices and elections, rather than locally.[40]

Local politics and elections went on much as before. Tombstone's political apparatus fielded a "fusion" ticket of Democrats and Republicans for the municipal election of 1888, and at the city political convention of 1890 "Democrats and Republicans were allowed a share of offices and no questions asked." In 1892 the two parties actually raised separate slates of candidates, but "harmony" tickets seem to have been more common.[41]

Jerome's political process produced no more innovation during that town's decline than did Tombstone's. In the municipal election of 1934, near the bottom of the Great Depression, thirty-nine voters reelected exactly the same council. In 1936—freed from United Verde's control, with active encouragement to vote by the chamber of commerce and the mining companies, in an era of political reform, and with a field of eleven candidates to choose from—Jerome's voters stormed to the polls in record numbers, and reelected four of the five councilmen then serving, the fifth losing his seat by two votes. Perennial councilman Carl E. Mills, manager of open-pit operations for United Verde and Phelps Dodge, numbered among those reelected by a comfortable margin. Mills went on to serve as the town's mayor for two years during the 1940s.[42]

Indeed, an examination of a roster of councilmen from the mid-1920s to the end of the 1930s shows only different combinations of the same cast, with occasional substitutions due to the retirement or election to higher office of those previously sitting. Much the same could be said of the town's nonelected officials. The most important of these individuals, town clerk R. E. Moore, served in that position for many years before and after the crash. As the *Verde Copper News* explained in 1930, "few changes are expected among employees in the various offices, inasmuch as most of the present office holders were reelected."[43]

Neither town's leadership, either inside or outside of formal politics, changed much as a result of bust. One sociologist maintains that disaster can be a catalyst for producing new leadership elites, but that did not happen in either of these towns. Certainly, a significant turnover occurred among Tombstone's leadership, both formal and informal, but that turnover resulted from the mobility of the population, rather than anyone's design to replace previous leaders. In Jerome the elite was much more stable, and the same names appeared over and over at the head of various civic and social endeavors.[44]

These elites held their places without much organized opposition, partly because of the nature of leadership in small communities. Local leadership in these two towns, as elsewhere, usually sought consensus rather than division when solving problems, and in neither case could the leadership class be blamed for the sad state of the local economy; in fact, its members often suffered as badly from bust as anyone. Further, though there might be occasional distractions in a small-town political environment, everyone realized that business was the engine that pulled the train. If that engine had to be rerailed, who better to do it than that combination of leading merchants and mining officials that composed the majority of the preexisting leadership?[45]

There was another reason for what might be regarded by outsiders as political apathy in these communities. Political contests are driven by a number of factors—including patronage and voter identification with candidates—but ideology is important. The fusion tickets and the apparent lack of debate came from the absence of significant ideological differences between Democrats and Republicans as to the conduct of local government, either before or during bust. Decline did not produce any fundamental changes in these two towns' political leaderships or ideas about local government because both parties and most of the leadership network agreed that the less government, the better— as long as it had enough power to provide the infrastructure and public safety services for which it had been organized. The only changes in government services in either town came from their reduction or elimination. That same conservatism about local government that fought for the smallest budgets possible also defeated disincorporation. "The city government is being economically administered," the *Tombstone Epitaph* opined during the disincorporation movement of 1890, ". . . and it would seem to be the part of wisdom to bear the evils we have than fly to those we know not of."[46]

Factions did appear in the disincorporation contests. The *Tombstone Prospector* charged that the disincorporation movement of 1887 was an attempt to blackmail the council into buying some mining claims within city limits. The *Tombstone Epitaph* held that the disincorporation movement of 1890 "appears to be a job on the part of someone to gain control of the city water works." Whatever the truth in these charges, this factionalism was not ideological; a general consensus favored reducing taxation by cutting the cost of local government. The *Verde Copper News* reported of one of Jerome's budget meetings in 1932 that every councilman favored "all the reduction possible . . . consistent with reasonable, economic government administration," and that each pledged to make further reductions whenever possible. The Jerome Taxpayers' Relief Association applauded the council for their efficiency and economy, but the watchdogs hardly needed to watch because the governors shared the thriftiness of the governed.[47]

This attitude extended beyond local governments during hard times. When a Democratic primary candidate for governor visited Jerome in the summer of 1932 he "criticized severely the needless expenditures" of the state's Democratic governor, and promised to cut costs. The *Verde Copper News* reported that his opinions were "well received in Jerome." R. E. Moore, seeking reelection as a county supervisor in 1932, considered the Depression to be an "opportunity for elected officials and heads of the various departments of the government to reduce expenses and lessen the burden of taxation and to carry on economical constructive work."[48]

No thought was given to using local government to provide relief or otherwise mitigate the bust in the 1880s, and none in the 1930s until there was New Deal money to distribute. Much like local churches, local governments were crippled financially and did not have enough money to maintain their own functions, let alone engage in relief efforts. In most cases town governors could have done little to control bust, even if so inclined. The Great Depression overwhelmed local, county, and state governments all over the country. The bust of a thousand frontier mining camps like Tombstone had to do with flooded mines, refractory ores, unprofitable markets, or other conditions well beyond the ability of a town's leadership to solve.[49]

But economic countermeasures were another question. If town government could not normally rally a town, the informal leadership of a community sometimes had that potential. An independent lead-ership class, as one sociologist observes, can shape and direct public opinion, and use its influence to promote the economic reorienta-tion that would guarantee the survival of a former mining community. Another sociologist believes that the caliber of a community's leader-ship "is a crucial determinant of the course of development," that is, in the possibility of a community recovering from bust. Whether a mining camp survived the decline of its mining depended upon the capability of its leaders and on the possibilities allowed by local cir-cumstances. If these two things combined fortuitously, they permitted a mitigation of bust.[50]

CHAPTER ELEVEN

Our Last Dollar
on a Sure Thing

By the beginning of the 1950s even Jerome's most optimistic resident could see that the end was near. The district's mines had limped along for twenty years, gradually curtailing their activities, but soon the United Verde Mine, the last of the district's great products, would suspend operations. Then what? Some of those who remained wanted to keep the town alive with tourism, but Jerome had no hotel and only two restaurants open. There appeared to be few other possibilities. The future held no certainties for Jerome or Clarkdale once mining ended. These communities' survival would depend upon the degree to which their residents could mitigate the economic hardship caused by the curtailment of mining.[1]

Departure offered the easiest way to dodge the catastrophe. In both Tombstone and Jerome a substantial percentage of the population abandoned the town very quickly after the onset of bust. Two weeks into Tombstone's lockout a witness recorded that more than 150 miners had departed the town within that week for other fields. Fewer opportunities existed elsewhere during the Great Depression; nevertheless, Jerome lost about half of its population in two years. How did those who remained in a busted community try to mitigate the damage caused by the loss of their keystone industry?[2]

Denial proved to be the first line of defense. Residents and especially newspaper editors—with their vested interest in community prosperity—simply would not admit that a serious problem existed. While the mining companies installed their pumps to fight Tombstone's water in 1883, the *Tombstone Republican*'s editor held the future to be "particularly bright." He predicted another boom in the district within a year "that will astonish many even of the most sanguine of those who have faith in its permanence."[3]

The failure of the pumps to solve the water problem, the labor dispute, and the wholesale depopulation of the camp over the following two and one-half years did not extinguish the *Tombstone Epitaph*'s optimism. The *Daily Tombstone*'s editor conceded in May 1886 that the destruction of the Grand Central's works by fire "was a very serious blow to Tombstone," but he also still believed that the camp would rebound. A little over a year later the *Tombstone Prospector* called Tombstone "the best town on the coast," and the year after that, in the summer of 1888, declared that "the mines in this camp never looked better."[4]

As Tombstone approached the tenth anniversary of the beginning of its decline, the *Prospector* had to admit that the Grand Central and the Contention pump works were beyond salvage and that the price of silver remained depressingly low. Still, in March 1892, the paper exhibited its perennial optimism—uninhibited by a mountain of unpleasant evidence—for all to read: "Tombstone is today the best town in the territory. Her future prospects were never brighter and her present prosperity is a gratification to all of us who have staked our last dollar on a sure thing." Even in Jerome, in the face of a much larger disaster, the local editor could see a resurrection forthcoming. "The normal condition of America is prosperity," he informed his readers a year into the Great Depression, "and America will get back to normal."

So it went. Somehow, a return to the "palmiest days" was always just around the corner.[5]

In addition to its persistent cheerleading, the home press conducted a relentless war against "croakers" and "calamity howlers." In September 1883 the *Prescott Miner* had the bad taste to suggest that Tombstone's mines might not be as valuable as previously thought. The *Tombstone Republican* reacted by calling the *Miner*'s article "a falsehood almost in its entirety," and charged the *Miner* with "total ignorance of the subject." The *Republican*'s editor admitted dull times, but only because the mining companies were installing pumps, in itself "evidence of their faith in finding ore bodies below water." Once the pumps began operating, he assured friend and foe, the riches extracted would "astonish these 'surface croppings' croakers like the *Miner*."[6]

Three years later, just after the Grand Central fire, the *Daily Tombstone* defended the honor of the silver camp from another challenge by a Prescott paper. This time former Tombstone resident Tom Farish told the *Prescott Courier* that unless the Grand Central works were rebuilt, Tombstone's glory days were gone. The *Tombstone*'s editor reprinted these comments and then told his readers that "if [Farish] was incited by malice . . . it would be impossible to compress more errors, to characterize them by no harsher name, into so brief a compass."[7]

The press did not monopolize this often aggressive denial of unfortunate facts, nor did metal-mining towns. One sociologist studying a fading coal town has found businessmen reluctant to even discuss the possibility of decline, believing such an acknowledgment derogatory to the town and harmful to future business. In another example, the Kansas Junior Chamber of Commerce passed a resolution in 1940 asking the state's newspapers to cease printing photos of dust-bowl scenes or "other pictures or material that are detrimental to Kansas or her people." Some of Jerome's residents even mocked the bad times with a "Hard Times Party," at which guests dressed in tattered clothing for a meeting of their contract bridge club.[8]

As bust wore on, a paranoid tinge began to creep into this denial, infecting the thinking and language of newspaper editors and residents alike. Some residents of both Tombstone and Jerome came to believe themselves thwarted by the machinations of powerful interests beyond their reach. In 1889 a writer for *Golden Era* magazine—doubtless primed by the locals—wrote that Tombstone's slowdown could be blamed on "the few autocrats whose grasp is only too great, and whose purpose is evidently to absorb the whole, if possible." Two years later the *Tombstone Prospector* grumbled that the work done to date had been the extraction of dividends and had barely touched the main ore bodies of the district. He went on to boost the prospects of a number of the district's glowing embers.[9]

After uvx shut down, some of Jerome's residents convinced a wpa writer that the company was up to something. The company had purchased much of the town's real estate, "ostensibly to forestall further suits," but the old-timers knew better. "It smells to the high heavens," wrote the wpa writer, who continued, "thinking of the possibilities for Jerome, a new Jerome, with a huge open pit mine where the town of Jerome now stands. That's the story one gathers on the streets of Jerome." A dozen years later, when Phelps Dodge wrapped up its operations in Jerome, the *Saturday Evening Post* informed its readers that "smart money seems to be betting that there still may be undiscovered metal along the fault."[10]

Visiting correspondents can hardly be faulted for incorporating the latest rumors into their stories. Rumors provided sustenance for residents with little else to consume. "You can pick up any sort of rumor you want in regard to the Tombstone mines, now-a-days," wrote the *Tombstone Epitaph*'s editor in 1886, "from a report that the companies are going to make a Christmas present of their properties to their foremen, to one that they have refused to exchange a tenth interest in the mines for the Southern Pacific Railroad." Almost any rumor gained adherents and might even become the basis for action, particularly the "washhouse rumors" that came from the mines.[11]

Local newspaper editors, though frequently vendors of rumors, did not intend merely to drift with them. Any self-respecting mining-town editor went beyond his role as cheerleader and defender of the faith to propose at least general antidotes to his town's decline. The *Tombstone Prospector* published "twenty-four reasons why Cochise County is destined to prosper," extolling the county's climate, agricultural products, natural resources, and infrastructure. The *Verde Copper News* published eight helpful hints on how residents could restore prosperity. The editor of that paper—along with many others—believed that the main problem during the Great Depression was a lack of confidence among consumers and investors. The *Prospector* opined in 1890 that Tombstone's problems could be blamed on a lack of initiative. It urged locals to abandon "the everlasting hobby of pumps and look after the trade that is going by without an effort to grasp any of it."[12]

All of these frantic efforts to ignore the obvious may seem ludicrous, even pathetic, in hindsight, but they were not wholly irrational. Most of these people knew from experience that mining—like other extractive industries—was cyclical. Its prosperity depended upon a host of factors, many of them beyond local control. This reality worked upon coal towns in the Midwest, timber communities in the Pacific Northwest, and farming towns on the Kansas prairie, as surely as it did upon mining camps in Arizona.[13]

Jerome experienced several depressions in its long history. Each, save the last, came before another boom. The collapse of 1919–1921, although not as long as that of the Great Depression, was almost as sharp. One of the great booms in the district's history followed. Even Tombstone, barely five years old when its bust began, had already been extensively rebuilt twice because of fires. One pioneer woman from the East, watching Tombstone's second reconstruction, observed that "it takes a great deal to discourage a Westerner." As long as the mines and markets justified it, residents would rebuild their town and forge ahead. Even in the depths of depression, fortunes could change with the arrival of new technologies, management, or economic conditions. Bust is no more normal or inevitable than boom.[14]

A second consideration also stemmed from experience. Civic leaders and newspaper editors lived in constant fear of bad publicity, which they were sure would frighten away investors. They expressed this concern often enough during the boom days, but bust magnified its urgency. The *Prospector* argued against disincorporation in 1887, in part because such a move would announce Tombstone's decline to the world. In contests between local economies, capital, like labor, is a highly mobile factor of production. Beyond the loss of future opportunities bad publicity about a mining district could retard present investment, which contributed to severe deflation, accelerating the bust.[15]

Jerome's editor warned his fellow citizens at the start of the Great Depression against hoarding their money, thereby producing the very conditions they sought to avoid. He believed that the best course was to "go ahead carefully and constructively" with spending and investments. But as bust wore on it became more and more difficult to gamble on a town's future. Joe Larson described Jerome's inhabitants in 1940 as "still leery about putting out their money for investments." The following year he confirmed that assessment, observing that "it looks like everybody is scared to death to invest anything in this town." The Larson family had certainly learned that lesson. When he finally paid off his long-standing debt to his friend Homer Nihell in June 1941, Larson wrote: "Believe me, our money is not going into mining stocks this time. We feel that government bonds are the safest." This loss of confidence, hence capital, contributed to bust. It was something that mining officials, merchants, and newspaper editors did their utmost to prevent.[16]

A third reason for unwarranted optimism among those who decided to remain was the lack of an alternative. Young America was optimistic in character to begin with, but during bust optimism becomes a survival skill. One researcher, sampling unemployed workers in Seattle in the 1970s, found among them "an uncanny denial of reality," as to both the state of the job market and their own chances of finding employment. Still, she found this self-delusion "functional," because her subjects had no other choice. Optimism may also have reduced social tensions in a busted community. Because recovery was just around the corner, no need existed for harsh or radical solutions to bust.[17]

Some evidence suggests that positive thinking may even have had tangible results. Several researchers have indicated that a loss of economic vitality in a community encourages people with talent or ambition to seek new opportunities elsewhere. A positive view of a community's future may stay that process and encourage reinvestment in the community through repairs to property, repainting, and similar activities. The belief in a town's viability can enhance its actual viability by causing residents to try other measures to mitigate bust.[18]

Residents tried many ideas as individuals and families and as a community to ease or reverse their economic situation. Perhaps the most widely practiced were returns to hunting and gathering and subsistence agriculture. Hunting and gathering activities included fishing and hunting, berry picking, and fruit gathering. Jerome's children helped by salvaging firewood from the town's demolished buildings. The people of the frontier or even of the 1930s had often been farmers themselves or were only a generation removed from that activity, and thus returned to it easily. They raised chickens, ducks, geese, and sheep, or even kept a cow. They smoked meat and canned fruits and vegetables when possible, although the inhabitants of mining camps often found gardening difficult in their alpine or desert habitations.[19]

Residents of these and other busttowns also engaged in what might be called subsistence crimes. Subsistence crimes were those directly tied to the survival of their perpetrators, as opposed to the economic crimes detailed previously that brought in money. In eastern coal communities people built shacks out of stolen wood on land that they did not own and raided coal mine dumps for fuel, or took it directly from passing trains. The Jerome Transfer Company suffered nocturnal raids on its coal bins in December 1930. In 1934 the local court convicted a group of residents of tapping bootleg electricity from the Upper Verde Public Utility. On several occasions men were arrested in Jerome for slaughtering beef cattle that they did not own. The Prescott forest ranger had to admonish Jerome's residents not to collect wood from the nearby forest without a permit. Although stories of poaching appeared in the papers only occasionally, the town's proximity to four national forests inhabited by thousands of deer must have offered enormous temptation. Even the bootlegging widely practiced in the Verde Valley and elsewhere at the beginning of the 1930s might be considered a subsistence crime, since moonshiners often bartered their product.[20]

All of these activities reduced household expenses, as did home gardening, canning, and sewing. Other tactics to reduce expenses included bartering for needed goods or services, postponing improvements to properties, and forgoing medical treatments when possible. Young adults deferred marriage, education, or separate households. Women and children sought work to increase household income. John Krznarich remembered that in busted Jerome children gathered recyclable materials, delivered newspapers, or helped local vendors or tradesmen as delivery boys or assistants. He recalled that the high school provided summer employment of cleaning and refurbishing to some students.[21]

During the Great Depression mining companies practiced—and newspapers endorsed—part-time "staggered" employment for as many workers as possible as the best and most equitable solution to mass unemployment. This concept has been practiced informally in busted towns in other eras as well. People used several part-time jobs rather than one full-time job to try to keep enough money coming in.[22]

At the level of daily living, these were often cash-poor economies even in the best times. In compensation, it did not take much currency to live decently. John Krznarich recalled that his family spent no money on automobiles because they walked everywhere, and that his father spent fifty cents a month to provide company medical insurance for his whole family and did not spend any money on life insurance. The family's entertainment, which included radio and newspapers, a phonograph, and cards and dominoes, did not cost very much. His father owned his own home, paid the mining company two dollars a year to rent the land upon which it sat, and paid very little for his utilities. One scholar of bust in the Oregon timber country has found that residents usually had very little money but that they ate well. He makes the valid observation that these resource-extractive economies frequently endured periods of boom or bust, so residents were well practiced in dealing with either.[23]

But there came a point when a family reached the limits of domestic economy. The victims of bust thereupon often resorted to kinship networks. A number of researchers have noted the importance of kinship in sustaining residents of a busted community. That sustenance might run from room and board at a relative's house to money received from relatives back east. This idea of kinship could extend beyond blood relatives to ethnic groups or even beyond to encompass an entire busted community. A Tombstone resident who grew up during the town's depression years recalled that "families felt a responsibility toward each other—if *they didn't help* one another then, there *was no help*."[24]

As time wore on, residents of a busted town tried to organize mitigation or relief at the community level. Sometimes that relief simply magnified what individual residents had already been doing. The local chapter of the Red Cross offered garden seed to Jerome residents who desired it. Local sportsmen, cooperating with United Verde, seined the company's Peck's Lake for carp and suckers, which the local charities board distributed throughout the district in the summer of 1932. That fall Jerome's charities board and several fraternal organizations sponsored a community canning project that produced more than twelve hundred quarts of canned fruit for the town's needy.[25]

Community organizations also did things that individuals or families were incapable of doing. Their efforts normally started slowly and modestly. As mentioned previously, churches were usually too involved in a struggle for their own survival to offer much assistance within the community. The record book of St. Paul's Church of Tombstone shows no evidence of any almsgiving or charity and relief work by the church's ladies' aid organization from 1881 to 1890. All of their fund-raising events supported the church itself. Excluding the Salvation Army, the same applies with rare exceptions to all of the religious organizations in Tombstone and Jerome. Local fraternal organizations did make modest attempts to provide relief at the beginning of bust. They targeted much of their assistance toward specific small-scale problems like uniforms for Tombstone's baseball team or eyeglasses for Jerome's impoverished students. But as the Great Depression persisted, fraternal and charitable groups of the Verde District devoted greater attention to poor relief and increasingly coordinated their efforts.[26]

The Great Depression did not introduce poverty to the Verde District, and Jerome had a preexisting system of poor relief. Local organizations contributed funds to a general pool, overseen by a board known as the Associated Charities. The manager of the Associated Charities referred reported cases of need to the city nurse for investigation. She, in turn, recommended the appropriate action in a particular case to the board.[27]

This was the system with which Jerome greeted the flood of relief cases that overwhelmed the town after 1929. In November 1930 the city nurse issued urgent calls for clothing, shoes, and bedding to distribute to the needy. At the same time community leaders began to further consolidate relief work through the office of the Associated Charities in order to eliminate redundant effort. By the end of the year that organization was spending three hundred dollars a month on relief in Jerome. An Associated Charities organized in the valley towns at the end of 1931 tried to do many of the same things.[28]

Jerome's fraternal and community organizations did their best to contribute to this effort. The Business and Professional Women's Club collected canned goods from its members and sewed garments for the needy. The ladies' auxiliary of the American Legion spent their meeting times sewing, and the legion paid for one-third of the three thousand school lunches that the Associated Charities furnished to Jerome schoolchildren in the spring of 1931. The Verde District Salvation Army also organized a sewing club at which district women made clothing for the poor. Local merchants donated clothing and groceries to the effort, with occasional contributions from individuals and United Verde. The town truck delivered firewood, split and sawn by town prisoners, to needy residents. Both the Salvation Army and the American Legion made efforts to find work for the district's unemployed or to serve as an exchange between employers and the jobless. Writing in the spring of 1932, Town Clerk R. E. Moore singled out the efforts of the Elks, American Legion, and Red Cross in the district and reported that the town had spent none of its own money on the relief effort to that date.[29]

As the Depression lengthened, the efforts of the Red Cross came increasingly to the fore. The local chapter drew upon the resources of the national organization to provide materials in great quantities to the district's destitute. In 1932 the Red Cross supplied thousands of yards of cloth for the district's seamstresses to render into clothing, and by October of that year the district had received over 75,000 pounds of flour from the agency. In 1933 the Red Cross provided the

Verde District with thousands of items of clothing, hundreds of blankets, and another 117,000 pounds of flour.[30]

Residents also sought to actively combat the decline of their town. Jerome revived its chamber of commerce in the spring of 1935, and within a year the organization had more than sixty members. The chamber publicized Jerome with maps, newspaper publicity, motion pictures, and a tourist and information bureau. The chamber conducted sightseeing tours, lobbied for road projects, and engaged in various civic improvements, including erecting signs at either entrance to town declaring Jerome "The Most Unique Little Town in America."[31]

In 1892 the *Tombstone Prospector* pushed for a town board of trade to assume many of the functions of a chamber of commerce. This suggestion seems better than two others the editor extended: statehood and a new name for the town. He contended that statehood would reassure capitalists waiting to invest in Cochise County and that Tombstone's name should be changed "to some other more appropriate one for this city. That it has had its drawbacks from having such a dismal sounding name is plain to everyone."[32]

The crown jewel of any busted town's revival-through-boosterism plan had to be its buy-at-home campaign. The *Prospector* opined that "if every dollar that is spent by our people in buying goods away from home was spent with our merchants, times would be far better in Tombstone." Merchants and editors witnessed the serious impact of the automobile on home trade in the Verde Valley for most of the 1920s, but bust increased pressure to support the home team. The Jerome Chamber of Commerce undertook a buy-at-home campaign as their primary objective. A close cousin to buy-at-home was an announced "bargain days," in which district merchants agreed to cut prices en masse to encourage home shopping.[33]

Advocates of buy-at-home tried something of the sort to revive their moribund industries. The *Prescott Courier* printed an editorial calling for the use of copper in roofing and household fixtures. These campaigns even reached the state level during the Great Depression. Proclamations from various Arizona governors declared a trade-at-home week in 1930, and another in 1931, asking citizens to take their summer vacations in the northern part of the state, "thus keeping as much Arizona money as is possible in circulation within our borders." One governor even considered it a "patriotic duty [to keep] Arizona dollars in Arizona."[34]

All of these voluntary exhortations and activities ultimately proved inadequate. The Great Depression would not go away no matter what the governor proclaimed, and the longer it lasted, the further local, district, and county charities fell behind. Not only was it impossible for them to meet demands, but they also had great difficulty raising any money to support their relief projects. The local chapter of the Red Cross, which joined Jerome's relief apparatus after demand overcame the Associated Charities in the spring of 1931, spent almost $3,800 in the year ending 1 July 1932. The chapter's fund-raising that fall netted less than $250, which its president declared as good as could be expected, considering the situation.[35]

By the summer of 1932 Governor George W. P. Hunt, who had been signing optimistic proclamations to the people, admitted the seriousness of Arizona's situation to the chairman of the federal Reconstruction Finance Corporation. "Our people have responded nobly to appeals for contributions to community chests and other charitable organizations," he wrote, but went on to report that "the funds of these charitable organizations are exhausted or approaching that condition. The legitimate demands for relief," he concluded, "are tragically greater than their ability to meet." Although charities continued to contribute—particularly toward direct relief—the battle to beat bust in both Tombstone and Jerome ultimately moved beyond the limits of those communities themselves.[36]

The earliest government action demanded in either case was not for direct relief, or even work relief, but for federal intervention to increase metal prices. Tombstone residents participated in the free-silver movement of the late 1880s and early 1890s. The *Tombstone Prospector* made no attempt to disclaim local self-interest when booming free silver in the summer of 1889; its campaign made no mention of inflationism or any of the other free-silver or populist rhetoric of the day. Even if Tombstone had immense reserves of silver wealth, the paper editorialized that "there will be no active operations in mining affairs until there is a change in the price of silver. There is no possibility of its taking a natural rise either, and the only hope for us is to agitate the free coinage of that metal with the hope that Congress will come forward and put a little artificial life into it." The *Prospector* saw in the free-silver movement an opportunity to subsidize the home industry.[37]

The paper spent the rest of that summer agitating Tombstone's citizenry, and in mid-October they responded with a public meeting attended by over 150 of Tombstone's mining, business, and opinion leaders. From that gathering came a call for a countywide meeting the following week to select delegates to a silver convention to be held in St. Louis at the end of November. Those delegates were duly nominated and dispatched to St. Louis, but ultimately nothing came of the convention or the free-silver movement. Except for occasional editorial musings, that seems to have been the end of the silver agitation in Tombstone. The price of the metal continued to slide.[38]

The movement to protect domestic copper during the Great Depression proved considerably more persistent and successful. The drive to impose a tariff on imported copper began almost as soon as Arizonans realized how serious the copper surplus had become. Petitions began to reach the state's United States Senate delegation by the end of 1930. Arizonans, from the governor on down, saw cheap foreign copper as the main cause of the copper glut, and thus believed a tariff to be an appropriate and rational solution to the problem.[39]

The chorus of voices expressing that opinion continued to grow as the Depression deepened. In April 1932 the governors of the twelve leading copper-producing states petitioned President Hoover, claiming unfair foreign competition, and asking for a copper tariff on the grounds of industrial distress, mass unemployment, and national security. Once again a conspiratorial view of circumstances crept into these petitions. One memorial to the Arizona State Senate in early 1933 saw the superabundance of foreign copper as "a well-organized effort [to permanently] destroy the copper production of this country."[40]

Protectionism carried the day in June 1932, when Congress passed a four-cent-per-pound tariff on imported copper. Protectionists quickly came to believe the four-cent duty inadequate, however, and in short order they began an effort to raise the duty to ten cents. Once more state legislatures passed resolutions; once more Arizona's governor enlisted other copper-state governors in the tariff crusade. Some people left their demand for federal intervention open-ended. A committee of United Verde miners and smeltermen petitioned President Roosevelt "to the end that foreign copper may be excluded from the domestic market, either by embargo or by a tariff sufficiently high that our domestic mines may operate at a profit."[41]

Governor B. B. Moeur had an even grander idea. At the end of 1933 he proposed to one federal official that the U.S. government declare an absolute embargo on the importation of foreign copper, then purchase the entire domestic copper surplus in the name of national security and full employment. The following March, Moeur wrote directly to President Roosevelt and unblushingly advocated that the U.S. government purchase the entire domestic surplus of 1.2 billion pounds of copper "at reasonable prices, say from nine to twelve and a half cents per pound." Even though none of these delicious dreams ever came true, copper-state officials succeeded in keeping the four-cent tariff intact for the remainder of the 1930s.[42]

The Great Depression also saw another attempt to artificially inflate the prices of silver and gold. This time the citizens of a bloc of western gold- and silver-producing states, including Arizona, lobbied Congress and the Roosevelt administration. They sought tariff protection, an increase in the government price of the metals, or, again, even the outright federal purchase of some percentage of the stocks of the two metals. Here, too, persistence paid off. Congress raised the price of silver almost one-third at the end of 1933, and almost another one-fifth in the spring of 1935, and raised the price of gold significantly in 1933 and again in 1934. In the latter year Roosevelt administration officials also urged Congress to put more silver into the monetary system and announced that they would seek a world bimetal agreement.[43]

Not everyone, then or since, considered these manipulations a good idea. Although at the time the copper tariff issue offered a bipartisan political opportunity—Arizona state and federal office seekers of both parties favored it—there was a surprising amount of doubt expressed about the efficacy of these measures, even in Arizona's mining and editorial circles. In the spring of 1935 the general manager of Phelps Dodge said that the increased gold and silver prices had not helped his company much, and he believed only a general improvement in the economy would return prosperity to mining.[44]

The *Prescott Courier* persistently criticized Depression-era federal interventions in the metal markets, warning during the original tariff drive of 1932 that a tariff would help only if demand for copper actually existed. The paper asked readers: "Is there any certainty that the buyers would respond and make large purchases at twelve or fourteen cents a pound when they fail to show an interest in the existing below-cost price[?]" The paper concluded that "a tariff . . . can have no immediate effect, and it takes no expert to see that." Two years later, during the movement to raise the copper tariff to ten cents, the *Courier* editorialized that "the assertion that a tariff will cure all [of copper's] ills is holding out false hope and accepting as fact a very flimsy illusion." The *Courier* warned six months later that currency manipulation through the federal Silver Purchase Act of 1934 would not produce recovery and might produce disaster.[45]

There has been general agreement since that era that the tariff redoubts constructed by nations in the early 1930s exacerbated the Great Depression. Less clear is what effect, if any, federal subsidy of various metals produced. One authority considers New Deal silver price support a reluctant concession by the Roosevelt administration to western mining interests and believes that it produced little long-term good, though it "paid handsome political dividends" to the administration. The jobs created thanks to federal metal price supports might be viewed in retrospect as another form of New Deal work relief.[46]

There was plenty of that, of course; Jerome lapped up its share of alphabet soup programs. But first came the Hoover administration's attempts to right the ship. That administration and Congress passed laws accelerating various federal construction programs, culminating in the Reconstruction Finance Corporation Act of July 1932, which loaned $1.5 billion to the states for physical improvements. Some of that money reached Arizona and Yavapai County, although the Hoover administration limited federal support to the states.[47]

After the Roosevelt administration took over, Jerome benefited from a variety of new federal programs designed to ease the Depression. The town's mining companies, businesses, and laborers did their part in the NRA. Hundreds of Jerome residents—citizens and aliens, men and women—found temporary employment on various New Deal projects. Residents of the Verde Valley labored under such agencies as the Emergency Relief Administration, the Civil Works Administration, the Public Works Administration, and a reinforced Reconstruction Finance Corporation. With all of its federal forestland, Arizona provided a natural theater for the Civilian Conservation Corps (CCC), and in the five years after 1933 that federal agency spent almost $40 million in the state. Yavapai County had four CCC camps, and dozens of Jerome youths participated in the program. The Verde District also benefited from massive distributions of federal food stocks beginning in the fall of 1933. By the end of that year Jerome's charity workers reported fewer families in need than for several years previously.[48]

Perhaps the WPA had the greatest impact on the Verde District itself, in terms of both immediate benefits and long-term physical improvements. Jerome's WPA workers built buildings, retaining walls, sidewalks, and drains, and repaired streets. Jerome's WPA sewing room employed some of the town's women. The town also had a WPA music project, which employed Frank Sonora to teach guitar, mandolin, banjo, or ukelele to anyone interested.[49]

Fig. 11.1. Identified as a "sewing class," this is probably a photograph of Jerome's WPA sewing room, which provided clothing for the destitute and employment for some of Jerome's women during the Great Depression. (Sharlot Hall Museum Photo, Prescott, Arizona, #BU-1 183P)

Fig. 11.2. Women, children, and a cyclist mustered to watch workers lay rail into Tombstone in 1903. Tombstone's business leaders and its newspaper editor saw railroad service as the solution to their town's decline. (Arizona Historical Society, Tucson, #28304)

Various of Jerome's prominent citizens helped study problems and organize solutions for county, state, and federal relief efforts throughout the 1930s. The residents of the busted copper camp administered as well as received federal largess. Former mayor J. P. Connolly represented the Home Owners' Loan Corporation in Jerome. Town Clerk R. E. Moore and United Verde executives Robert Tally and Val DeCamp held important positions in various county and state relief activities. In both Tombstone and Jerome residents also focused their attention on specific projects—in each case improvements to infrastructure—that they thought would help bring about their salvation.[50]

Many people in Tombstone convinced themselves that if they only had a railroad, they would also have prosperity. The district came tantalizingly close to welcoming the iron horse on several occasions. In the summer of 1882 a gang graded a roadbed to within two miles of town. Rumors had it that the town might become a division point on the Atchison, Topeka, and Santa Fe line. But the grading stopped, and nothing further came of it. Once the depression set in, many in Tombstone believed it especially important to get a railroad of their own.[51]

That opportunity presented itself in 1888, when the directors of the Atchison, Topeka, and Santa Fe determined to extend their line south from Fairbank to the burgeoning copper camp of Bisbee. When the *Tombstone Prospector* interviewed one of the Santa Fe's agents in April, the agent warned that the railroad to Bisbee would not pass through Tombstone "unless her people wish it." That was all the invitation the paper needed to sound the call to arms. The *Prospector* called on Tombstone's "go-ahead men" to leave the "mossbacks" behind and seize this opportunity to end the town's isolation. It warned that the people of Tombstone and Cochise County "must not sit down and wait for fortune to shine upon them. They must assist and foster all of the industries which will spring up in the county." The paper was not adverse to publishing dire predictions about the town's fate without a railroad. It claimed that Tombstone would lose the county seat, that mining companies would relocate their headquarters to Bisbee, and that Tombstone would "dwindle to an insignificant cow camp," all because "the large amount of trade which naturally comes to this city at present from Sonora and Bisbee will be entirely cut off if the road runs around us."[52]

All of the editor's agitation culminated in a town meeting at the beginning of June. The meeting by itself apparently got the railroad's surveyors started, and residents worked to ensure the railroad's arrival by creating a subscription list and appealing to residents and the town council for support. The day after beginning their drive proponents of the project had raised almost $3,000 from twenty-six subscribers. At its meeting of 6 June 1888 the council answered popular demand by resolving to "aid said railroad company, financially and otherwise ... as an inducement to build the road by way of Tombstone." Nine days later the council granted the railroad a right-of-way through town. By the middle of June the *Arizona Weekly Citizen* reported that Tombstone's residents were waging a hard fight to gain the railroad, which that newspaper thought would do much to enhance the silver camp's prosperity. The *Prospector* concurred, saying that "the first spike will be the last nail in the coffin of hard times."[53]

But bust would not stay buried. Only a week later the *Prospector* conceded a most bitter defeat. Although the subscription drive among Tombstone's citizens produced more than $7,000 for the railroad, that amount was insufficient to offset the added cost of constructing the line through Tombstone. On 22 June 1888 the railroad's directors announced their decision to build into Bisbee along the easy grades of the San Pedro River. Tombstone had been jilted again. The *Prospector*'s editor was left with nothing for his efforts, save the unpleasant task of congratulating Fairbank and Bisbee "on their good fortune in securing the great advantages which will accrue to them by the building of the road." Within a week he had recovered his spirits, however, and had returned to printing his mining schemes and railroad dreams.[54]

Tombstone had one more crack at its own railroad in the nineteenth century. That effort was a do-it-yourself project. In the fall of 1891 townspeople went through much the same process as they had three years before. They held another town meeting, attempted to gather enough subscription money to build the railroad—this time a branch line from Fairbank to Tombstone—and read more prosperity sermons from the town's editor. The Tombstone and Fairbank road, the *Prospector* assured them, would restore to Tombstone "as good times as we ever had." The paper admitted that the town had not recovered from the last railroad debacle, and urged everyone to invest 5 percent of their assessed valuation in railroad stock. It informed residents that this was their chance to regain lost ground and reminded them that this railroad did not ask for tribute, only interested investors. "That the road will pay from the start," the *Prospector* assured readers, "is plain to anyone who will take the trouble to look at the matter in a fair light."[55]

This home-built railroad was another scheme of Tombstone's ever active magnate R. W. Wood. He capitalized his company at $200,000, gained a fifty-year franchise and twenty-year tax exemption from the town, and brought in another survey party to study the terrain. Wood figured that this time he would need to raise $25,000 locally, as the new company would have to provide its own equipment and facilities. By the time the survey party arrived in late November 1891 he had raised roughly half of that amount, but this scheme also died aborning. Tombstone would have to wait another decade for a railroad's arrival.[56]

In contrast to Tombstone's failures to secure a railroad, Jerome's pet project succeeded—as far as it went. Jerome's plan, also formulated prior to the bust, called for the completion of State Highway 79 from Flagstaff to Prescott. That highway passed down the beautiful Oak Creek Canyon, through the Verde Valley and Jerome, over Mingus Mountain Pass, and through Lonesome Valley to Prescott. As early as 1927 the *Verde Copper News* boomed the proposed highway as "an incomparable stimulant to business," and predicted the finished road would flood Oak Creek Canyon—and presumably towns nearby—with "a golden stream of automobiles." Only a month after the stock market crash, Yavapai County editorialists and residents began clamoring for funds to build the highway and thereby hire the unemployed.[57]

This combination of road building and relief was a natural pairing. Relief projects needed to be started quickly, and in many cases road projects had already been planned and surveyed, and the equipment stood ready. All that remained was to hire laborers, and busttown had plenty of those. By May 1930 the Oak Creek Canyon road had made it onto the federal relief plan, and that November it received an appropriation of $275,000. Over the next two years various Yavapai County road projects benefited from federal and state emergency funding.[58]

Boosters of Highway 79 saw more potential in the road than just a make-work project for the area's unemployed, although they usually presented their pleas in terms of relief of economic distress. United Verde's smelter superintendent, C. R. Kuzell of Clarkdale—one of Highway 79's most vociferous proponents—believed that the road would "open up a tourist wonderland" in Oak Creek Canyon, which he immoderately compared to the Grand Canyon. He also maintained that the completion of Highway 79 presaged great commercial development in

Fig. 11.3. A work crew improving the Jerome-Prescott highway at the first curve above Jerome in Hull's Canyon in the latter 1930s. Road construction projects provided much of the relief work sponsored by New Deal agencies in Arizona during the 1930s. (Arizona Collection, Arizona State University Libraries, CP HD3890.a7.d9x, v1/b)

the area. Kuzell wrote to the regional Forest Service office in Albuquerque, New Mexico, urging the managers of that agency to pour all possible money into the project, much of which passed through national forests. Kuzell believed that locals viewed the completion of Highway 79 "as about the most desirable thing in our community life outside of the resumption of the copper industry. Highway 79," he explained, "means easy egress both north and south for our own people, and access to, we believe it safe to say, tens of thousands of tourists and visitors annually." Other persons besides Kuzell bombarded the relevant agencies with pleas for support, but until 1933 theirs remained a piecemeal effort.[59]

In July of that year various interested parties moved to create an association to boost the Highway 79 project. The organization came into being at a meeting in Flagstaff on 16 August 1933. Forty-three people attended the meeting, including seven from Jerome. In two years membership swelled to a peak of about three hundred individuals and organizations, seventy-one of them from Jerome, which had the largest representation of any town on the route. A number of the town's leading lights participated in the Highway 79 Association, including Town Clerk R. E. Moore, who served as president; Town Attorney Perry Ling; Postmaster Ross Cunningham; and T. A. Miller of Jerome's principal business house, the T. F. Miller Company. The association's letterhead proclaimed Highway 79 to be "the Southwest's Most Beautiful, Scenic and Necessary Commercial Highway." The organization's constitution stated its purposes to be to obtain the construction of a permanent surfaced highway from Flagstaff to Prescott, and "to advertise in every practicable manner the beauties and advantages of Highway 79, to the end that it may be used by companies operating busses as well as the motoring public at large, as generally as possible."[60]

The organization thereupon set about booming the project—holding annual meetings to coincide with the state highway budgeting process and lobbying the state highway commission, the U.S. Forest Service, Arizona's congressional delegation, and others concerned with funding road construction. To fulfill the organization's second stated purpose members used a variety of publicity devices designed to draw tourists to their highway. The most important, a pamphlet titled *Highway 79: "The Greatest Scenic Drive in America,"* was produced in co-operation with the Jerome Chamber of Commerce. Eighteen Jerome businesses purchased advertising in the pamphlet, permitting its publication in 1937, before completion of the highway. The association saw the first tangible results of its efforts by the end of 1933. Four years later the state highway plan for 1936–1937 budgeted $583,418 for Yavapai County, of which almost $416,000—71 percent—went to Highway 79 projects. The completed highway was dedicated in Sedona on 8 October 1939, and Jerome's KCRJ broadcast the ceremonies across the Verde Valley.[61]

While it might appear that seeing or seeking a busted town's salvation in rails or roads was unrealistic, these community leaders did understand their situation and acted somewhat correctly. One scholar of California's Mother Lode country declares transportation to be the most important element in the survival of any particular portion of the region. After Bisbee got the railroad it did supercede Tombstone as the supply and mining center of southeastern Arizona, and it also picked up a considerable portion of the area's cattle trade. And again appearances entered the equation. A railroad's arrival in itself signaled prosperity in a district and might encourage persistence and investment.[62]

Yavapai County's busted citizens discovered that their Highway 79 Association did gain attention and did command a certain amount of respect. The state highway budget passed in July 1932 spent only $3,000 on all of the road, while in 1935 a single contract for five miles of the highway between the summit at Mingus Mountain and Jerome was let for almost $120,000. This increase certainly reflected the

massive infusion of New Deal money, but states and counties determined how that money was spent. The Highway 79 Association made
sure that the relevant agencies kept its project among the well-financed
routes.[63]

Leaders of both Tombstone and Jerome understood that in order to
survive the collapse of mining, their economies had to diversify. Tombstone's home-built railroad of 1891 would be more than a cheaper way
to move ore; it would also restore Tombstone as "the jobbing center of
southern Arizona and northern Sonora, just as it was before railroads
took it away from us." It would restore or even increase the town's
previous vitality. It might even gain Tombstone a U.S. district court, at
least according to rumor.[64]

The validity of these particular promises aside, case after case indicates that economic diversification makes the difference between a
live town—however robust—and a ghost. Neither Tombstone nor Jerome
became a ghost town, partly because both sought to diversify their
economies even before their busts began. Tombstone had the county
seat and remained a supply center for the area. Jerome had already
begun to explore the potential of a growing tourist trade brought in
on improved roads. Both towns quickly attempted to enhance these
sources of revenue and to take advantage of other opportunities that
fortune cast their way. Tombstone tried to cash in on southeast Arizona's burgeoning cattle business of the late 1880s by becoming the
farming and ranching center for the region. The former proprietor of
Tombstone's Occidental Hotel, Joe Pascholy, was among those who invested in local cattle-ranching enterprises, and interested parties had
organized a Tombstone Stock Association by 1888 to further the cause.[65]

Tourism was another alternative—one from which a number of former mining towns have since made a living. Tombstone has developed
into one of the leaders in this trade, though for its outlaw legends
rather than its mining history. While the process of capitalizing on
Tombstone's past did not begin until almost fifty years after its boom,

Jerome converted from mining minerals to mining tourists almost immediately. Tourism provided the incentive for the road-building campaigns and the road signs and pamphlets advertising the community, and those efforts to attract tourists paid off. The WPA guide to Arizona, published in 1940, reported that while mineral production in Jerome had fallen by half in the preceding decade, tourism had doubled.[66]

Once the last of Jerome's big mines closed in the early 1950s residents realized that tourism alone stood between their town's prosperity and oblivion. The few people who remained redoubled their efforts to attract visitors. The same summer that the United Verde Mine suspended operations for the last time, residents organized the Jerome Historical Society and opened a museum in the heart of downtown. As the years went by, the town tried to attract artists and craftsmen; hosted music festivals, barbecues, and celebrations; started a historic-homes tour; and for a few years sponsored an annual automobile race up the hill from the Verde Valley. The town sought and received National Historic Landmark status from the federal government and enlisted state support, which ultimately took the form of Jerome State Historic Park. The transformation from heavy industry to retail trade is now so complete that a Depression-era miner returning to Jerome today would hardly recognize the place. Jerome's current collection of curio and gift shops, antique dealers, art galleries, craft shops, confectioneries, coffee houses, and cafes bears very little relation to the gritty working-class town of a half century and more ago.[67]

So too in Tombstone. Both towns took advantage of what historical geographer Richard Francaviglia has termed *technostalgia*. The gift shops and restaurants might make the sales, but it is frontier history and nostalgia that lure the customers. Tombstone was one of the first communities to figure this out. To mark the fiftieth anniversary of the town, residents inaugurated "Heldorado Days" in October 1929. This celebration featured a reenactment of the O.K. Corral fight and a number of other events designed to evoke nostalgia for the frontier days. Tourism gained additional momentum in Tombstone during the 1950s thanks to publicity garnered from televised westerns and the establishment of Tombstone Courthouse State Park. Although Jerome lacks frontier mystique, it does have World War I and the Roaring Twenties to play with, as well as two first-rate bonanza kings in William Andrews Clark and Jimmy Douglas.[68]

Pensioners have also abetted the survival of both Tombstone and Jerome. People began to settle in Tombstone because of its climate as early as the 1890s. In 1892 some thought was given to establishing a sanitarium there to take advantage of the mild, dry air. A number of retirees have since taken up residence in the town, particularly in recent years. Jerome has been even more fortunate in this regard. The dawn of what has been called the pension age coincided with the end of that town's big mining operations after World War II. When the United Verde Branch closed, a number of its employees retired and remained in town. As bust took hold and real estate prices collapsed, Jerome and Clarkdale—and busted towns like them—attracted pensioners who had fixed and limited incomes and desired a quiet, small town in which to live. By 1965, a dozen years after the final shutdown of its major mining operations, Jerome consisted mainly of newcomers, many of them retirees.[69]

Busted towns had these and other ways available to diversify their economies, depending upon their locations, resources, and circumstances. The object in every case was to establish multiple sources of income for a community, both to recover from current economic difficulties and to preclude their repetition. The editor of the seventy-fifth-anniversary edition of the *Tombstone Epitaph* demonstrated that he had learned that lesson. "Gradually people have come to realize that everything cannot be lashed to a mining venture," he wrote. "Thinking has changed, and other economic factors have been exploited. Today we are not dependent solely upon any one factor. Yes, failure of any of a number of factors to produce anticipated revenue would hurt, [but] it would not spell financial disaster in the sense that closing the mines did years ago."[70]

Not all schemes panned out. The railroad to Bisbee and the later construction of Douglas, Arizona, ended Tombstone's chances to be the entrepôt and mining headquarters for northern Mexico. Southeastern Arizona ranchers suffered the same hard times in the late 1880s as did those in the cattle business in the rest of the West. And some schemes were frankly unrealistic from their inception. The *Tombstone Prospector*'s booming of a mescal distillery for Tombstone in 1891, on the grounds that "someday mescal will become a popular drink all over the United States," did not help much. A call by the same paper the following year to plant sugar beets in the arid Sulphur Springs Valley nearby also came to nothing.[71]

As the great number of western ghost towns attests, no single strategy guaranteed the ultimate survival of a busted town. Those who decided to remain during bust did their best to minimize its impact by seeking alternate sources of income or willingly accepting a reduced standard of living. If they succeeded, their town survived. The degree to which they succeeded depended upon local circumstances, some of which they could control; upon outside forces that they could not; and upon their willingness to fight for the life of their town. The decision to stand and fight or to take flight also often involved differences between the nineteenth- and twentieth-century West.[72]

CHAPTER TWELVE

But a Monument to the Glories of the Past?

One hundred and twenty years have passed since Tombstone first started to decline, and more than sixty since bust descended upon Jerome. Both communities faced lean times—even extinction—but both have survived. Although neither has regained its former significance in either population, affluence, or influence, both have recovered a measure of vitality, and both offer clues to understanding the nature of boom and bust in the American West.

The year 1893, the tenth anniversary of the beginning of Tombstone's decline, opened with the usual professions of optimism. Surely, this would be the year when things began to turn around. But the national depression that followed the panic of 1893, coupled with the repeal of the Sherman Silver Purchase Act by Congress and the cessation of silver purchases by India's mints, plunged the price of silver down the shaft, and pulverized any hope of a mining revival in Tombstone for the rest of the nineteenth century. Tombstone's mines operated fitfully, if at all, and the end finally seemed to have arrived when the stalwart Tombstone Mill and Mining Company shut down in 1896. Tombstone had become, as one writer eloquently put it, "all that its name implied, it was but a monument to the glories of the past."[1]

But the new century saw some reasons for hope. In February 1900 E. B. Gage, former superintendent of the Grand Central, returned to Tombstone with capitalist Frank Murphy. That October, together with W. F. Staunton, the former superintendent of the Tombstone Mill and Mining Company, Gage and Murphy visited the district's mines. Gage had long dreamed of consolidating Tombstone's mines under one company and dewatering them with one massive effort.[2]

To that end the three men organized the Tombstone Consolidated Mining Company in July 1901, with Gage as president, Staunton as superintendent, and Murphy as financier and liaison to a parent company called the Development Corporation of America. Gage returned

Fig. 12.1. An overview of Tombstone from the Tombstone Hills, ca. 1890. The Cochise County Courthouse is visible at the far left and part of Schieffelin Hall at the extreme right. (Arizona Historical Society, Tucson, #29270)

to town in September 1901, and Tombstone Consolidated began acquiring claims and sinking a combination pumping and haulage shaft.
The company purchased the rights to over seventy claims, covering
about two square miles of area. Its collection included the holdings of
the old Grand Central, the Contention Consolidated, the Tombstone
Mill and Mining Company, and the Head Center and Tranquility Mines,
as well as their buildings and machinery. The *Engineering and Mining
Journal* estimated in 1902 that Tombstone Consolidated's properties
had yielded twenty-four of the twenty-five million dollars that Tombstone had produced to that time.[3]

The company began to sink a one thousand–foot-deep four-
compartment shaft on the old Flora Morrison claim. Two compart-
ments moved ore and personnel, while the other two pumped over
1,750 gallons of water per minute, more than half again the capacity of

Fig. 12.2. The Tombstone Consolidated Mines, ca. 1910. (Arizona State Library,
Archives, and Public Records, Archives Division, Phoenix, #97-0436)

the Grand Central and the Contention works combined. Profiting from experience, this company built its headframe and shaft house entirely of steel. By 1902 Tombstone's pulse had quickened to such an extent that outsiders began to notice. *Harper's* magazine reported that "old buildings are being cleaned out, and people are moving in.... The prospects are good for a large and prosperous camp.... Its new prosperity will be built upon a more solid foundation, and it will become greater than ever." The *Engineering and Mining Journal* informed its readers that same year that Tombstone Consolidated's expertise and resources gave "reason for anticipating a brilliant outcome."[4]

This time it did look as if dreams might come true. Tombstone finally received its railroad. Attracted by the renewed activity, the El Paso and Southwestern began building into Tombstone from Fairbank in 1902, inaugurating its line on 5 May 1903. The initial efforts of Tombstone Consolidated also boded well. The company refitted the old Tombstone Mill, introduced the cyanide leaching process to the district, and reworked waste dumps and old drifts. The arrival of the railroad ended the town's transportation problem, and by 1906 the company was treating 250 tons of ore per day, and shipping 2,500 tons of concentrates per month for smelting at El Paso, Texas. By the beginning of 1904 Tombstone Consolidated's pumping and haulage shaft passed seven hundred feet in depth, well below the previous water table, and the company had raised 675 million gallons of water in less than a year. By 1907 Tombstone Consolidated was excavating a pumping station on the one thousand–foot level, averaging about five hundred thousand dollars in production per year, and had made a profit for four years.[5]

By then, however, cracks in the company's foundation had begun to show. In April 1907 General Manager Staunton informed stockholders that "the heavy expenses of pumping and advance development work in opening the ground below water level, from which we are not yet in shape to derive any returns, have continued to absorb the earnings, and this condition may continue for some little time." Once it began operating on the one thousand–foot level, the company discovered that pumping 5 million gallons of water per day was not enough. At the end of that September the *Tombstone Epitaph* reported that Tombstone Consolidated planned to increase its pumping capacity to 10 million gallons per day. It took the better part of the next year to get the massive new pumps installed, but by the end of 1908 the mine lifted close to 7 million gallons of water per day. The *Epitaph* had to admit that 1908 had been filled with "vexatious delays and obstacles almost insurmountable," but assured its readers that 1909 would find Tombstone hailed as the greatest mining camp in the West.[6]

Fig. 12.3. Because of the district's flooding problems, miners worked in rain gear in the Tombstone Consolidated, ca. 1900. (Arizona State Library, Archives, and Public Records, Archives Division, Phoenix, #97-0429)

Fig. 12.4. The river of water being pumped out of the Tombstone Consolidated Mines, ca. 1910. In the thirteen months before October 1909, Tombstone Consolidated removed more than one billion gallons of water from Tombstone's mines. (Arizona State Library, Archives, and Public Records, Archives Division, Phoenix, #97-2850)

But 1909 brought disaster. Tombstone Consolidated had struggled through the year with its herculean task. From the beginning of September 1908 to the end of September 1909 the company lifted 1,828,215,012 gallons of water, averaging over 5 million gallons per day. In May 1909 the pumping system failed temporarily due to a contaminated fuel supply. By the time it could be resurrected, the pumps on the one thousand–foot level were submerged and inoperable, and the company was forced to retreat to the eight hundred–foot level. Tombstone Consolidated ordered more dewatering equipment and called on the stockholders for salvation—again. On 8 November 1910 Murphy told the stockholders bluntly that they had not been forthcoming enough and that "unless funds are made available ... operations will have to come to a close, to be followed by liquidation, which would ... result in almost a total loss to those who have invested in this great enterprise." A month later the *Epitaph* reported that company officials sought to amend Tombstone Consolidated's articles of incorporation in order to double its debt limit to six million dollars. In January 1911 the paper reported that the company would reorganize, but those efforts came to nothing. The Tombstone Consolidated Mining Company abandoned pumping and suspended operations for good that same month.[7]

In 1914 Phelps Dodge tried its hand in the Tombstone District. That company purchased all of the holdings of Tombstone Consolidated, but hopes that the great copper concern would resurrect Tombstone and its mines proved ill-founded. The company did ship manganese and fluxing ores from the district for a number of years but never mined on a large scale. The district produced more than twenty-eight million dollars from 1879 to 1907, but only eight million dollars from 1908 to 1934. In 1933 Phelps Dodge rendered its opinion of the potential of the old bonanza camp by selling out to the Tombstone Development Company. That firm explored extensively, especially after 1965, in an unsuccessful attempt to find previously undiscovered mineral bodies. Tombstone Development also leached some tailings before passing control of its properties to Tombstone Exploration, Inc., in 1979. Tombstone Exploration brought mining back to Tombstone with an extensive exploration program and an open-pit and leaching operation. For a time the operation became the largest open-pit silver mine in the United States. The pit consumed the sites of the Contention and Grand Central, and employed about two hundred people. The company leached ore from tailings, underground works, and the pit and anticipated that its reserves would last until 2005.[8]

But as often happened in Tombstone's history, things did not quite work out as planned. By 1983 Tombstone Exploration had moved 6 million tons of material from the pit, including 1.5 million tons of ore, but the following June the company suspended operations and laid off most of its workers "due to the economics of the operation." In 1985 Tombstone Exploration, Inc., entered bankruptcy proceedings and abandoned its Tombstone operation. Another company, Cowichan Resources, met the same fate in 1990 after it tried to work the property. Some of Tombstone's residents were frankly glad to see the mining companies go. Times have changed, and the town now has another industry to replace mining.[9]

The year 1929 proved a pivotal one for the drowsy former mining town. In November the voters of Cochise County elected to relocate the county courthouse—one of Tombstone's few remaining sources of income—to Bisbee. The citizens of the smelter town of Douglas had instigated this move, seeking the county seat for their town. They had tried to achieve the same goal in 1917, but with Douglas the only other

Fig. 12.5. Tombstone's first "Heldorado Days" celebration in October 1929 included this parade. (Arizona Historical Society, Tucson, #880)

option on the ballot, Bisbee's voters joined those of Tombstone to keep the courthouse in Tombstone. In 1929 all three towns were eligible, and Tombstoners, realizing that they would be outvoted in any case, threw their support to Bisbee to keep the seat from being relocated all the way to Douglas.[10]

Losing the county seat delivered another hard blow to the already-groggy Tombstone. One witness reported in 1930 that "only the purchases of a few miners and stockmen keep the town from falling into ruin." With mining suspended and the courthouse removed, the former silver queen of the Southwest began to drift downward again. By 1940 the town's population had dwindled to 822, its lowest level since 1900.[11]

Fig. 12.6. Since the 1920s Tombstone's residents have understood that their town's survival depends upon celebrating its frontier past—a celebration that has almost nothing to do with the area's mining history. The sign (*right*) states: HERE IN THIS STREET WYATT EARP PLAYED HIS LONE HAND AGAINST A LYNCH MOB OF 300 AND SAVED HIS PRISONER "JOHNNY-BEHIND-THE-DEUCE" WHO WAS HIDDEN IN VOGAN'S BOWLING ALLEY, JAN. 15, 1881. (CP SPC 63-56, Arizona Collection, Arizona State University Libraries)

Fig. 12.7. A billboard beside a road approaching Tombstone.

The key to Tombstone's salvation had been discovered less than a month before the vote to remove the courthouse. In the last week of October 1929 Tombstone celebrated the fiftieth anniversary of its founding with its first "Heldorado Days." The hundreds of people who attended watched a mock stagecoach robbery and a reenactment of the gunfight at the O.K. Corral. The idea for the pageant largely belonged to Bill Kelly, the *Tombstone Epitaph*'s editor. That paper's founder, John Clum, attended as the guest of honor, accompanied by a number of other pioneers. Residents had discovered the reason for Tombstone's continued existence and have since staged a celebration every October.[12]

Although the poverty of the Great Depression and the travel restrictions of World War II certainly pinched Tombstone's fledgling tourism business hard, the town survived to prosper in the 1950s. By 1960 Tombstone's population had increased more than one-third again from that of 1940. The age of the automobile vacation and the booming popularity of the legends of the Old West—of which Tombstone held a hard hand to beat—combined to create a steady income for the town. In 1962 Tombstone became Arizona's first National Historic Landmark. By 1980 the town had a population of 1,632, its largest since the 1880s, and tourism had long since replaced mining as the town's principal source of income. Today even an off-season weekday finds a few dozen tourists wandering the streets, shopping in the curio stores, or eating Earp burgers in the replica saloons. You too can visit the old courthouse (now a state park), take a ride in a Concorde coach, or—for three dollars—"Walk Where They Fell." Tombstone is the fortunate one, the only survivor of the Tombstone District. All of the San Pedro River towns that milled and transported Tombstone's ores have long since returned to the desert dust.[13]

Although it came even closer to death than Tombstone, Jerome has also survived. As 1939 turned into 1940, it looked as if the copper

Fig. 12.8. Present-day Tombstone, looking north from the base of the Tombstone Hills with the former Cochise County Courthouse, now the state park's headquarters, at center.

Fig. 12.9. Jerome photographed from the hogback, ca. 1940. Cleopatra Hill domi-
nates the town on the left, with United Verde's works appearing in the center and to
the right. Some subsidence damage is visible in the center of the town. Compare this
photograph to the modern view on page 3. (SPC 150:14.39x, Arizona Collection, Arizona State University Libraries)

camp might boom again. Even though United Verde's open-pit operation ceased in 1940, things began to pick up as the storm of World
War II blew ever harder upon the American shore. Jerome's population rose during the first two years of the decade as Phelps Dodge
imported miners into the district to meet the increased demand for
copper. One publication estimated that the company had a twenty-
year supply of ore with which to meet this latest emergency.[14]

The new war did not bring another bonanza to Jerome, however.
World War II may have actually accelerated the town's decline. Even
before the United States entered the war, Jerome suffered a chronic
labor shortage. Men left Jerome's mines for better-paying, easier, and

safer jobs in the burgeoning defense industries of California. Joe Larson noticed this trend as early as mid-1941. He reported that the younger men and good mechanics were leaving for the shipyards and aircraft plants in southern California and wrote that Jerome's businessmen "complain that we have such a floating population here now; they stay only long enough to make enough to take them to the next place." Four months later Larson described Jerome's population as "a floating one, and not the desirable kind." Larson's own wife and daughter joined the migration after the war began; they spent the war years building aircraft in California.[15]

The Phelps Dodge United Verde Branch won commendation from the federal government for its efforts in 1942, but the company could not stop hemorrhaging labor. The town's population may have declined by half between mid-1942 and 1944, and United Verde spent much of the war working only two shifts because of its labor shortage. The government and the company tried several measures to solve this problem—without much success—including freezing wages, restricting the movement of laborers, and importing soldiers and other amateurs to mine. John Muretic, in charge of an underground crew at the United Verde during the war, regarded most of the civilian replacements as draft dodgers of dubious value to the mine. Once copper production ceased to be a government priority in 1944, attrition accelerated, and most of those who left never returned. In the spring of that year Perry Ling believed the end to be in sight for United Verde and probably for Jerome as well.[16]

In order to meet its war quota while operating shorthanded, the company had stripped the mine of its remaining high-grade ores and could not then go back and economically recover much of the low-grade material left underground. Its twenty-year reserve of ore vanished in a half dozen. In October 1946 a mine official reported that "mining operations are now essentially confined to the cleaning up of ore . . . and the completion of a few open stopes. . . . This vast ore zone,

with over sixty years of active productivity, can now be said to be effectively bottomed." A journalist visiting the town two years later noticed
that "the majority of Jerome citizens are pessimistic about their town's
future. . . . The life of Jerome seems to be dwindling with its copper
deposits; a fatalistic attitude . . . seems to be all-pervasive."[17]

The early 1950s found the mine down to a skeleton crew, the Clarkdale smelter and most of Jerome's businesses closed, and Phelps Dodge
renting company housing to anyone who would have it. It had been
quite a run. In seventy-five years the district had produced almost one
billion dollars in minerals from works that extended almost a mile
underground. But in January 1953 Phelps Dodge announced that the
United Verde Branch would be closed for good. The company removed
its last ore from the mine in March and shipped its last concentrate in
May 1953.[18]

Just as in Tombstone, mining has proceeded fitfully since the suspension of the last major operation. One of the United Verde Extension's successor companies, Verde Exploration, Ltd., did considerable
sampling in the district in the 1940s and 1950s, but found nothing
to justify further development. Phelps Dodge also searched for ores
as late as the 1980s and engaged in some leaching operations in the
late 1970s and early 1980s. The independent Big Hole Mining Company leased the United Verde pit from Phelps Dodge for a number of
years, employing a handful of men to ship upward of one thousand
tons of ore per month recovered from the walls of the pit. In the late
1980s and early 1990s a dozen men worked underground at the uvx
extracting fluxing ores, but that operation ceased by the mid-1990s.[19]

The 1950s saw Phelps Dodge disengage from the district and the
town battle for survival through the establishment of its historical
society and museum, and other activities used to attract tourists and
residents. Jerome's population may have dropped below 100 in the
mid-1950s, but at the end of the decade the town remained alive, with
about 250 residents and a toehold on its future.[20]

During the 1960s the town solidified its position with the formation of the Jerome State Historic Park, headquartered in the old Douglas mansion deeded to the state by the family in 1962. Civic leaders and citizens poured effort into this endeavor, crowned when Lewis Douglas, son of James Douglas, presided at the dedication ceremony for the new state park on 14 October 1965. Residents' efforts to preserve and interpret Jerome's historic sites and events gained even more legitimacy in 1967 when the town won designation as a National Historic Landmark.[21]

By 1975 almost every inhabitable residence and commercial space had an occupant. At the end of 1978, *U.S. News and World Report* found that "more homes are occupied than vacant," reported rising real estate prices, and claimed that the former copper camp averaged about 10,000

Fig. 12.10. Although Jerome lacks Tombstone's Wild West pedigree, the town's serpentine roads, spectacular setting, bars, and boutiques attract motorcyclists and other tourists.

Fig. 12.11. Contemporary Jerome photographed from Sunshine Hill. Compare this to figure 2.3 on page 52. While much of the central business district remains, the "Mexican Colony" below the town has disappeared, as has the company housing above Clark Street.

visitors per month. In 1980 Jerome had a population of 420, and a number of thriving specialty stores, art galleries, and restaurants. A community profile conducted that year revealed that roughly one-third of those employed worked in service occupations and a little over one-quarter in wholesaling or retailing and that the town had an unemployment rate under 8 percent. Clearly, Jerome had moved beyond bust.[22]

As Jerome has recovered a measure of prosperity, officials and residents have devoted even more attention to historic preservation. In 1965 residents established a community service organization to promote and assist community projects. Eleven years later the town created a planning and zoning commission to preserve historic structures and regulate growth. In 1980 and 1981 Jerome conducted a historic site survey, covering 265 of the town's buildings. Today Jerome seems secure in a future populated by artists, shopkeepers, retirees, and tourists, and proud of its industrial heritage.[23]

Tombstone and Jerome are abnormal in having survived their mining days and attained some measure of postmining prosperity. Most mining towns disappear once mining ends, as their populations drift away. A career in mining has always meant mobility, but this was especially true in the period before World War I, when many of the people who roamed the mining West possessed a boomer mentality. *Boomer* is railroad slang for an employee whose wanderlust causes him to change locales frequently. The term applies to most of those drawn to the Tombstone excitement in the 1880s.

Some people remained after Tombstone's boom and in many other busted mining camps. They may have lost the boomer mentality as they got older; the mean age in Tombstone's voting registers rose from 36.7 in 1882 to 41.3 in 1892. Police Chief Gage, ensnared by circumstances to the point of suicide, was fifty-nine years old and had a wife and five children. Other people stayed because vested interests meant that they had more to gain by remaining than by starting over or because they had gained a place of importance in the community. The *Tombstone Prospector* described people who worked against the railroad in 1888 as "those persons who have made their all in this camp, and now wish to deprive others from doing anything to better their condition." Still others remained because they had those unrealistic expectations of a revival of the town's fortunes. One old-timer—a resident since 1880—wrote to the *Prospector* in 1920, forty years after the glory days, stating "that within three years the real, permanent boom in Tombstone will be here in full bloom."[24]

But most people moved on. They understood that any nineteenth-century mining venture was a gamble and that the most regular outcome in gambling is losing. The argonauts had their fortunes to find and their adventures to experience—or at least their ends to meet.

They left it to others to examine the meanings of boom and bust, and that examination began almost immediately. Both Mark Twain and John Muir pondered the meaning of the ghost town in the 1860s and 1870s, long before historians took up the issue. Twain claimed in *Roughing It* that "in no other land, in modern times, have towns so absolutely died and disappeared, as in the old mining regions of California." One of his contemporaries concurred, writing of the Sierra towns of the late 1860s that "now half of these are wholly deserted, and the others, with few exceptions, are in a decaying condition, with many houses and stores unoccupied, and often with only a small proportion of their old populations." The ghost town quickly became a symbol of the West's shortcomings, with the blame for same being variously assigned, usually elsewhere. Populist rhetoric blamed the busts of the 1880s and 1890s on goldbugs, railroads, or nonproducers, rather than on arid climate or overproduction of silver.[25]

Bernard DeVoto was the first historian to confront boom and bust in the West. In reaction to Frederick Jackson Turner's understanding of westward expansion as the victorious march of progress, DeVoto saw failure in the West's cycles of boom and bust. He articulated his views in a 1934 *Harper's* magazine article titled "The West: A Plundered Province." DeVoto—like the Populists—blamed the West's busts on eastern capitalists and industrialists, who made a colony of the West, plundering its resources and exploiting its labor to build an industrial colossus in the East. Western mineral development, he believed, "has not made the West wealthy. It has, to be brief, made the East wealthy. Very early the West memorized a moral: the wealth of a country belongs to the owners, and the owners are not the residents or even the stockholders but the manipulators."[26]

DeVoto believed that the crux of the problem was that the West needed capital more than it needed labor. It needed capital to defeat the region's vastness and aridity, to overcome the high cost of industrial mining. But capital could come only from the East and could come only at the price of western independence. He believed that this debt peonage meant that the West was "the one section of the country in which bankruptcy, both actuarial and absolute, has been the determining condition from the start."[27]

Nineteen years later DeVoto sounded the same theme in a 1953 article for *Collier's* magazine, "Our Great West: Boom or Bust?" in which he argued that hand in hand with the hope and promise of the Golden West came heartbreak and despair—and ghost towns. The boomtown might be a utopia, "doing a brisk business in corner lots. But always there were the Utopias that went broke. It was the country of the false color, the exhausted placer claim, the bobtail straight, the lost chance, the road that ran out to nowhere." To DeVoto, and others since, the ghost town is a brand of failure upon the West—failure wrought from either exploitation or myopia.[28]

This belief in the colonial victimization of the West—sustained by such leading western historians as Bernard DeVoto and Walter Prescott Webb—lasted through the Great Depression and World War II. After the war revisionists like Earl Pomeroy and Gerald Nash argued that the colonial period had ended, its demise brought about by World War II, which had ended the West's absolute dependence upon the extractive industries. Pomeroy believed that the West had been becoming more like the East since 1900 anyway, even without the revolutionary transformations in the western economy brought about by the war.[29]

Colonialism as an explanation returned to dominance and probably reached the apex of its expression in the "sagebrush rebellion" of the 1980s. The emotions of the sagebrush rebels are perhaps best displayed in Richard D. Lamm and Michael McCarthy's *Angry West: A Vulnerable Land and Its Future*. Lamm, a former governor of Colorado, and McCarthy, a journalist, echoed the idea of the plundered province. "Left behind," after the boom days end, they wrote, "is a wasteland, its skeletal boomtowns and cratered-out landscape a graphic reminder of days past. Western people, pawns in an ugly and endless war, regroup and rebuild. . . . The history of the West, in many respects, is a history of exploitation." Again, eastern magnates harvested the wealth, and western townspeople paid the price. While Lamm and McCarthy certainly did not let industrial capitalists off the hook, they directed most of their arrows at the federal government, which they believed had replaced the robber barons of old as the exploiter of the region. They contended that federal policy—through domination by powerful lobbies or incompetence or inattention—favored eastern interests. The West bore the costs, whether from silver pricing in the nineteenth century or atomic testing in the twentieth.[30]

In recent years the New West school of historical interpretation has gone beyond regionalism and has returned some of the blame to the people of the West. The ghost town fits the New Western interpretation of the region's exploitation—in this version by both residents and outsiders. New Western history finds the ghost town of the American West to be a symbol of the failure—through greed, hubris, or man's inhumanity to nature—of the American frontier experience, capitalism, or even American civilization itself. According to the New West's most prominent historian, Patricia Nelson Limerick, boom and bust reveal the ethos of the West: "the attitude of extractive industry—get in, get rich, get out." Ghost towns were simply the physical remains of that creed. The invading culture failed to understand the West's different conditions, and that failure produced the abandoned buildings—the headstones of abandoned dreams. She believes that the ghost town symbolized failure in the westering experience and that the historians of the West who preceded her "have not taken failure seriously."[31]

If one takes the ghost town seriously, Limerick argues, "one has to notice that often enough, this process [of western expansion] could more accurately wear the opposite label—western contraction. Expansion, construction, and growth are certainly part of the region's history, but so are contraction, retreat and abandonment." These ghost towns show, further, that the West is "the region where we can most profitably study the interplay of ambition and outcome, the collision between simple expectation and complex reality, and the fallout from optimistic efforts to master both nature and human nature." She concludes that we can learn from all of this failure and construct for ourselves "a less imperial and more permanent society," instead of one that sits "on the narrow and precarious foundation of extractive industry and unrestrained economic ambition." Limerick, like others before her, sees the ghost town as a teacher of hard lessons that, if heeded, will lead to a better life in the West.[32]

While the bleached bones of a former mining town might encourage such thinking, to define as a "failure" a settlement that was organized for a purpose, has served that purpose, and then has ceased to exist is certainly a curious use of that term. By that definition the Apollo space program was a disaster, since what it did was establish a half-dozen ghost towns on the moon. Indeed, one could define all life as a failure, as it inevitably ends in death. A more reasonable perspective understands that land and resources have different uses during different periods of history. Yesterday's mining camp becomes today's wilderness, farm, or town, just as surely as yesterday's wilderness, farm, or town becomes today's mine. Nobody placers along Cherry Creek these days, but that hardly denotes the failure of Denver—or the West.[33]

Those who are ready to understand the West in terms of its victimization often use the ghost town, or more generally boom and bust, as the cornerstone of their argument. They certainly have some valid points to make. The West has always been beholden to absentee investment, and those eastern and European capitalists—or the immigrants who invested their labor, for that matter—have often sought to profit from the West and then retire from the region. The extractive industries—mining, agriculture, lumbering, and fishing—have always labored under the burden of market prices that they did not control and could barely influence. The histories of both Tombstone and

Jerome offer compelling examples. Given these motives and circumstances, many believe that the West has been a playground for unrestrained capitalism, heedless of consequences. A prominent historian of the timber industry has concluded that "the market prevailed and the social and economic stability of the region was never a major consideration in private and public decision making." Granted, the West has had a more volatile economy than the East. But it has been more fortunate than the South, most of which was chronically impoverished and depressed from the end of the Civil War until the last generation—busted without boom for a century.[34]

Some more zealous defenders of the West have gone beyond the mere assertion of economic disadvantage, and in that other great western tradition have identified the black hats and white hats, the exploiters and exploited. Lamm and McCarthy call the people of the West "pawns in an ugly and endless war" of boom and bust. External forces, including federal policy, wars, international economic conditions, "and the colonial whims of big business," have battered the West, and "the ultimate victims were the towns' people. History has treated the mineral frontiers as rich, romantic chapters in the story of America's march west. But there was nothing romantic about their instability or about the suffering of their people."[35]

Westerners have always craved the monies invested by corporations and the federal government, then resented it when those same entities tried to manage their investments in the region. At the same time that Tombstone's newspapers were constantly bewailing the lack of attention the area received from outside capitalists, the author of the *Tucson and Tombstone General and Business Directory* complained that Tombstone "is, unfortunately, mainly owned by Eastern capitalists, who show little, if any, interest in the permanent or lasting prosperity of the city." But absentee control is the price one pays for absentee investment.[36]

Those inclined to see westerners as the victims of eastern capitalist domination must also overlook the legion of easterners and Europeans who sank their money into mining ventures that never showed a profit or were swindled outright by western confidence men. When the great bubble of stock speculation in Goldfield, Nevada, burst in early 1907, as much as $150 million in American investment may have gone with it—a loss that substantially exceeded the entire value of the district's production in its first fifty years. Even the paying properties did not produce unlimited wealth. Though it certainly enriched William Andrews Clark and Phelps Dodge, the Arizona Department of Mineral Resources estimated in 1952 that United Verde did not average a 10 percent return on investment over its life. In 1980 the Phelps Dodge Corporation made a profit of $91 million—on $2.1 billion in total assets. A 4.33 percent annual return on investment hardly qualifies as unfettered exploitation.[37]

If one blames the federal government or corporations for bust, it seems only fair to credit them for boom. The roads, railroads, industrial enterprises, supply towns, and financial centers established all over the West are the legacies of resource booms long past. Neither San Francisco, Seattle, Denver, Wichita, Albuquerque, nor Anchorage owes its origins or development to its textile mills or gently rolling farmlands. All derived from extractive industries of one sort or another, as did most of the rest of the West. The mineral booms at Tombstone and Jerome certainly did as much or more to develop their sections of the Southwest as anything else.[38]

The oft-expressed notion that these "failures" of the West "broke individual lives prodigally, [and] turned hope into despair wholesale," as DeVoto put it, also bears examination.[39] One certainly should not minimize the suffering bust inflicted on individuals like Police Chief Gage, but at the same time one should ask whose dreams were demolished and how. The dreams of speculators may have died when the stamps ceased to drop, but what about most of the people who actually lived in these towns? They did not have mortgages or health and pension plans to keep them shackled to their jobs. Those things did not exist for most people prior to World War I. What kept you at a job in a nineteenth- or early-twentieth-century mining town was that you liked the people, the place, or its prospects. When those conditions ceased there was usually a job very much like this one somewhere up the road.

Tombstone pioneer William Miller, who spent forty years drifting about the West, remembered later that he "never worked long at any one thing, made money easily and spent it freely as I earned it. When I had $10.00 I had all the money I wanted and was ready for new sights and places." In Miller's case, as in most, this boomer mentality was not especially remunerative, but it had other compensations. Miller reminisced at age seventy-six that he had "never found a fortune, but I have made many friends and had a glorious time." People like Miller, who followed the mining rushes across the western frontier, understood the transient nature of the business better than those who have since sought to explain it. Any mineral deposit, no matter how vast, is finite and ultimately depleted. The frequent use of the word *camp* to describe substantial nineteenth-century cities like Leadville, Virginia City, or Tombstone is certainly suggestive.[40]

Even if we consider only the successful mining towns, their finite, unrenewable mineral bodies were often their one redeeming feature. This is the signal difference between mining towns and other resource-extractive communities whose farm products, timber, or fish can be regenerated—even if such conservation has not always been practiced. Once they had exhausted their ores, many mining towns—located on forbidding desert mesas or in forbidding alpine meadows—lost their only reason for existence. To be disillusioned by this process of settlement and abandonment, one had to have illusions about it in the first place. The people who named places like Suckertown, California, and Small Hopes, Colorado, certainly understood the nature of nineteenth-century mining.[41]

We might also ask how many of these people really meant to live out their days in places like the Wood River District, "the darkest part of

darkest Idaho," or Goldfield, Nevada, "where the desolations meet"—
even if given that choice. How about Coldfoot, Alaska, or Bedbug, California? Russell Elliott wrote of his hometown of McGill, Nevada, that
"the only possible motive for anyone intruding into this hostile environment was a mining boom." One of Tombstone's female pioneers
wrote that "the attractions of Tombstone . . . are easily summed up in
one word—money. . . . People do not come here to live because it is
an agreeable abiding place, but because wages and business are good.
The expense of living is high, but one is able to save something toward
a pleasenter [*sic*] home elsewhere if one practices economy." When
that money dried up it was time to move on—caterwauling of the camp
newspaper notwithstanding—and most people understood that.[42]

When the Tombstone excitement occurred, they came from everywhere. One newspaper reporter exploring the silver boomtown wrote
that "there is no mining camp from Placerville to White Pine, nor from
Frazer River to Muleje they have not visited. I think I can learn more
about California, Nevada, Idaho, Utah and Montana by a sojourn of
a single week in Tombstone than I could by a two years' tour through
the Pacific States and Territories." William Miller had "no reactions"
to the primitive camp when he arrived in Tombstone in 1879 because
he "had seen just such places from my earliest recollections. A 'strike,'
a stampede, and a new town."[43]

When the excitement died, they went everywhere. There was always
news of another excitement—in nearby Pearce, Bisbee, or Sonora, or
in places farther removed. Former residents of Tombstone ended
their lives in Alaska, Australia, and New York City. After Tombstone
businessman Sol Israel relocated to Seattle, he listed by name nine
other former Tombstone residents living in that city. After her days in
Tombstone, Nellie Cashman went elsewhere in the American Southwest; to Kimberley, South Africa; and then to Canada during the Klon-
dike rush, where, in Dawson, Yukon Territory, she ran into her old
friend John Clum, Tombstone's former mayor.[44]

But for the most striking example of the boomer mentality we must return to Edward Schieffelin, discoverer of the Tombstone District. Schieffelin—a prospector of twenty years' experience—realized something like four hundred thousand dollars out of his Tombstone discoveries, a fantastic sum of money in the 1880s. Clearly, he could have spent the rest of his life in genteel retirement. He tried; he did everything a bonanza king was supposed to do. He moved to California, had a mansion built in San Francisco for his mother, and married a proper wife. But boomer Schieffelin could not stick it out. Before long he was off prospecting in Alaska and elsewhere in the West. He died in 1897, fifteen years after his big bonanza, alone in his prospector's cabin twenty miles into the Oregon woods. At his request he was buried in

Fig. 12.12. Ed Schieffelin asked to be buried at a site where he camped while making his Tombstone discoveries. His monument, in the shape of a miner's cairn, lies about two miles west of Tombstone, which appears on the horizon at the center of this photograph.

prospector's garb, not among the wealthy in a San Francisco cemetery, but among the boulders in Tombstone's hills. Nine years before his bonanza he had written, "I am restless here and wish to go somewhere that has wealth for the digging of it. I can't say that I care to be rich—it isn't that. . . . But I like the excitement of being right up against the earth trying to coax her gold away to scatter it." For Schieffelin, and those like him, what mattered most was the chase, not the catch.[45]

By the twentieth century the transience and individualism of nineteenth-century boomtowns were being replaced by the order and stability of great corporate mining communities. The debate among historians over whether the history of the West should be understood as the history of westward expansion or the history of the western region—of process or place—has obscured the possibility that it might be both. Mining in the nineteenth-century West was a process of settlement: discovering ore bodies, building towns, stringing telegraph wires, and laying rails. By World War I most of the major discoveries had been made and the districts populated. By then western mining had become a matter of place—more precisely, places: fewer and larger mineral districts, many of them towns of long duration. As these places matured they became, in the words of one mining historian, "modern industrial communities, not greatly different from eastern cities of similar size."[46]

The corporate welfare increasingly offered after World War I by companies such as United Verde to reduce turnover of employees also increased the attachment of workers and their families to particular places. John Krznarich's father worked as a timberman at United Verde from 1916 until he retired with the mine in 1953. Joe Larson, who arrived in Jerome in 1899, was still seeking his destiny there fifty-four years later. Franklin Roosevelt's copper-tariff petitioners noted that "many of us have been with the company for twenty years or more, few of us have been employed by it for less than ten years. Our children have been born here and have been educated in schools [here]. . . .

Some of our sons have followed us into the mine and smelter, and others of our families will probably be furnished employment if operations can be resumed." John Muretic estimated that 70 percent of the men on his crew during World War II were born and educated in Jerome and were the sons of miners. With the cooperation of the mine's employment office, he hired the sons and brothers of men already underground.[47]

These are generalizations, of course. The nineteenth century had its home miners and the twentieth century had its boomers. While one might argue that people stayed in Jerome and other towns during the 1930s only because of the magnitude of the Great Depression—which eliminated opportunities elsewhere—lack of opportunity did not prevent people from tramping during the 1930s. Hoards of people migrated throughout the decade in search of a better chance somewhere else, and that tendency applied to mining camps like Jerome. There was little chance to grow foods at most mining sites and few alternative employers once the big mines shut down. However bad things got in Phoenix and southern California, those places provided more opportunities than Arizona's copper camps, and many people chose to migrate thither. Still, the urge to stay kept many behind, and researchers of the timber and steel busts of the 1970s noticed the same reluctance to relocate during an era when general economic conditions were far better than they were during the 1930s.[48]

Even members of groups at the lower end of the economic or social order, usually regarded as transient and assumed to have no loyalty to localities, often had a great stake in the Verde District. By the time Juana Sanchez died in the mid-1930s, she had resided in the district for almost thirty years and had borne nine children, all of whom lived in Jerome. Rafaila Rodriquez remembered later that "there were [Latino] families that the father, the grandfather ... and the great-grandfather had all worked in the mines, so ... their roots were really deeply embedded there." Further, the demographics of Jerome more closely resembled those of stable, long-lived communities than they did the young, male, footloose boom camps of the frontier.[49]

Thus, the dislocation caused by the bust of these settled mining towns during the twentieth century was more traumatic for the people involved than was that of the frontier towns of the nineteenth century. Ray Consolidated Mines of the Ray District of Arizona recruited many workers in Mexico with the assurance that, in the event of a company shutdown, they would be returned home at company expense. But the free transportation did not ease the departure. "It was a sight to see," wrote a biographer of that district, "as trainload after trainload took their departure and headed southward as people shed tears with their farewells." That sort of attachment to the community remained for many in the Verde District even after the great copper mines finally played out and Jerome was all but abandoned after 1953.[50]

For twenty years after United Verde's final shutdown, a ghost town reunion brought hundreds of former townspeople back to Jerome annually. At a class reunion in 1974 Arthur Duran may have articulated the difference between the nineteenth- and twentieth-century West: a difference between a process orientation of exploration, discovery, and settlement and a place orientation attained by living in a long-standing community, with well-established institutions and amenities. "I don't know just what it is that makes people from Jerome feel so close to each other . . . ," Duran said. "Maybe it was all the hard times that brought us together, and maybe now we depend on each other."[51]

Introduction

1. *Arizona Daily Star*, 10 Aug. 1884.

2. For example, Malcolm J. Rohrbough devotes three hundred pages of *Aspen: The History of a Silver Mining Town, 1879–1893* to the history of that camp up to the silver panic of 1893, and a ten-page epilogue to the bust period from 1893 to the 1930s. Likewise, Robert L. Spude's twenty-two-page article on its boom, "Swansea, Arizona: The Fortunes and Misfortunes of a Copper Camp," ends with one paragraph on the town's decline. Neither of these authors has been singled out because he did something unusual. Dozens of other examples would have served as well.

3. *Arizona Weekly Citizen*, 24 July 1886. In the interests of clarity, the spelling, capitalization, and punctuation in all quotations used in this book have been modernized. In no case have these changes altered meaning.

4. *Jerome Chamber of Commerce News Bulletin*, 24 Dec. 1935. Most sources place Jerome's total value of product at around $900 million.

5. Page Smith, *As a City upon a Hill: The Town in American History*, 33; Thomas J. Noel, Paul F. Mahoney, and Richard E. Stevens, *Historical Atlas of Colorado*, pl. 23; Patricia Nelson Limerick, "Haunted by Rhyolite: Learning from the Landscape of Failure," 21. The search for books on nonmining ghost towns is not completely hopeless. The ghost town guides to Kansas, Texas, and the Pacific Northwest, cited in the following notes, deal mainly with agricultural, forestry,

and transportation ghost towns, though coal, oil, and other types of mining get their fair share of attention. The ghost town guides themselves are an interesting study. They are uniform in style and content, even in their titles. These books usually consist of brief biographies of the towns, in alphabetical order, with an accompanying photograph of the boom days or of the present site. The biographies discuss the foundation and growth of the new community and prominent citizens or notorious crimes, if any, and perhaps conclude with a paragraph explaining why the town busted. Limerick is correct in her observation that analysis, of either booms or busts, is not their strength.

6. E. Gorton Covington, "Safford Flared but Briefly"; David Rich Lewis, "La Plata, 1891–93: Boom, Bust, and Controversy," 5–21; Rex Myers, "Boom and Bust: Montana's Legacy of High Hopes . . . and Lost Dreams."

7. Michael Malone, "The Collapse of Western Metal Mining: An Historical Epitaph"; Sandra Dallas, *Colorado and Utah Ghost Towns and Mining Camps*, 139; Donald C. Miller, *Ghost Towns of Nevada*, 145; Bruce Ramsey, *Ghost Towns of British Columbia*, 215–26; Raye Carleson Ringholtz, *Uranium Frenzy: Boom and Bust on the Colorado Plateau.*

8. Roger M. Olien and Diane D. Olien, *Oil Booms: Social Change in Five Texas Towns*; Daniel Fitzgerald, *Ghost Towns of Kansas: A Traveler's Guide*, 154; John S. Spratt Sr., *Thurber, Texas: The Life and Death of a Company Coal Town*; Donald C. Miller, *Ghost Towns of Wyoming*, 13, 87; T. Lindsay Baker, *Ghost Towns of Texas*, 96; James E. Sherman and Barbara H. Sherman, *Ghost Towns and Mining Camps of New Mexico*, 90, 221; Donald C. Miller, *Ghost Towns of Idaho*, 72; Donald C. Miller, *Ghost Towns of Montana*, 169; personal observations and conversations with mining historian David Wolff.

9. Donald Worster, *Under Western Skies: Nature and History in the American West*; Robert R. Dykstra, *The Cattle Towns*; Donald Worster, *Dust Bowl: The Southern Plains in the 1930s*; R. Myers, "Boom and Bust"; Baker, *Ghost Towns of Texas*, 36, 73; D. Fitzgerald, *Ghost Towns of Kansas*, 152, 232, 237; Nancy Burns, "The Collapse of Small Towns on the Great Plains: A Bibliography," 23; personal observations.

10. William G. Robbins, *Hard Times in Paradise: Coos Bay, Oregon, 1850–1986*; Ramsey, *Ghost Towns of British Columbia*, 140–41; personal knowledge derived from employment in the Alaskan fishing industry, summer 1993.

11. D. Fitzgerald, *Ghost Towns of Kansas*, 31–33, 114; Baker, *Ghost Towns of Texas*, 114, 116; D. C. Miller, *Ghost Towns of Idaho*, 36; Ramsey, *Ghost Towns of British Columbia*, 25; D. C. Miller, *Ghost Towns of Wyoming*, 8; James E. Sherman and Barbara H. Sherman, *Ghost Towns of Arizona*, 125; D. C. Miller, *Ghost Towns*

of Nevada, 125; Donald C. Miller, *Ghost Towns of California*, 86; Dykstra, *The Cattle Towns*, 51–53; Norman D. Weis, *Ghost Towns of the Northwest*, 38; Sherman and Sherman, *Ghost Towns of New Mexico*, 166.

12. Terry Bacon, "Steins, New Mexico: A Railhead Town Abandoned by Progress," Special Collections, Northern Arizona University, Flagstaff; D. C. Miller, *Ghost Towns of Nevada*, 92; Baker, *Ghost Towns of Texas*, 123; W. F. Cottrell, "Death by Dieselization"; David F. Myrick, "A Chapter in the Life of Raso: A Railroad Ghost Town"; Richard O. Davies, *Main Street Blues: The Decline of Small Town America*, 169–70; Baker, *Ghost Towns of Texas*, 110; D. Fitzgerald, *Ghost Towns of Kansas*, 220–21.

13. Bacon, "Steins, New Mexico," Special Collections, Northern Arizona University; Davies, *Main Street Blues*, 167–69; Burns, "Collapse of Small Towns," 9–12; Weis, *Ghost Towns of the Northwest*, 25; D. Fitzgerald, *Ghost Towns of Kansas*, 185.

14. D. C. Miller, *Ghost Towns of Nevada*, 70; Baker, *Ghost Towns of Texas*, 125; D. Fitzgerald, *Ghost Towns of Kansas*, 254, 269, 299.

15. D. Fitzgerald, *Ghost Towns of Kansas*, 275–76.

16. Ibid., 109–11, 249; Baker, *Ghost Towns of Texas*, 11; Sherman and Sherman, *Ghost Towns of New Mexico*, 52.

17. Rohrbough, *Aspen*; Lambert Florin, *Colorado and Utah Ghost Towns*, 13; Baker, *Ghost Towns of Texas*, 130–32; Dallas, *Colorado and Utah Ghost Towns*, 127.

18. U.S. Strategic Bombing Survey, *The Effects of Strategic Bombing on German Morale*; U.S. Strategic Bombing Survey, *The Effects of Strategic Bombing on Japanese Morale*.

19. Allen H. Barton, *Communities in Disaster: A Sociological Analysis of Collective Stress Situations*, 128, xliv.

20. J. C. Raines, Leonora E. Berson, and David M. Gracie, *Community and Capital in Conflict: Plant Closings and Job Loss*; Terry F. Buss, *Mass Unemployment: Plant Closings and Community Mental Health*; Terry F. Buss and F. Stevens Redburn, *Shutdown at Youngstown*; Katharine H. Briar, *The Effect of Long-Term Unemployment on Workers and Their Families*.

21. Lewis E. Atherton, *Main Street on the Middle Border*, xv; Robert Wiebe, *The Search for Order, 1877–1920*. For a brief introduction to this subfield, see Davies, *Main Street Blues*, 4–5, 87–88.

22. Oliver Knight, "Toward an Understanding of the Western Town."

23. P. Smith, *As a City Upon a Hill*, viii–ix. See Duane A. Smith, *Rocky Mountain Mining Camps: The Urban Frontier*, for a discussion of the importance of these communities in the settlement and development of the West.

24. W. Turrentine Jackson, *Treasure Hill: Portrait of a Silver Mining Camp*, 1–2.

25. Another very useful and entertaining book in this vein is Paul Fatout's *Meadow Lake: Gold Town*.

26. Ronald M. James, *The Roar and the Silence: A History of Virginia City and the Comstock Lode*, xxi.

27. Andrew Gulliford, *Boomtown Blues: Colorado Oil Shale, 1885–1985*, 223, 14 (emphasis his).

28. Ibid., 4–5 (emphasis his).

29. Sherman and Sherman, *Ghost Towns of Arizona*, 127; Nancy Cook, "Cleary"; Mark Twain, *Gold Miners and Guttersnipes*, 113.

30. Sherman and Sherman, *Ghost Towns of Arizona*. I extracted the mean and median figures by tabulating their dates. See D. Smith, *Rocky Mountain Mining Camps*, for the best discussion of the urban nature of the "frontier" mining experience.

31. A dilemma when writing about any community is to balance its individuality against its representativeness as a member of a class. I shall endeavor herein to understand the ways in which Tombstone and Jerome typified communities of their type and era, while not forgetting that these were singular places built and inhabited by real people.

Chapter One. Future Growth and Prosperity Is Assured

1. John Pleasant Gray, "When All Roads Led to Tombstone," Arizona Historical Society, Tucson, 2–3.

2. Adolphus Noon, "Visit to Tombstone City," 7 Feb. 1880. This was a published letter dated Tucson, 5 Jan. 1880.

3. Jeanne Devere, "The Tombstone Bonanza, 1878–1886," 17, 19.

4. Lonnie E. Underhill, *The Silver Tombstone of Edward Schieffelin*, 39.

5. *Arizona Weekly Citizen*, 5 Oct. 1878, 6 June, 13 June, 1 Aug., 22 Aug. 1879; Underhill, *Silver Tombstone*, 39.

6. *Arizona Weekly Citizen*, 25 Jan. 1879, 17 Jan., 18 Dec. 1880, 13 Mar., 27 Nov. 1881; Underhill, *Silver Tombstone*, 47–48.

7. William P. Blake, *Tombstone and Its Mines: A Report upon the Past and Present Condition of the Mines of Tombstone, Cochise County, Arizona*, 64; *Arizona Weekly Citizen*, 1 Aug. 1879, 27 Mar., 20 Nov. 1880; "Dividends Paid in the Tombstone District."

8. Patrick Hamilton, *The Resources of Arizona* (1883), 77; *Arizona Weekly Citizen*, 30 May 1879; Blake, *Tombstone and Its Mines*, 66.

9. Blake, *Tombstone and Its Mines*, 63; *Arizona Weekly Citizen*, 7 Jan. 1883. The *Tucson and Tombstone General and Business Directory, for 1883 and 1884* says the monthly production average for 1882 was $433,155.44 and for 1883 was $452,487.29 (110).

10. *Arizona Weekly Citizen*, 28 Sept. 1878, 11 Oct. 1879.

11. *Arizona Weekly Citizen*, 24 July 1880.

12. "Tombstone," 22.

13. Karen Thure, "Tombstone 1881," 4; Blaine Peterson Lamb, "Jewish Pioneers in Arizona, 1850–1920," 160; Donald MacNeil and Bessie MacNeil, Personal and Business Correspondence, 1884–1894 (notes), Arizona Historical Society, Tucson; Alice Emily Love, "The History of Tombstone to 1887," 21. Of course, opulence is relative. Although the *Tombstone Epitaph* grandly reviewed the Grand Hotel in September 1880 (Douglas D. Martin, *Tombstone's "Epitaph,"* 49), an eastern visitor wrote home over a year later that it "owes its name to its being a house of two stories or perhaps even two-and-a-half, but is otherwise unpretentious[. I] was shown to a pretty wretched little room" ("Peabody Tells of Tombstone," Arizona Historical Society, Tucson, 6).

14. *Arizona Weekly Citizen*, 27 Mar. 1880, 4 Mar. 1883.

15. John A. Rockfellow, Papers, Arizona Historical Society, Tucson, 135; Ben T. Traywick, "Tombstone's Hell's Belles," *Bisbee Gazette*, 23 Oct.–4 Nov. 1987, Arizona Historical Society, Tucson.

16. *Arizona Weekly Citizen*, 4 Dec. 1880, 27 Feb. 1881; D. D. Martin, *Tombstone's "Epitaph,"* 49–52.

17. Love, "History of Tombstone," 72; James G. Wolf, "Story of James G. Wolf," Arizona Historical Society, Tucson, 18.

18. "Peabody Tells of Tombstone," 6; *Arizona Weekly Citizen*, 4 June 1882.

19. *Arizona Weekly Citizen*, 21 Feb. 1880; William Vincent Whitmore, Biographical Sketch of Dr. Goodfellow of Tombstone, 1880–91, Arizona Historical Society, Tucson.

20. *Arizona Weekly Citizen*, 5 Sept. 1879, 31 Jan. 1880. One author indicates that by then lots in the central business district were fetching one to two thousand dollars each (Henry P. Walker, "Arizona Land Fraud: Model 1880, the Tombstone Townsite Company," 9).

21. In mid-1880 flour cost six dollars per hundred pounds, hay twenty dollars per ton, bacon fifteen to twenty cents per pound, whiskey two to eight dollars per gallon, and imported beer five dollars per dozen bottles. Nails cost fourteen dollars per keg, as did horse or mule shoes. Blasting powder ran five and one-half

dollars per keg, with blasting caps at one dollar and a half per hundred ("Price of Labor, Provisions, Etc., at Tombstone"; *Arizona Weekly Citizen*, 27 Nov. 1881). Retlef reported in February 1880 that water cost one cent per gallon and wood cost from three to four dollars per cord in summer and five to seven dollars in winter (*Arizona Weekly Citizen*, 21 Feb 1880). A year later, another witness recorded that while water had stayed one cent per gallon, the price of wood had gone up to eight to nine dollars per cord (*Arizona Weekly Citizen*, 27 Feb. 1881).

22. "Price of Labor, Provisions, Etc."; Love, "History of Tombstone," 74; *Arizona Weekly Citizen*, 27 Feb. 1881.

23. Walker, "Arizona Land Fraud," 11; *Tucson and Tombstone Directory*, 112.

24. *Arizona Weekly Citizen*, 16 June 1883.

25. U.S. Bureau of the Census, *Statistics of the Population of the United States at the Tenth Census*, 99; Cochise County, Ariz., Census, 1882. The federal census of 1880 gave the incorporated town of Tombstone a population of 973, but, as there is no way to separate town from precinct in the manuscript census, we will have to use the precinct figure for analysis. George W. Parsons estimated the population upon his arrival in February 1880 at 2,000 ("Early Days in Tombstone"), as did William Henry Bishop in 1881 (*Across Arizona in 1883: Including Glimpses of Yuma, Tombstone, Tucson*, 9). Two years later Patrick Hamilton put Tombstone's population at 6,000 (*The Resources of Arizona* [1883], 50).

26. *Arizona Weekly Citizen*, 30 Aug. 1878, 1 Feb. 1879; "Peabody Tells of Tombstone," 7.

27. *Arizona Weekly Citizen*, 1 Nov. 1879; *Statistics at the Tenth Census*, 99; Cochise County, Census, 1882.

28. D. D. Martin, *Tombstone's "Epitaph,"* 40; *Statistics at the Tenth Census*, 99; Cochise County, Census, 1882.

29. *Statistics at the Tenth Census*, 99; Cochise County, Census, 1882. Throughout this book the term *Latino* designates those persons of Latin American nativity and those of U.S. nativity whose fundamental identification is with the Spanish language and a Latin American culture. The term *white* designates non-Spanish European and American Caucasians. People in both Tombstone and Jerome used the terms *Mexican* and *American* to designate Latinos and whites. Both groups called the Latino quarter of Jerome the "Mexican Colony" ("Colonia Mexicana"), for example. The problem with those labels is that many of the "Mexicans" were American citizens, either by birth, treaty, or naturalization, while many of the "Americans" were European resident aliens. While the term *white* has its weaknesses, it is leagues better than the currently fashionable but terribly misleading *Anglo*.

30. *Arizona Journal Miner,* 30 July 1880; *Statistics at the Tenth Census,* 99; Cochise County, Census, 1882; Ben T. Traywick, *The Chinese Dragon in Tombstone,* 5, 7, 14.

31. D. D. Martin, *Tombstone's "Epitaph,"* 39, 41.

32. Love, "History of Tombstone," 77; Hamilton, *The Resources of Arizona* (1883), 51; Thure, "Tombstone 1881," 6; *Tucson and Tombstone Directory,* 118, 121.

33. *Arizona Weekly Citizen,* 15 Jan. 1882; *Tucson and Tombstone Directory,* 120, 122.

34. Quoted in *Arizona Weekly Citizen,* 17 Jan. 1880.

35. Odie B. Faulk, "Life in Tombstone," 504, 507; *Arizona Weekly Citizen,* 4 Dec. 1880, 10 Dec. 1882, 8 Apr. 1883; Thure, "Tombstone 1881," 5. The smaller venues included the Turnverein Hall, Ritchie's Hall, the Sixth Street Opera House (in reality, a dance hall next to the Bird Cage that was also called the Free and Easy), the Oriental and Crystal Palace saloons, and Danner and Owen's saloon (which hosted the first religious service in Tombstone) (Pat M. Ryan, "Tombstone Theater Tonight! A Chronicle of Entertainment on the Southwestern Mining Frontier," 51).

36. Clair Eugene Willson, *Mimes and Miners: A Historical Study of the Theater in Tombstone,* 10, 18, 45.

37. D. D. Martin, *Tombstone's "Epitaph,"* 48; Love, "History of Tombstone," 76; Odie B. Faulk, *Tombstone: Myth and Reality,* 113; Faulk, "Life in Tombstone," 507; Thure, "Tombstone 1881," 7; *Arizona Weekly Citizen,* 4 Dec. 1880.

38. "Peabody Tells of Tombstone," 3; *Arizona Weekly Citizen,* 25 Dec. 1880.

39. D. D. Martin, *Tombstone's "Epitaph,"* 106, 109; Willson, *Mimes and Miners,* 30–31; *Arizona Weekly Citizen,* 27 Nov. 1880, 1 Jan. 1881, 27 Oct. 1883. The *Arizona Weekly Citizen,* discussing another match, reported that "betting on the result is quite lively" (28 Apr. 1883).

40. *Arizona Weekly Citizen,* 14 May 1882, 4 Aug., 11 Aug. 1883.

41. Certainly, things were rambunctious in the early days. By the fall of 1879 Tombstone had a "reputation in cutting and shooting," at least in the *Arizona Weekly Citizen* (12 Sept. 1879), which derided a shooting incident two years later with the title "Traditional Tombstone Amusements" (20 Nov. 1881). There was certainly some justification for the district's rowdy reputation. In October 1879 the *Tombstone Nugget* editorialized: "We would very much dislike to see another such disgrace upon our camp as that which occurred on Sunday night last—the promiscuous firing of revolvers by lewd women, and men who seem to spend most of their time with them. Law and order was at a discount, and riot ran

rampant" (quoted in *Arizona Weekly Citizen*, 11 Oct. 1879). The *Tombstone Epitaph* reported "a number of vigorous nose mashings" on a payday in February 1882 (quoted in *Arizona Weekly Citizen*, 19 Feb. 1882).

Others discounted the town's unsavory reputation, however. Arthur Laing wrote in February 1881 that Tombstone's residents were "for a mining community . . . quite well behaved. Not once have I heard a pistol-shot in town, and it seems to me that a few 'drunks' (they are pretty heavily fined for this) constitute the sum of the offenses committed" (*Arizona Weekly Citizen*, 27 Feb. 1881). A year before that observation, Retlef had noted that part of the reason lawyers were struggling to find work was that there had been no crime in the preceding three weeks (*Arizona Weekly Citizen*, 21 Feb. 1880). The missionary Endicott Peabody, newly arrived from the East in January 1882, and predisposed to notice unchristian behavior, recorded of his first Sunday morning stroll that "the streets were thronged with loafers, chiefly miners and gamblers, but all were quiet and there was a total absence of rowdyism" ("Peabody Tells of Tombstone," 6). A visitor to Tombstone in May 1883 wrote that "there is as good order in Tombstone as in any town in the West, and less street drunkenness than in most Eastern cities" (*Arizona Weekly Citizen*, 26 May 1883). Robert M. Boller, a teamster in Tombstone from 1880 to 1882, recalled later that "the tougher element was but a small part of Tombstone" (Reminiscence of Early 1880s, Arizona Historical Society, Tucson, 4). William N. Miller, a resident of the camp from 1879 until 1883, seconded that assessment in his later years, writing that "the idea there was a 'man for breakfast' every morning is a stretch of imagination" (Manuscript, Arizona Historical Society, Tucson, 6–7).

42. Wolf, "Story of James G. Wolf," 5; Mike Greg, "History of the Religious Growth of Tombstone, Arizona," Arizona Historical Society, Tucson, 25; Faulk, "Life in Tombstone," 497; Hamilton, *The Resources of Arizona* (1881), 27.

43. Greg, "Religious Growth of Tombstone," 16–18, 25.

44. *Tucson and Tombstone Directory*, 107; Walker, "Arizona Land Fraud," 6; Hollis Cook, "Too Little—Too Much: Water and the Tombstone Story," in *History of Mining in Arizona*, ed. J. Michael Canty and Michael N. Greeley, 229; *Arizona Weekly Citizen*, 21 Mar. 1879.

45. *Arizona Weekly Citizen*, 7 Mar., 18 July, 27 Sept. 1879.

46. *Tucson and Tombstone Directory*, 107; "Tombstone"; *Arizona Weekly Citizen*, 27 Feb. 1881.

47. *Arizona Weekly Citizen*, 19 Sept. 1879; *Tucson and Tombstone Directory*, 130; Walker, "Arizona Land Fraud," 14; John Myers, *The Last Chance: Tombstone's Early*

Years, 37. The town-site controversy did add to the tumult of the early days, and Walker goes so far as to blame it for most of the town's famous violence. The controversy certainly embittered some of those directly involved. Among them was John Gray, who claimed that his father, Mike, was cheated out of a fortune rightfully his, and "became the object of much abuse, unjustly so, by the lot jumpers" ("When All Roads Led to Tombstone," 11–12). Others did not take Mike Gray's claims so seriously. William Miller recalled Gray's techniques: "If you didn't buy a lot he asked you to move. If you didn't do either, he gave you a deed to your lot for $5.00 to pay for recording, if you didn't recognize his claim he run [sic] a bluff as long as he could. All in good part, no serious trouble. Though if you left your property unguarded you were likely to fin[d] [it] in the street" (Manuscript, 5). By 1883, the year our study begins, the town-site controversy had largely played itself out, and the lawsuits, though "very vexatious, [were] practically decided in favor of the settlers" (*Tucson and Tombstone Directory*, 110).

48. D. D. Martin, *Tombstone's "Epitaph,"* 123; *Arizona Weekly Citizen*, 7 Aug. 1881.

49. *Arizona Weekly Gazette*, 1 June 1882, Newspaper and Ephemera File, Arizona Historical Society, Tucson; Grace McCool, *Sunday Trails in Old Cochise: A Guide to Ghost Towns, Lost Mines, and Buried Treasure in Cochise County, Arizona*, 92.

50. D. D. Martin, *Tombstone's "Epitaph,"* 128; Hamilton, *The Resources of Arizona* (1883), 51.

51. *Arizona Weekly Citizen*, 15 Aug., 22 Nov. 1879, 14 Feb., 21 Feb. 1880, 27 Feb. 1881, 12 May 1883; *Tombstone Epitaph*, 15 July 1882.

52. *Tombstone Daily Nugget*, 10 July 1881; *Arizona Weekly Citizen*, 4 Sept. 1881; *Tombstone Republican*, 20 Oct. 1883; Hal LaMar Hayhurst, *Hardpan: A Story of Early Arizona*, 106.

53. *Arizona Weekly Citizen*, 11 Oct. 1879, 21 Feb. 1880, 10 Apr. 1881, 17 Nov. 1883; *Tombstone Republican*, 6 Oct., 20 Oct. 1883.

54. *Arizona Weekly Citizen*, 7 Dec. 1878, 11 Jan., 22 Nov. 1879; Love, "History of Tombstone," 55.

55. *Arizona Weekly Citizen*, 31 Jan. 1880, 22 Jan. 1882, 5 May 1883.

56. McCool, *Sunday Trails*, 14; Ben T. Traywick, *Some Ghosts along the San Pedro*, 10–11, 27, map; *Arizona Weekly Citizen*, 20 Mar. 1881.

57. Love, "History of Tombstone," 18; *Arizona Weekly Citizen*, 17 Jan. 1880, 6 Aug. 1882; *Tucson and Tombstone Directory*, 112, 126–27; Town of Tombstone, Minutes of Town Council, 14 Dec. 1883.

58. *Arizona Weekly Citizen*, 18 Sept. 1880, 25 June 1882.

59. *Arizona Weekly Citizen*, 11 Oct., 29 Oct. 1879, 7 Feb., 21 Feb. 1880; *Tucson and Tombstone Directory*, 110, 113–15.

60. D. D. Martin, *Tombstone's "Epitaph,"* 127.

61. Hamilton, *The Resources of Arizona* (1881), 27; *Tombstone Arizona Daily Star*, 30 May 1883; *Arizona Weekly Citizen*, 16 June 1883. Tombstone's glorious future certainly seemed a credible assumption in December 1882, when the *Arizona Weekly Citizen* reported that the Tombstone Mill and Mining Company had made sufficient money in the previous month to pay its operating expenses, plus its interest payments and $25,000 on the principal of its loans (17 Dec. 1882).

Chapter Two. Apartments or Houses Are Impossible to Obtain

1. *Verde Copper News*, 21 Dec. 1928.

2. Lewis J. McDonald, "The Development of Jerome: An Arizona Mining Town," 43, 49, 51. For Jerome mines' leadership in output, McDonald cites *Mineral Resources of the United States* (Washington D.C.: United States Government Printing Office, 1932). See also the *Verde Copper News* indicating that the UVX was the seventh-leading copper producer in the state in 1929, with state production "far in excess of the record of 1918" (24 Jan. 1930).

3. WPA Writers' Program, *The WPA Guide to 1930s Arizona*, 89; F. E. Doucette, comp., *Arizona Year Book, 1930–1931*, 52; McDonald, "Development of Jerome," 49. The *Verde Copper News* carried copper prices on its masthead.

4. Arizona State Mine Inspector, *Annual Report* (for the year ending 30 Nov. 1929), gives the employment figure for Jerome's three big mines as 1,983 for the second half of 1929—1,240 at United Verde, 556 at UVX, and 187 at Verde Central (3). McDonald, "Development of Jerome," 43, gives slightly higher numbers—1,585 at United Verde, 575 at UVX, and 185 at Verde Central, for a total of 2,345, though he does not indicate from what source his figures were derived. Yavapai County, Ariz., Yavapai County Assessor's Office, Automobile Registrations, 1920–1932; John Carl Brogdon, "History of Jerome, Arizona," 142; *Verde Copper News*, 4 Jan. 1929.

5. McDonald, "Development of Jerome," 16–17; Brogdon, "History of Jerome, Arizona," 13–14.

6. Herbert V. Young, *Ghosts of Cleopatra Hill: Men and Legends of Old Jerome*, 30–33.

7. Brogdon, "History of Jerome, Arizona," 20; James Brewer, *Jerome: A Story of Mines, Men, and Money*, 4; Aliza Caillou, ed., *Experience Jerome and the Verde Valley*, 128; McDonald, "Development of Jerome," 27.

8. U.S. Bureau of the Census, *Thirteenth Census of the United States: Abstract of the Census ... with Supplement for Arizona*, 573–74; Thomas Refermat, "Transportation in Jerome," Hayden Library, Arizona State University, Tempe, 5; Herbert V. Young, *They Came to Jerome: The Billion Dollar Copper Camp*, 33, 51; Brogdon, "History of Jerome, Arizona," 33; "Verde Mining District," 23 Mar. 1901; Nancy Lee Pritchard, "Community: A Combination of Tradition and the Individual," 18.

9. U.S. Bureau of the Census, *Thirteenth Census of the United States*, 573–74.

10. For details on the United Verde Extension's development and discovery, see McDonald, "Development of Jerome," 32–38; H. V. Young, *Ghosts of Cleopatra Hill*, 123–24; and Brogdon, "History of Jerome, Arizona," 39–40.

11. Brogdon, "History of Jerome, Arizona," 42–43.

12. Ibid., 43–46; Highway 79 Association, Minute Book, Yavapai County Chamber of Commerce, Grace M. Sparkes, Collected Papers, 1904–1953, Hayden Library, Arizona State University, Tempe; *Verde Independent*, 15 Apr. 1965.

13. C. E. Mills, "Ground Movement and Subsidence at the United Verde Mine," 171.

14. Lewis J. McDonald, *Jerome, Arizona: The Unique Town of America*, 15–16. Herbert Young wrote: "I was employed in the administrative department of the United Verde Copper Company and its successor for four decades ... and had opportunity to closely observe their operations. The Clarks were generous to their towns and employees—too generous in the eyes of other mining men, who were astonished and a bit dismayed at the amounts spent not only for the needs but for the comforts and pleasures of their employees. To those who had to account to watchful stockholders for all expenditures this was an example they could not match" (*Ghosts of Cleopatra Hill*, 65).

15. McDonald, "Development of Jerome," 81–83; H. V. Young, *They Came to Jerome*, 54–55; Brogdon, "History of Jerome, Arizona," 60–63. Brogdon also believes that heavy investment by miners in the mining stocks of the district probably did much to dampen labor unrest.

16. Brogdon, "History of Jerome, Arizona," 74–75; H. V. Young, *They Came to Jerome*, 56; Pritchard, "Community," 24.

17. L. A. Parsons, "The New Surface-Plant for the United Verde Copper Company," 873.

18. *Verde Copper News*, 1 Jan., 16 Feb. 1926, 28 Sept. 1928, 1 Mar., 26 Mar. 1929.

19. *Verde Copper News*, 2 Apr., 19 Apr., 20 Sept. 1929, 24 Jan. 1930.

20. "Arizona: Greatest Producer of Copper in the United States"; McDonald, *Jerome, Arizona*, 10.

21. Mills, "Ground Movement and Subsidence," 161–64; *Verde Copper News*, 5 July 1929; "United Verde Extension Annual Report for 1929 and Quarterly Report for the Third Quarter of 1929," 1 Nov. 1929, in Joe Larson, Collection, Special Collections, Northern Arizona University, Flagstaff.

22. Nancy R. Smith, "Some Early History of the Verde Central Property"; Manager's Report, Verde Central Mines, Inc., 22 Jan. 1930, in "Mining Reports, Arizona, 1901–1930," Hayden Library, Arizona State University, Tempe; *Verde Copper News*, 1 Jan. 1929.

23. Historian Russell R. Elliott, who grew up in the company town of McGill, Nevada, records that company towns were discernable by their rows of well-built homes of similar types and absence of hotels, saloons, and red-light districts. He notes that the company owned the land on which the town sat, and that independent businesses existed at company pleasure, if they existed at all. Company towns were unincorporated and lacked local government; thus, residents lived under the legal jurisdiction of the county. While Jerome had company-built houses and neighborhoods, none of the other conditions mentioned above applied to the town (*Growing Up in a Company Town: A Family in the Copper Camp of McGill, Nevada*, 29; and *Nevada's Twentieth Century Mining Boom: Tonopah, Goldfield, Ely*, 220–22).

24. H. V. Young, *They Came to Jerome*, 30, 86–87.

25. "To keep employees in a contented frame of mind," Robert Tally wrote, "they must have good wages, kind and just treatment from their employers, good working and living conditions, plentiful opportunity for physical and mental recreation, and good churches and schools.... We believe in taking an active interest in worthy civic movements, and are always willing to cooperate in any constructive program for the benefit of the community, county, state or nation. We do not believe that our responsibilities cease with the comfort and contentment of our employees" ("Relations with Employees"). In a similar vein, the United Verde Hospital's chief surgeon, A. C. Carlson, believed that the hospital, though "a heavy cost in dollars and cents" to the company, "is actually a sound investment to both employee and employer.... It is the soundest foundation for maximum production of any industry, for aside from its economic value it fosters a sense of loyalty as well as making the men's work more attractive" (Walter V. Edwards, M.D., "Jerome Takes Its Medicine II," in *They Came to Jerome: Proceedings of the Seventh Annual Historic Symposium, August 25, 1984*, by the Jerome Historical Society, 4; United Verde Mine, Miscellaneous News Clippings, file folder "United Verde 2," Sharlot Hall Museum and Archives, Prescott.

26. *Verde Copper News*, 30 Apr. 1926, 21 June, 7 Sept. 1927, 23 Mar., 12 June, 10 July, 16 Oct. 1928; H. V. Young, *They Came to Jerome*, 121–22; Noel Pegues, "Recreation in the Verde District"; J. C. Harding and C. L. Guynn, "Employment and Welfare at the United Verde," 103. That assessment was confirmed years later by Herbert Young, secretary to Robert Tally, who noted that labor-management relations "remained on a fairly friendly footing" throughout the 1920s (*They Came to Jerome*, 56).

27. Pegues, "Recreation in the Verde District," 108–9; *Verde Copper News*, 10 June, 20 Dec. 1927, 10 Feb. 1928, 19 Mar., 19 Apr., 19 Nov. 1929; Brogdon, "History of Jerome, Arizona," 79; Caillou, *Experience Jerome*, 252.

28. "Jerome," file folder 27, Sharlot Hall Museum and Archives, Prescott; H. V. Young, *They Came to Jerome*, 30; Caillou, *Experience Jerome*, 242.

29. Charles F. Willis, "Jerome Business Houses."

30. *Verde Copper News*, 27 Dec. 1927, 4 Jan. 1929.

31. *Verde Copper News*, 5 Nov. 1926, 20 Mar. 1928; boxes 7 and 9, Perry Ling Collection, 1920s–1940s, Special Collections, Northern Arizona University, Flagstaff; Yavapai County, Ariz., Yavapai County Assessor's Office, Automobile Registrations, 1920–1932.

32. *Verde Copper News*, 26 Aug. 1927, 12 June 1928, 12 Mar. 1929, 18 Apr. 1930; First National Bank of Arizona, Ledgers and Account Books, 1860s–1948, Northern Arizona University, Flagstaff.

33. H. V. Young, *They Came to Jerome*, 59, 109; Brogdon, "History of Jerome, Arizona," 96–99.

34. Caillou, *Experience Jerome*, 228–32; *Verde Copper News*, 25 Oct. 1927.

35. Brogdon, "History of Jerome, Arizona," 96–99; Caillou, *Experience Jerome*, 226–27; *Verde Copper News*, 2 Feb. 1926.

36. H. V. Young, *They Came to Jerome*, 12, 91; Brogdon, "History of Jerome, Arizona," 77.

37. H. V. Young, *They Came to Jerome*, 12; Brogdon, "History of Jerome, Arizona," 75.

38. U.S. Bureau of the Census, *Fifteenth Census of the United States*, 156. In 1930 the government—which listed Latinos as "white" in every other census in the first half of the twentieth century—decided to list them as "other." Because "other" never had more than 35 members in any census after 1920, I have estimated 2,800 Latinos from the 2,853 "others" of 1930, which, when taken from the total 1930 census figure of 4,932, produces the 57 percent figure. *Verde Copper News*, 24 May, 16 Sept. 1927, 4 May, 25 May, 7 Aug. 1928, 14 Mar. 1930.

39. Charles L. Mann, "The Mexicans of Jerome: Impressions of 1929," in WPA Writers' Program, Manuscripts, box 8, Arizona State Department of Library and Archives, Phoenix.

40. John Krznarich, Personal Recollections of Jerome, Ariz., 1930s and 1940s, oral history tape 862, 29 July 1991, Sharlot Hall Museum and Archives, Prescott; *Verde Copper News*, 25 Sept. 1928, 7 May 1929, 4 Mar. 1932, 11 Jan. 1935. Russell Elliott believed that athletic contests "proved to be a major force in breaking down the ethnic barriers that had earlier divided the community [of McGill, Nevada] into ethnic towns" (*Growing Up*, 109).

41. *Verde Copper News*, 16 Sept. 1927, 4 May 1928, 12 July, 29 Nov. 1929; Krznarich, oral history tape 863, 29 July 1991.

42. H. V. Young, *They Came to Jerome*, 121; Brogdon, "History of Jerome, Arizona," 56–57; *Verde Copper News*, 18 Dec. 1928. After the shooting of one Latino by another, the paper, reporting the failure of police to find the assailant, noted that "the not unusual feature of Mexican troubles is that stone-wall protection is often given one of their clan in trouble [which] is hindering police in their attempt to arrest the man" (3 Feb. 1928). When a Latino resident was convicted of assault in 1929, the newspaper commented that "this is one of the very few convictions found . . . in the Mexican cases arising from this district" (8 Feb. 1929).

In *Gunfighters, Highwaymen, and Vigilantes: Violence on the Frontier*, Roger D. McGrath observes of Aurora, Nevada, and Bodie, California, that "Mexicans and especially Chinese often preferred to avoid dealing with the justice system. Mexicans . . . sought to personally avenge what they thought to be wrongs. Chinese let the secret societies handle most of Chinatown's problems. The authorities quickly learned that if both the victim and the perpetrator of a crime were Chinese, they could expect little or no cooperation, even from the victim himself" (253).

43. Caillou, *Experience Jerome*, 246; Brogdon, "History of Jerome, Arizona," 87.

44. *Verde Copper News*, 23 Feb. 1926, 19 July, 9 Aug., 30 Aug. 1927.

45. Brogdon, "History of Jerome, Arizona," 87–88; *Verde Copper News*, 9 Apr. 1929.

46. Brogdon, "History of Jerome, Arizona," 78–80; Caillou, *Experience Jerome*, 243; *Haven Methodist Visitor*, "Jerome, Arizona" File, Special Collections, Northern Arizona University, Flagstaff; *Verde Copper News*, 10 May, 3 June 1927, 15 Nov. 1929.

47. *Verde Copper News*, 24 Sept., 10 Oct. 1926, 5 Aug. 1927, 28 Mar., 3 Apr. 1928, 19 Nov. 1929; Caillou, *Experience Jerome*, 256.

48. *Verde Copper News*, 20 Dec. 1927, 29 June 1928; Caillou, *Experience Jerome*, 246; wpa Writers' Program, *WPA Guide to Arizona*, 335.

49. *Verde Copper News*, 2 Mar., 29 Oct., 3 Dec., 17 Dec. 1926, 11 Jan. 1927, 20 Nov. 1928, 16 Apr. 1929.

50. *Verde Copper News*, 29 Jan. 1926, 22 Apr. 1927, 27 May, 13 July 1928, 16 July 1929; H. V. Young, *They Came to Jerome*, 117.

51. *Verde Copper News*, 23 Aug. 1927; Charles F. Willis, "Special Edition on the Verde District."

52. *Verde Copper News*, 21 Jan. 1927, 20 Aug., 29 Oct. 1929; H. V. Young, *They Came to Jerome*, 46.

53. R. K. Duffey, "Housing and Service for Employees," 99; Arizona State Mine Inspector, *Annual Report* (1930), 3; *Verde Copper News*, 18 Oct. 1929.

54. Brogdon, "History of Jerome, Arizona," 14.

55. McDonald, "Development of Jerome," 75; Brogdon, "History of Jerome, Arizona," 110–13, 126; Richard Martin, "One Hundred Years of Government in Jerome," in *Bring It All Back to Jerome: Proceedings of the Sixth Annual Historic Symposium . . . August 27, 1983*, by the Jerome Historical Society, 3–4; Wesley W. Stout, "Want to Buy a Ghost Town?"

56. H. V. Young, *They Came to Jerome*, 43, 45; Brogdon, "History of Jerome, Arizona," 56; *Verde Copper News*, 5 Mar. 1926, 4 May, 13 Nov., 16 Nov. 1928, 24 Sept. 1929.

57. *Verde Copper News*, 12 Nov., 17 Dec. 1926, 16 Nov. 1928, 24 Sept., 8 Oct. 1929.

58. *Verde Copper News*, 11 Jan. 1927, 16 Nov. 1928.

59. *Verde Copper News*, 23 Apr., 19 Oct. 1926, 27 Apr., 2 Nov. 1928, 4 Apr. 1930.

60. McDonald, *Jerome, Arizona*, 18; Brogdon, "History of Jerome, Arizona," 81–82; H. V. Young, *They Came to Jerome*, 134; Maynard Davenport, "An Analysis of the Capacity and Utilization of the Jerome School Plant," 18–20, 25.

61. Davenport, "Jerome School Plant," 25, 33; Brogdon, "History of Jerome, Arizona," 84.

62. *Verde Copper News*, 5 Feb., 29 Nov. 1929, 24 Jan. 1930.

Chapter Three. This Ill-Omened City

1. D. D. Martin, *Tombstone's "Epitaph,"* 255–57; "Mining Reports, Arizona, 1901–1930," 256.

2. *Arizona Weekly Citizen*, 17 Apr. 1881, 28 Apr., 5 May 1883; D. D. Martin, *Tombstone's "Epitaph,"* 257; Blake, *Tombstone and Its Mines*, 17; Devere, "Tombstone

Bonanza," 20. The Head Center hit water at 545 feet in April 1881, and the same thing happened at about the same time in the Contention, Grand Central, West Side, and Empire Mines.

3. *Arizona Weekly Citizen*, 23 June 1883; Blake, *Tombstone and Its Mines*, 17; *Tombstone Republican*, 5 Jan. 1884.

4. *Arizona Weekly Citizen*, 22 Dec. 1883, 19 Jan., 16 Feb. 1884; *Tombstone Republican*, 19 Jan., 22 Mar. 1884; Blake, *Tombstone and Its Mines*, 17.

5. *Tombstone Republican*, 29 Dec. 1883; *Arizona Mining Index*, 19 Jan. 1884.

6. Blake, *Tombstone and Its Mines*, 17; "Mining Reports, Arizona, 1901–1930," 257; *Arizona Daily Star*, 7 June 1883; *Tombstone Republican*, 5 Jan. 1884.

7. Tombstone Mill and Mining Company, *Annual Report on Mines and Mills, with Production and Expenses for the Year Ending 31 March 1883*, i, ii, 3–5, 7.

8. *Arizona Weekly Citizen*, 29 Apr., 3 May, 10 May 1884; *Arizona Daily Star*, 2 May 1884.

9. *Arizona Weekly Citizen*, 10 May, 17 May 1884.

10. *Arizona Mining Index*, 21 June 1884; George W. Parsons, Diaries, 6, 7, 8, 9 May 1884, Arizona Historical Society, Tucson.

11. *Arizona Weekly Citizen*, 24 May 1884; *Arizona Daily Star*, 11 May, 18 May 1884; *Arizona Mining Index*, 21 June 1884.

12. *Arizona Weekly Citizen*, 24 May 1884; *Arizona Daily Star*, 12 June 1884.

13. G. W. Parsons, Diaries, 14 May, 29 May 1884; *Arizona Daily Star*, 2 June 1884.

14. *Arizona Daily Star*, 5 June 1884.

15. *Arizona Daily Star*, 6 June, 17 June, 3 July, 26 July 1884; *Arizona Weekly Citizen*, 12 July, 26 July, 2 Aug., 23 Aug. 1884.

16. *Arizona Daily Star*, 1 Aug., 3 Aug. 1884; *Arizona Weekly Citizen*, 9 Aug 1884.

17. *Arizona Daily Star*, 6 Aug. 1884.

18. *Arizona Daily Star*, 10 Aug. 1884; *Arizona Weekly Citizen*, 16 Aug. 1884.

19. *Arizona Daily Star*, 10 Aug. 1884.

20. *Arizona Daily Star*, 14 Aug., 16 Aug., 17 Aug., 21 Aug., 29 Aug., 10 Sept., 13 Sept. 1884.

21. *Arizona Daily Star*, 6 Aug. 1884; D. D. Martin, *Tombstone's "Epitaph,"* 257; Hamilton, *The Resources of Arizona* (1884), 157.

22. *Arizona Weekly Citizen*, 2 Aug., 13 Dec. 1884, 8 July, 15 Aug., 29 Aug., 12 Sept., 19 Sept., 26 Sept., 7 Nov., 12 Dec., 19 Dec. 1885.

23. "The Tombstone Mill and Mining Company, Arizona."

24. *Arizona Weekly Citizen*, 3 Apr. 1886; *Arizona Daily Star*, 2 Mar. 1886.

25. *Arizona Weekly Citizen,* 29 May 1886; *Arizona Daily Star,* 28 May, 29 May 1886.

26. *Arizona Weekly Citizen,* 29 May 1886; *Daily Tombstone,* 27 May 1886; *Arizona Daily Star,* 28 May 1886; G. W. Parsons, Diaries, 26 May, 27 May, 30 June 1886.

27. *Arizona Daily Star,* 8 June 1886; G. W. Parsons, Diaries, 19 July, 24 July 1886; *Arizona Weekly Citizen,* 14 Aug. 1886; J. H. Young, "Tombstone, Arizona," 486.

28. *Arizona Weekly Citizen,* 11 Aug. 1888, 4 Jan., 15 Feb., 1 Mar. 1890; *Tombstone Prospector,* 7 Oct. 1888; Blake, *Tombstone and Its Mines,* 18.

29. Ben T. Traywick, *The Chronicles of Tombstone,* 10; *Tombstone Prospector,* 1 Apr., 24 June 1888; *Arizona Weekly Citizen,* 7 Apr. 1888, 4 Jan 1890.

30. *Arizona Weekly Citizen,* 25 May 1889; *Tombstone Prospector,* 7 Dec. 1889.

31. D. Smith, *Silver Saga,* 180, 216; *Tombstone Prospector,* 4 Jan., 11 Jan., 18 Jan., 8 Feb., 7 June 1890; *Arizona Weekly Citizen,* 18 Jan. 1890; Traywick, *The Chronicles of Tombstone,* 11; Ben T. Traywick, *The Mines of Tombstone,* 16–17.

32. *Arizona Daily Star,* 30 July 1891; *Tombstone Prospector,* 28 Dec. 1891.

33. *Arizona Daily Star,* 8 Jan., 23 Jan., 23 Mar. 1892; *Tombstone Prospector,* 16 Jan., 20 Jan., 23 Jan. 1892.

34. *Tombstone Prospector,* 28 July 1892.

Chapter Four. Employment by Mines Practically Nil

1. Young reports that Tally, worth over one million dollars, was nearly wiped out in the crash (*Ghosts of Cleopatra Hill,* 113). *Verde Copper News,* 14 Feb. 1930.

2. *Prescott Courier,* 8 Dec. 1929, 17 Jan., 25 Feb. 1930; *Verde Copper News,* 28 Mar. 1930.

3. *Prescott Courier,* 19 Apr. 1930, 11 Feb. 1933; McDonald, "Development of Jerome," 104. Ten days after the stock market crash, the *Prescott Courier* (9 Nov. 1929) reported that "the leaders in the copper industry are very confident that the 18 cent price will be maintained at least well into next year and are very cheerful over the outlook in spite of stock market reactions," and that the eighteen-cent price "is here to stay," with good domestic volume in the coming weeks. The newspaper also printed the more ominous opinion of "foreign metal brokers" that the producers "'present policy may keep up prices at a fictitious level for some time, but the longer the price is kept at an unjustifiable level, the more the markets will suffer subsequently, and the price must then of necessity go lower than it would have done if rationalization had never taken place.'"

4. *Prescott Courier*, 3 Apr., 15 Apr., 27 Nov. 1930; Margaret Maxwell, "The Depression in Yavapai County," 210.

5. *Verde Copper News*, 29 Aug., 30 Sept., 17 Oct. 1930; *Prescott Courier*, 30 Sept., 18 Oct. 1930; Christina Ellen Dickson and Robert Henry Dickson, *Dickson Saga: Story of Our Married Life*, 67–69; John Muretic, *To Heaven in the West*, 85.

6. *Verde Copper News*, 12 Dec. 1930, 13 Jan. 1931; folder 6, Larson Collection.

7. *Verde Copper News*, 16 Jan., 27 Jan. 1931; *Prescott Courier*, 18 June, 29 Sept. 1930, 6 May 1931; United Verde Extension, "Quarterly Reports for the First and Second Quarters of 1931," Larson Collection; Arizona State Mine Inspector, *Annual Report* (1929–1931).

8. *Prescott Courier*, 8 May, 28 May 1931; *Verde Copper News*, 8 May, 3 July 1931.

9. United Verde Extension, "Report for the Second Quarter of 1931," Larson Collection; *Verde Copper News*, 12 June, 25 June 1931; *Prescott Courier*, 18 June, 27 June 1931.

10. *Verde Copper News*, 27 Mar., 31 Mar., 8 Sept., 29 Sept., 3 Nov. 1931; P. G. Spilsbury, Arizona Industrial Congress, to Governor Hunt, 8 July 1931, box 5, Governor's Office, Arizona, Governors' Files, Arizona State Department of Library and Archives, Phoenix; Mills, "Ground Movement and Subsidence," 169, 170; E. M. J. Alenius, "A Brief History of the United Verde Open Pit, Jerome, Arizona"; Muretic, *Heaven in the West*, 86.

11. *Verde Copper News*, 18 Aug., 4 Sept., 22 Sept. 1931, 15 Jan. 1932; *Prescott Courier*, 12 June, 17 Sept. 1931, 12 Mar. 1932.

12. *Prescott Courier*, 1 Mar., 7 Dec. 1932, 21 Feb. 1933, 24 Feb., 5 Apr. 1934; *Verde Copper News*, 16 Feb., 15 Mar. 1932. A United Verde Extension official put the cost of copper production before taxes at 7.43 cents per pound in February 1931 (*Verde Copper News*, 27 Feb. 1931). With credits for gold and silver values in the UVX ore, the cost of copper production dropped to less than seven cents per pound. Although production costs varied somewhat because the price of labor fluctuated, clearly the industry could not operate with copper at five cents per pound.

13. Gov. G. W. P. Hunt to Chairman of Reconstruction Finance Corp., 11 Aug. 1932, Governor's Office, Arizona, Governors' Files; *Prescott Courier*, 17 Mar., 24 Mar., 27 May, 3 Sept. 1932, 21 Feb., 1 Sept. 1933; Maxwell, "Depression in Yavapai County," 210; Brogdon, "History of Jerome, Arizona," 76; *Verde Copper News*, 29 Apr. 1932; folder 6, Larson Collection.

14. *Verde Copper News*, 14 July, 11 Aug., 25 Aug. 1933; *Prescott Courier*, 19 Jan., 11 Feb., 5 Aug., 29 Dec. 1933, 24 Feb., 5 Apr. 1934; Gov. Moeur telegram to President Roosevelt, n.d.; NRA administrator Gen. Hugh Johnson to Gov. Moeur,

16 Dec. 1933; Gov. Moeur telegram to Gen. Johnson, 26 Dec. 1933, all in Governor's Office, Arizona, Governors' Files; Arizona State Mine Inspector, *Annual Report* (1929–1933).

15. *Prescott Courier*, 4 Jan., 29 Jan., 5 Apr., 2 May, 28 Sept., 26 Oct. 1934; *Verde Copper News*, 7 Dec. 1934; WPA Writers' Program, *WPA Guide to Arizona*, 92.

16. *Verde Copper News*, 28 Dec. 1934; *Prescott Courier*, 27 Dec. 1934, 5 Feb. 1935; H. V. Young, *Ghosts of Cleopatra Hill*, 114–15.

17. *Prescott Courier*, 13 Feb., 4 Mar. 1935; Brogdon, "History of Jerome, Arizona," 34, 147; Wayne Johnson, "Jerome, Arizona: The Billion Dollar Copper, Mineral, and Silver Camp," 63; "Jerome," file folder 27, Sharlot Hall Museum and Archives.

18. *Prescott Courier*, 11 Sept., 18 Sept., 20 Sept., 14 Oct. 1935, 24 Jan. 1936. The *Courier* reported on 11 Sept. 1935 that the Yavapai County Welfare Board had experienced a 35 percent reduction in the number of people on relief over the previous year, "and nine-tenths of this reduction is traceable directly to clients having found employment in or through mining."

19. *Jerome Chamber of Commerce News Bulletin*, 18 May 1936; *Prescott Courier*, 31 Aug., 18 Sept., 2 Dec., 14 Dec., 29 Dec., 31 Dec. 1936, 5 Jan., 16 Jan., 16 Feb. 1937.

20. United Verde Extension, "Report for the Third Quarter of 1930," folder 6, Larson Collection; *Prescott Courier*, 16 Aug. 1934, 25 June, 26 Dec. 1935, 15 Jan., 26 Feb. 1937, 17 June, 15 Sept. 1938; *Verde Copper News*, 9 Nov. 1934; *Jerome Chamber of Commerce News Bulletin*, 31 Dec. 1935; Brogdon, "History of Jerome, Arizona," 147; Arizona State Mine Inspector, *Annual Report* (1938).

21. *Prescott Courier*, 30 Mar., 12 May, 24 May, 31 May, 22 Oct., 11 Dec. 1937; Larson to Nihell, 27 Mar., 27 Sept., 26 Oct., 23 Nov., 23 Dec. 1937, Larson Collection.

22. *Prescott Courier*, 14 Feb., 8 Mar., 19 May, 9 June 1938; Larson to Nihell, 25 May 1938, Larson Collection.

23. Larson to "Friend Harry," 27 Apr. 1939, Larson Collection; *Prescott Courier*, 29 Sept., 21 Oct., 25 Oct., 18 Dec., 30 Dec. 1939.

Chapter Five. The Drawbacks Incident to Mining

1. Tombstone Mill and Mining Company, *Annual Report*, ii.

2. Jonathan Dembo, "The Pacific Northwest Lumber Industry During the Great Depression," 51; Mann, *After the Gold Rush*, 71; Allen D. Edwards, "Influence

of Drought and Depression on a Rural Community: A Case Study of Haskell County, Kansas," 36.

3. Larson to Whiteley, 12 Oct. 1936, Larson Collection; *Arizona Weekly Citizen*, 1 June 1889.

4. Charles H. Shinn, *The Story of the Mine*, 163–71; Cochise County, Ariz., Index to Deeds of Mines; *Tombstone Republican*, 17 Nov. 1883; *Arizona Daily Star*, 24 Aug. 1886; *Arizona Weekly Citizen*, 2 June 1888.

5. Traywick, *The Mines of Tombstone*, 17; Cochise County, Ariz., Index to Deeds of Mines.

6. *Verde Copper News*, 10 Feb., 3 Mar., 19 July 1931.

7. William A. Evans, *Two Generations in the Southwest*, 86; *Prescott Courier*, 13 Feb. 1935, 24 May 1937; Brewer, *Jerome*, 9, 11.

8. Leonor Lopez, *Forever, Sonora, Ray, Barcelona: A Labor of Love*, 17; Andrea Yvette Huginnie, "Strikitos: Race, Class, and Work in the Arizona Copper Industry, 1870–1920," 340; *Prescott Courier*, 9 Feb. 1935. The WPA Writers' Program's *WPA Guide to Arizona* claims that of the copper companies, "none were as aggressive and successful as the Phelps Dodge interests" (91), but in the same era Kennecott Copper Corp. was slowly gaining control of the Guggenheim properties in Nevada, Utah, Arizona, and New Mexico (Elliott, *Growing Up*, 139).

9. Shinn, *Story of the Mine*, 168; James, *Roar and the Silence*, 78–79, 105; "Mining Reports, Arizona, 1901–1930"; Tombstone Consolidated Mines, Bankruptcy Papers, 1911, Hayden Library, Arizona State University, Tempe.

10. Mann, *After the Gold Rush*, 9. Unfortunately, there is no statistical verification for the newspaper reports of the return of lessees. The Cochise County Index to Leases contains only a few entries for mines in the district in the years following the bust. One evidently did not have to register a lease of a mining property with the county as one did a sale.

Leasing or tribute mining also sometimes occurred during the early days of a district. The system was well suited to Goldfield, Nevada's small pockets of high-value gold ores, and the leasing system required little capital, an advantage in remote locations in their early days. Fading districts like Goldfield or the Wood River District of Idaho returned to leasing, often with the encouragement of the local newspaper. See Sally S. Zanjani, *Goldfield: The Last Gold Rush on the Western Frontier*, 42–43, 233; Elliott, *Nevada's Mining Boom*, 13, 160; and Clark Spence, *For Wood River or Bust: Idaho's Silver Boom of the 1880s*, 216–19.

11. Otis E. Young Jr., *Western Mining: An Informal Account of Precious-Metals Prospecting, Placering, Lode Mining, and Milling on the American Frontier from Spanish*

Times to 1893, 25; Tombstone Mill and Mining Company, *Annual Report*, 12; *Arizona Daily Star*, 2 Mar., 9 Mar. 1886; *Daily Tombstone*, 22 Feb., 14 Apr. 1886.

12. *Arizona Daily Star*, 2 Apr. 1891, 4 Aug., 20 Aug. 1892; *Tombstone Prospector*, 3 Apr. 1891, 8 Sept. 1892.

13. Duane A. Smith, *Mining America: The Industry and the Environment, 1800–1980*, 116; *Verde Copper News*, 18 Jan. 1935; *Prescott Courier*, 2 Jan., 10 Apr. 1931; Maxwell, "Depression in Yavapai County," 218.

14. WPA Writers' Program, Manuscripts, box 8; *Prescott Courier*, 17 Aug., 8 Sept. 1932, 7 Oct. 1938, 10 July 1939; *Verde Copper News*, 18 Jan. 1935.

15. *Prescott Courier*, 20 Dec. 1934, 10 July 1939.

16. Larson to "Friend Harry," 10 Sept. 1938; Larson to Nihell, 25 Mar. 1939, both in Larson Collection; Walton to Gov. Jones, 3 Feb. 1939; and Willis to Jones, 22 Dec. 1938, both in Governor's Office, Arizona, Governors' Files; *Prescott Courier*, 7 Apr. 1938.

17. *Prescott Courier*, 14 Oct., 15 Oct., 17 Oct. 1932, 22 Dec. 1933, 30 Oct. 1934; D. Smith, *Mining America*, 116; O. E. Young Jr., *Western Mining*, 22; Maxwell, "Depression in Yavapai County," 218.

18. Donald E. Miller and Richard E. Sharpless, *The Kingdom of Coal: Work, Enterprise, and Ethnic Communities in the Mine Fields*, 299, 319, 321, 332.

19. WPA Writers' Program, *WPA Guide to Arizona*, 91; *Prescott Courier*, 24 Jan. 1936, 19 Jan. 1938; Larson to Nihell, 28 Nov. 1939, Larson Collection; Record Book of Small Mines, Yavapai County, in Sparkes, Collected Papers.

20. "Jerome, History," vertical file, Special Collections, Northern Arizona University, Flagstaff; "Jerome," file folder 27, Sharlot Hall Museum and Archives; Larson to Nihell, 28 Nov. 1939, Larson Collection.

21. *Tombstone Prospector*, 6 Sept. 1892; WPA Writers' Program, *WPA Guide to Arizona*, 89.

22. Governor Moeur to Illinois man, 1 Mar. 1935, Governor's Office, Arizona, Governors' Files; WPA Writers' Program, Manuscripts, box 7; *Prescott Courier*, 7 Oct. 1938; Maxwell, "Depression in Yavapai County," 218.

23. *Tombstone Prospector*, 27 July 1889; "Mining Reports, Arizona, 1901–1930," 256; Yavapai County Chamber of Commerce, "Mine Production ... Yavapai County, 1880–1936," Hayden Library, Arizona State University, Tempe; WPA Writers' Program, Manuscripts, box 8; Richard Lowitt, *The New Deal and the West*, 115.

24. Jackson, *Treasure Hill*, 220; D. Smith, *Silver Saga*, 178–79; Yavapai County, Ariz., General Index to Mines, records 150 locations in the Verde District in the

period 1925 to 1929, inclusive; seventy-one locations in the period 1930 to 1934, inclusive; and ten locations in the period 1935 to 1937, inclusive. *Arizona Weekly Citizen,* 10 Jan. 1890.

25. Rufus Kay Wyllys, *Arizona: The History of a Frontier State,* 291. James C. Foster discusses mobility as an expression of discontent in the late nineteenth and early twentieth centuries ("Western Miners and Silicosis: The Scourge of the Underground Toiler, 1890–1943"). See Stephen M. Voynick, *The Making of a Hardrock Miner,* for a discussion of this phenomenon among modern miners. Mann, *After the Gold Rush,* 89–91; *Arizona Weekly Citizen,* 10 May 1884.

26. *Arizona Daily Star,* 6 May, 20 May 1884; *Arizona Weekly Citizen,* 10 May 1884.

27. *Arizona Daily Star,* 29 May 1884.

28. *Arizona Daily Star,* 29 May, 24 June, 25 June, 30 July 1884.

29. *Arizona Weekly Citizen,* 24 May, 16 Aug. 1884; *Arizona Daily Star,* 1 Aug, 6 Aug. 1884.

30. Jackson, *Treasure Hill,* 128–38; *Arizona Mining Index,* 10 May 1884; Florin, *Colorado and Utah Ghost Towns,* 112; James, *Roar and the Silence,* 140; Elliott, *Nevada's Mining Boom,* 144; Spence, *Wood River or Bust,* 199–205. John Spratt describes the post–World War I strike and shutdown in the coal town of Thurber, Texas, as very much fitting the same pattern (*Thurber, Texas,* 116).

31. *Arizona Daily Star,* 10 July, 1 Aug. 1884.

32. Jackson, *Treasure Hill,* 137–38; D. Smith, *Silver Saga,* 239.

33. *Prescott Courier,* 7 May, 9 May, 17 July, 21 Aug., 6 Sept. 1934; *Verde Copper News,* 6 July 1934. In a famous example, Goldfield, Nevada, had a notorious labor war in 1907 and 1908, while Tonopah, less than thirty miles away, experienced no significant labor dispute until after World War I (Elliott, *Nevada's Mining Boom,* 144–45).

34. Richard Melzer, *Madrid Revisited: Life and Labor in a New Mexican Mining Camp in the Years of the Great Depression,* 33–35; Dembo, "Pacific Northwest Lumber Industry," 58; *Verde Copper News,* 15 Sept. 1933.

35. *Verde Copper News,* 15 Sept. 1933, 16 Mar., 23 Mar. 1934.

36. *Verde Copper News,* 1 Apr., 11 Apr., 18 Apr. 1930, 29 Sept., 3 Nov. 1931, 16 Feb. 1932; *Prescott Courier,* 19 Apr. 1930, 14 Apr. 1931, 5 Aug. 1933.

37. H. V. Young, *Ghosts of Cleopatra Hill,* 65; Muretic, *Heaven in the West,* 85; United Verde Mine and Smelter Committees, Petition to President F. D. Roosevelt, 7 Sept. 1933, Governor's Office, Arizona, Governor's Files.

38. *Prescott Courier,* 9 May, 14 June, 26 June, 27 June, 29 June, 1 July 1935, 12

Oct. 1936, 11 Jan., 16 Jan., 2 Feb., 3 Feb., 9 Mar., 20 Apr. 1937; Joshua Freeman et al., *Who Built America? Working People and the Nation's Economy, Politics, Culture, and Society*, 353, 378–80; WPA Writers' Program, *WPA Guide to Arizona*, 100. Elliott recorded that union locals from the International Union of Mine, Mill, and Smelter Workers revived at McGill and Ruth, Nevada, in 1935 (*Growing Up*, 139).

39. Brogdon, "History of Jerome, Arizona," 76; *Prescott Courier*, 22 Apr., 14 June, 15 Oct. 1938.

40. Larson to Nihell, 19 Dec. 1938, 25 Apr., 28 Nov. 1939, Larson Collection. Larson is not a disinterested source, he being engaged in a long-running and bitter (at least on his side) dispute with Phelps Dodge over water rights at his mining properties. He also had a long-standing indebtedness to the United Verde Extension. There is other evidence, however, to support the idea that management-employee relations cooled after the Phelps Dodge takeover and that the company was certainly less involved in the community than United Verde had been. Kingdon to Larson, 27 May 1932; United Verde Extension to Larson, 19 June 1939; and Larson to Notman, 7 Jan. 1948, all in Larson Collection.

41. *Prescott Courier*, 19 Nov., 21 Nov. 1938.

42. *Arizona Weekly Citizen*, 10 Nov. 1888, 1 June 1889; D. Smith, *Silver Saga*, 210, 239. McGrath reports that by 1881 investors had spent one-half million dollars developing the Mono Mine in Bodie, California, without finding the Bodie vein (*Gunfighters, Highwaymen, and Vigilantes*, 108). And these were only the legitimate ways investors could lose their money; for the illegitimate ways, see Clark C. Spence, "I Was a Stranger and Ye Took Me In," and George Graham Rice, *My Adventures with Your Money*.

43. *Verde Copper News*, 2 Dec. 1932; Olien and Olien, *Oil Booms*, 85; Muretic, *Heaven in the West*, 34; *Verde Independent*, 15 Apr. 1965.

Chapter Six. Owing to the Dullness of Trade

1. G. W. Parsons, Diaries, 10, 11, 12, 13 May 1884.

2. Wallace E. Clayton, "Fortune-Seeker's Diary Notes Turbulent Events: Tombstone's Early Years Depicted by Eye-Witness," *Tombstone Epitaph* (Jan. 1992), 14; *Arizona Daily Star*, 11 May 1884; *Arizona Weekly Citizen*, 17 May 1884.

3. *Arizona Weekly Citizen*, 17 May 1884; *Arizona Daily Star*, 11 May, 15 May, 29 May, 4 June 1884; *Tombstone Prospector*, 26 Sept. 1889.

4. *Arizona Daily Star*, 13 May, 29 May 1884; *Arizona Weekly Citizen*, 10 May 1884, 24 Jan., 10 Oct. 1885, 31 Dec. 1887; *Tombstone Prospector*, 24 Oct. 1891. Jackson

describes almost exactly the same bust in Hamilton, Nevada, in 1869, as occurred in Tombstone fifteen years later. The collapse of the local bank constricted the money supply, which brought on deflation, depression, and mass out-migration within six months (*Treasure Hill,* 154–59).

5. *Prescott Courier,* 10 Apr. 1930, 17 Nov. 1933, 3 Jan. 1936; *Verde Copper News,* 28 Mar. 1930, 8 Sept., 24 Nov., 29 Dec. 1933; *Jerome Chamber of Commerce News Bulletin,* 17 Dec. 1935.

6. D. Smith, *Rocky Mountain Mining Camps,* 7, 75; Hums (pseud.), "The Attractions of Tombstone," Arizona State Department of Library and Archives, Phoenix; Rockfellow, "Log of an Arizona Trailblazer," Rockfellow Papers, 57.

7. G. W. Parsons, Diaries, 18 May 1884, 24 Sept. 1886; Ling to Wooldridge, 16 Sept. 1931, Ling Collection.

8. *Arizona Daily Star,* 12 June 1884; *Verde Copper News,* 29 Apr. 1932; Lary Michael Dilsaver, "From Boom to Bust: Post Gold Rush Patterns of Adjustment in a California Mining Region," 26; Shinn, *Story of the Mine,* 163; *Prescott Courier,* 22 Sept. 1931; Yavapai County Chamber of Commerce to Governor Moeur, 18 Aug. 1933, Governor's Office, Arizona, Governors' Files; Ben T. Traywick, *The "Tombstone Epitaph" and John Philip Clum,* 46; "Price of Labor, Provisions, Etc."; *Daily Record-Epitaph,* 12 Nov. 1885; *Tombstone Prospector,* 15 Feb. 1892; Willson, *Mimes and Miners,* 77; Larson to "Friend Harry," 23 Jan. 1939, Larson Collection; Stout, "Want to Buy a Ghost Town?"

9. Herman R. Lantz, *The People of Coaltown,* 193; *Arizona Weekly Citizen,* 5 Sept. 1879, 20 Nov. 1881; "Price of Labor, Provisions, Etc."; Bishop, *Across Arizona in 1883,* 9.

10. *Arizona Daily Star,* 25 June 1884; *Tombstone Prospector,* 22 July 1889, 30 Nov. 1892; Fatout, *Meadow Lake: Gold Town,* 83–85. Ramsey reports that when the placer camp of Granite Creek went under in 1886, one house that cost $600 to build sold for two glasses of whisky, and another that cost $1,500 sold for firewood for $15 (*Ghost Towns of British Columbia,* 192).

11. Home Owners' Loan Corp. application by Candelario Beltran, Apr. 1934, John P. Connolly, Collection, Special Collections, Northern Arizona University, Flagstaff; Larson to Nihell, 28 Feb., ? May 1936, 23 Mar., 25 May, 19 Dec. 1938; and Nihell to Larson, 30 Nov. 1935, 1 Apr. 1937, 2 July 1941, all in Larson Collection. After Aurora, Nevada, began its rapid decline in 1864, its town lots lost 95 percent of their value, and one witness reported a two-story brick building, which could not have been purchased for five thousand dollars during the boom, offered for five hundred dollars (McGrath, *Gunfighters, Highwaymen, and Vigilantes,* 10).

12. D. E. Miller and Sharpless, *Kingdom of Coal*, 85; Mann, *After the Gold Rush*, 74; *Daily Record-Epitaph*, 4 Sept. 1885; *Tombstone Prospector*, 29 Jan 1889; box 4, Ling Collection.

13. *Prescott Courier*, 10 Apr. 1931; Ling to Woolbridge, 11 Oct. 1932; and Amster to O'Keefe, 31 Aug. 1935, both in Ling Collection. Farmers had a significant advantage over miners when it came to personal subsistence; if unable to trade their product with local merchants for other goods and services, they could easily take up subsistence agriculture (Pamela Riney-Kehrberg, "In God We Trusted, in Kansas We Busted ... Again: A Social History of Dust Bowl Kansas, 119, 282; Edwards, "Drought and Depression," 52).

14. *Tombstone Prospector*, 22 Mar. 1888, 23 Sept. 1889, 8 Apr. 1891; Larson to Nihell, 8 Dec. 1941, Larson Collection.

15. Amster to O'Keefe, 31 Aug. 1935, Ling Collection; Larson to Nihell, 19 Dec. 1938, Larson Collection.

16. Sparkes, Collected Papers; Annette Marie Atkins, "Everything but the Mortgage: The Response to Agricultural Disaster in Minnesota, 1873–1878," 57; Lantz, *People of Coaltown*, 191; Liberty Garage and Popular Store cases, Ling Collection; Edwards, "Drought and Depression," 61; *Prescott Courier*, 26 July 1939; Larson to "Friend Harry," 3 Mar. 1939, Larson Collection. James reports that when the Big Bonanza began to dry up on the Comstock, miners and construction workers suffered first, followed by restaurants, saloons, stores, and other businesses (*Roar and the Silence*, 236).

17. *Verde Copper News*, 1 Aug. 1930; J. Myers, *Last Chance*, 38; Lantz, *People of Coaltown*, 91, 192–93; First National Bank of Arizona, Insurance Accounts Receivable, 1929–1932; Riney-Kehrberg, "In God We Trusted," 120–21; Krznarich, oral history tape 863; Larson to Nihell, 24 Feb. 1939, Larson Collection; Briar, *Effect of Long-Term Unemployment*, 53–54; Edwards, "Drought and Depression," 7.

18. Krznarich, oral history tape 862; Yavapai County, Ariz., Yavapai County Assessor's Office, Automobile Registrations, 1920–1932.

19. Agreement between Shea Family and Reese and Amster Motors, 26 Mar. 1928; Coleman to Ling, 12 Mar. 1928; and Ling? to D. J. Shea, 13 Mar. 1928, all in Ling Collection; Yavapai County, Ariz., Yavapai County Assessor's Office, Automobile Registrations, 1920–1932.

20. Liberty Garage Case, box 4, Ling Collection; *Verde Copper News*, 21 Aug. 1931.

21. Liberty Garage Case, box 4, Ling Collection.

22. ibid.; and Ling to Duffey, 21 Feb. 1933, both in Ling Collection.

23. *Verde Copper News*, 3 Dec. 1929; Amster to O'Keefe, 31 Aug. 1935; Ling to Standard Oil, Phoenix, 28 Dec. 1931; and Amster to Lyons, 12 Mar. 1937, all in Ling Collection; *Prescott Courier*, 25 Nov. 1937.

24. D. D. Martin, *Tombstone's "Epitaph,"* 6; *Arizona Daily Star*, 12 Feb. 1886. A biography of the *Tombstone Prospector* on its microfilm leader (Arizona State Department of Library and Archives, Phoenix) indicates that the paper changed owners in October 1887, February 1888, February 1889, April 1895, and August 1913. *Tombstone Prospector*, 26 Apr. 1888.

25. *Arizona Weekly Citizen*, 24 Nov. 1888, 6 Dec. 1890; *Arizona Daily Star*, 25 Dec. 1910.

26. *Verde Copper News*, 1 Apr., 10 June 1930, 4 Sept. 1931, 6 Dec. 1932, 1 Feb. 1935; *Prescott Courier*, 2 Feb. 1935, 31 May 1939; *Jerome Chamber of Commerce News Bulletin*, 3 Dec. 1935, 18 May 1936.

27. Cochise County, Ariz., Great Registers of Cochise County, 1882 and 1892; Sanborn Map Company, "Sanborn Fire Insurance Maps"; *Tombstone Prospector*, 14 Apr. 1887; Muretic, *Heaven in the West*, 8; *Verde Copper News*, 30 June 1931, 12 Apr. 1932. Nevada's Comstock had a similar bootlegging industry in its hills in the early 1930s (James, *Roar and the Silence*, 252).

28. D. Smith, *Rocky Mountain Mining Camps*, 225; Town of Tombstone, Minutes of Town Council, 17 Dec. 1887; *Tombstone Prospector*, 17 Oct. 1887, 28 Oct. 1889; U.S. Bureau of the Census, Territory of Arizona, Tenth Census Enumeration Volume; Krznarich, oral history tape 862; Larson to Nihell, 27 May 1939, Larson Collection; Town of Jerome, Police Court Dockets, 1932–1937; Mountain States Telephone and Telegraph Co., Telephone Directories, 1930–1940, Arizona State Department of Library and Archives, Phoenix; Caillou, *Experience Jerome*, 230, 232; *Verde Copper News*, 11 Sept. 1931, 21 July 1933; Ben T. Traywick, *Eleanora Dumont, Alias Madam Moustache*, 7; J. Myers, *The Last Chance*, 70, 75, 77; *Arizona Daily Star*, 16 Jan. 1938.

29. Ryan, "Tombstone Theater Tonight," 52, 68, 72–73; Traywick, *"Tombstone Epitaph" and Clum*, 44; *Arizona Weekly Citizen*, 18 Aug. 1883; Willson, *Mimes and Miners*, 22–23, 30–31, 60–61, 67, 69–70.

30. George M. Blackburn and Sherman L. Ricards, "The Prostitutes and Gamblers of Virginia City, Nevada, 1870," 257; Anne M. Butler, *Daughters of Joy, Sisters of Misery: Prostitutes in the American West*, 150; Allan G. Bird, *Bordellos of Blair Street: The Story of Silverton, Colorado's Notorious Red Light District*, 183. Blackburn and Ricards believe that prostitutes and gamblers did not remain long in Virginia City after the boom because they were parasites of mining. Butler argues that the

periods of consolidation after booms or of recession just before depression were the times most likely to see attempts by town reformers to restrict or prohibit prostitution.

31. *Tombstone Prospector*, 18 Dec. 1888, 9 Aug. 1890; *Daily Record-Epitaph*, 9 Sept. 1885; "Publisher's Department," *Golden Era* (May 1889): 239; Moore to MacNeil, 29 Apr., 21 Aug. 1888, MacNeil Correspondence.

32. "Blinn's Lumber Yard"; and Blinn to MacNeil, 8 Sept. 1888, both in ibid.

33. Moore to MacNeil, 17 Sept. 1888; and Blinn to MacNeil, 9 Oct. 1888, both in ibid.

34. Moore to MacNeil, 26 Dec. 1888, 12 Apr., 27 May 1889, ibid.

35. Moore to MacNeil, 27 May, 11 June, 29 Sept. 1889, 9 Jan. 1890, ibid.; *Tombstone Prospector*, 3 July 1889, 22 Nov. 1890.

36. *Tombstone Prospector*, 1 Sept. 1891, 25 Mar. 1892; MacNeil Correspondence; *Tucson and Tombstone Directory*; Charles F. Willis, "T. F. Miller Company Store Started in Year 1890"; Box 10, Ling Collection.

37. *Arizona Daily Star*, 26 July 1891; *Verde Copper News*, 18 Feb., 16 Dec. 1930, 17 Apr. 1931; *Prescott Courier*, 8 Aug. 1938; Riney-Kehrberg, "In God We Trusted," 128–31.

38. *Tombstone Prospector*, 30 Nov. 1892; *Verde Copper News*, 20 May 1932, 24 Mar. 1933; Don Harrison Doyle, *The Social Order of a Frontier Community: Jacksonville, Illinois, 1825–70*, 90. Doyle has discovered that merchants in Jacksonville, Illinois, benefited from the panic of 1857, which drove smaller county firms out of business, and made Jacksonville the trade center of the county. Those who survived the panic emerged stronger.

39. "Publisher's Department," *Golden Era* 38 (May 1889): 239–41; *Tombstone Prospector*, 24 May 1887, 20 Feb. 1889, 28 Jan., 29 Jan., 5 July, 25 Dec. 1890, 23 Sept. 1891, 14 Jan., 23 Jan., 11 Aug. 1892.

40. Popular Store Case, Ling Collection; *Verde Copper News*, 27 Oct. 1933, 2 Mar. 1934.

41. *Daily Record-Epitaph*, 9 Sept. 1885; Blinn to Moore, 27 May 1889, MacNeil Correspondence; William Hattich, *Tombstone*, 30; Evelyn Cooper, "C. S. Fly of Arizona: The Life and Times of a Frontier Photographer," 45–46. Lantz observes that trade also migrated, as customers started shopping in nearby towns with lower prices, unable to pay the higher prices charged by local merchants—who then had to charge still higher prices to cover lost business (*People of Coaltown*, 204). Today Jerome has no supermarket or gas station; one must drive to Cottonwood or Prescott.

42. Mann, *After the Gold Rush*, 23, 27; "Mining Reports, Arizona, 1901–1930," 260; Tombstone Mill and Mining Company, *Annual Report*, 14; *Arizona Weekly Citizen*, 26 Feb., 20 Aug., 12 Nov. 1882; boxes 7 and 10, Ling Collection; *Arizona Daily Star*, 12 Jan 1938. Zanjani reports that civil suits decreased during the 1907 lockout at Goldfield, Nevada (*Goldfield*, 194).

43. Ling to Wooldridge, 16 Sept. 1931; Bankruptcy Referee to Ling, 24 Nov. 1937; and Popular Store Case, all in Ling Collection; "Jerome," file folder 27, Sharlot Hall Museum and Archives; Stout, "Want to Buy a Ghost Town?"; Hayhurst, *Hardpan*, 222; *Arizona Weekly Citizen*, 24 July 1886; Hattich, *Tombstone*, 49–51.

44. Riney-Kehrberg, "In God We Trusted," 120–21; Charles Leland Sonnichsen, *Billy King's Tombstone: The Private Life of an American Boom Town*, 196; "Publisher's Department," *Golden Era* (May 1889): 239; Mann, *After the Gold Rush*, 83; D. Smith, *Silver Saga*, 193; Cochise County, Ariz., Great Registers, 1882, 1884, 1886, 1890, 1892; H. V. Young, *They Came to Jerome*, 168.

45. Buss and Redburn, *Shutdown at Youngstown*, 116; Krznarich, oral history tape 862; Doyle, *Social Order*, 97–103.

46. Data for these conclusions were extracted from Cochise County, Ariz., Great Registers, 1882, 1884, 1886, 1892. One piece of evidence suggests greater mobility among miners than nonminers. In both communities each mine industry worker sustained a greater number of persons after the bust than before. In 1882 each Tombstone miner supported a little less than 2 persons and in 1890 supported 2.5. In Jerome the figure went from one miner supporting 2.6 persons in 1930 to 3.15 in 1940. This indicates a slightly lower persistence among miners than among other classes. For a discussion of the relationship between "basic" and "nonbasic" laborers in a one-industry town, see L. W. Cassaday, "The Economics of a One-Industry Town."

47. Sanborn Map Company, "Sanborn Fire Insurance Maps," Jerome, Arizona, 1924, 1938; Mountain States Telephone and Telegraph Co., Telephone Directories, 1930–1940; *Arizona Weekly Citizen*, 31 Dec. 1887; Stuart M. Blumin, *The Urban Threshold: Growth and Change in a Nineteenth-Century American Community*, 85–95; Mann, *After the Gold Rush*, 66–67, 83; D. Smith, *Rocky Mountain Mining Camps*, 104; D. Smith, *Silver Saga*, 191, 283.

48. *Arizona Daily Star*, 17 June 1884.

49. Moore to MacNeil, 11 Sept. 1888, MacNeil Correspondence.

50. Moore to MacNeil, 31 Oct. 1889, ibid.; Barton, *Communities in Disaster*, 40; Blumin, *Urban Threshold*, 207. Blumin notes the extreme vulnerability of all

nineteenth-century merchants—one bust or blunder could ruin years of work. D. Smith (*Rocky Mountain Mining Camps,* 59–61) and Jackson (*Treasure Hill,* 125, 220) both remark upon the speculatory nature of running a business in a frontier mining camp.

Chapter Seven. Two Chinese, an Irishman, a Frenchman, and a Negro

1. *Verde Copper News,* 16 Feb. 1934.

2. Olien and Olien, *Oil Booms,* 82; Jackson, *Treasure Hill,* 63. While many local publications criticized the census takers, the publishers of the *Tucson and Tombstone Directory* sympathized with those beleaguered officials: "The large proportion of the population is composed of single male adults, men who 'are here today and gone tomorrow'—men who will change their place of abode or lodging two, three or four times in a single month. They sleep two, three, or five or six in the same cabin. Again, the compiler, or 'Directory man,' may apply during the day time from door to door through whole blocks of the same street without being able to get the names of the occupants, or the necessary information from those miners who happen to be working on the 'day shift,' while those who are working on the 'night shift' are naturally averse to being disturbed from their hard earned slumbers to answer to their names and occupation[s. . . . And] there are many miners coming or going, and working from time to time, either within the limits of the city proper . . . or elsewhere in the immediate vicinity, and spending most of their spare time and wages in the city itself" (103–4). *Arizona Weekly Citizen,* 26 May, 16 June 1883; Love, "History of Tombstone," 72; *Arizona Republic,* 28 June 1957. One notices that population estimates for these towns tend to grow with the passage of time.

3. *Verde Copper News,* 28 Feb., 1 Apr., 25 Apr. 1930.

4. Estimates placing Tombstone's peak population at around 6,000 may be found in Cochise County, Census, 1882; *Tucson and Tombstone Directory,* 103–4; Hamilton, *The Resources of Arizona* (1883), 50; and Bollar, Reminiscence of Early 1880s, Arizona Historical Society, Tucson.

5. Estimates putting Jerome's peak population at ten thousand or below may be found in U.S. Bureau of Census, *Sixteenth Census of the United States,* 91; Sparkes, Collected Papers; *Prescott Courier,* 31 May 1930; *Verde Copper News,* 6 May 1930; and Duffey, "Housing and Service," 99.

6. Whitmore, Biographical Sketch of Dr. Goodfellow of Tombstone; Hamilton, *The Resources of Arizona* (1884), 77; Jules Baumann, "Tombstone, Arizona," 399; Sanborn Map Company, "Sanborn Fire Insurance Maps," Tombstone, 1889; U.S.

Bureau of Census, *Thirteenth Census of the United States*, 573–74; *Tombstone Prospector*, 14 July 1890, 24 Oct. 1891.

7. Ling to Watson, 2 Feb. 1932, Ling Collection; Home Owners' Loan Corp. application, Candelario Beltran, Apr. 1934, Connolly Collection; Allen, "Yavapai County . . . Jerome," 4 Apr. 1936, in WPA Writers' Program, Manuscripts, box 7; U.S. Bureau of Census, *Sixteenth Census of the United States*, 91; McDonald, "Development of Jerome," 87; Davenport, "Jerome School Plant," 13.

8. Doyle, *Social Order*, 95, 111–12; Don Harrison Doyle, "Social Theory and New Communities in Nineteenth-Century America," 155; Dean Esslinger, *Immigrants and the City: Ethnicity and Mobility in a Nineteenth Century City*, 42–43; Stephen Thernstrom and Peter R. Knights, "Men in Motion: Some Data and Speculations about Urban Population Mobility in Nineteenth-Century America," 2, 4–5, 12–13, 16, 22–24; Mann, *After the Gold Rush*, 141. In taking a sample from the 1880 federal census, then trying to find the same people in the Cochise County Census of 1882, I could positively identify in the county census 58 of 217 (26.7 percent) from the original sample from the federal census of 1880. Adding the probables increased that number to 71 of 217 (32.7 percent). Olien and Olien, *Oil Booms*, 38; Atkins, "Everything but the Mortgage," 50; Edwards, "Drought and Depression," 5, 14, 19–20, 23.

9. *Arizona Daily Star*, 5 June, 12 June 1884; *Prescott Citizen*, 28 Jan. 1931; J. C. Brodie to Governor Moeur, 13 Dec. 193[3?], Governor's Office, Arizona, Governor's Files.

10. *Daily Tombstone*, 9 Feb., 12 Feb., 16 Feb., 17 Feb. 1886; *Arizona Weekly Citizen*, 10 Apr. 1886. In spite of the anti-Chinese sentiment common in the West, declining placer camps were often abandoned to Chinese miners, who were willing to work deposits for lesser returns.

11. *Daily Tombstone*, 17 Feb., 20 Apr. 1886.

12. Town of Tombstone, Minutes of City Council, 11 June, 16 June 1886, 2 Feb., 18 Feb., 20 Apr. 1887; *Tombstone Prospector*, 21 Apr. 1887; *Arizona Weekly Citizen*, 7 May 1887; *Daily Tombstone*, 25 Feb. 1886.

13. *Daily Tombstone*, 27 Feb., 10 Mar., 9 Apr., 30 Apr., 4 Aug., 20 Aug. 1886; *Arizona Weekly Citizen*, 17 Apr. 1886.

14. *Daily Tombstone*, 1 Mar., 23 Aug. 1886.

15. *Verde Copper News*, 18 July, 29 July, 22 Aug. 1930, 17 Apr., 3 July, 14 Sept., 15 Dec. 1931, 24 Mar. 1933; *Prescott Courier*, 13 Apr. 1932; Nancy R. Smith, "Jerome: Man's Changes to One Mountain," 7; U.S. Bureau of the Census, *Fifteenth Census of the United States*, 153.

16. *Arizona Daily Star*, 9 Feb. 1886; *Daily Tombstone*, 17 Feb. 1886; *Arizona Weekly Citizen*, 20 Feb., 17 Apr., 15 May 1886; T. A. Larson, *History of Wyoming*, 141; *Arizona Journal Miner*, 30 July 1880; Traywick, *Chinese Dragon*, 6; Shin-Shan Henry Tsai, "American Exclusion against the Chinese," 39; Florence C. Lister and Robert H. Lister, "Chinese Sojourners in Territorial Prescott," 57, 60, 62, 64.

The Wood River anti-Chinese movement, eerily similar to Tombstone's, featured a district in decline, included a meeting to organize an anti-Chinese committee, an economic boycott, an attempt to establish a white-run steam laundry, and the failure, by the end of 1886, to force the Chinese to leave the area. But unlike Tombstone, the Wood River agitation did feature some violence. Bodie, California's anti-Chinese movement fragmented over the question of violence. Zanjani maintains that anti-Chinese expulsion and exclusion succeeded at Goldfield, Nevada, in 1904 and after because an anti-Chinese riot at Tonopah the previous year had killed one Chinese resident and injured many. Goldfield's Chinese residents took the threat seriously and departed en masse. See Spence, *Wood River or Bust*, 90–96; McGrath, *Gunfighters, Highwaymen, and Vigilantes*, 137–39; and Zanjani, *Goldfield*, 96–97.

17. *Verde Copper News*, 31 May, 7 June, 16 July, 18 Oct. 1929, 30 June 1931; *Prescott Citizen*, 23 June 1939. J. C. Brodie to Governor Hunt, 9 Jan. 1931; and Brodie to Governor Moeur, 13 Dec. 193[3?], both in Governor's Office, Arizona, Governors' Files; Spratt, *Thurber, Texas*, 118; Lopez, *Forever*, 16–17; Heather Hatch, Karen Dahood, and Al V. Fernandez, "Clifton: Photography and Folk History in a Mining Town," 334.

18. G. W. Parsons, Diaries, 3 Apr. 1886; Town of Tombstone, Minutes of Town Council, 6 Nov. 1886, 20 Apr., 10 May, 21 May 1887.

19. The secretary of the Arizona Board of Public Welfare wrote to the secretary to the governor on 5 Feb. 1935, reporting that although 16.5 percent of the state's welfare caseload consisted of aliens—who constituted 20 percent of the total of individuals on relief—37.5 percent of the people in alien families were American-born (Governor's Office, Arizona, Governors' Files). *Prescott Courier*, 30 Apr. 1930, 13 Apr. 1932.

20. *Tombstone Epitaph*, 5 June 1930; *Arizona Republic*, 13 Jan. 1938; *Arizona Daily Star*, 16 Jan. 1938; *Verde Copper News*, 14 June 1930; Lawrence Michael Fong, "Sojourners and Settlers: The Chinese Experience in Arizona," 243; Cochise County, Census, 1882; U.S. Bureau of Census, Territory of Arizona, Twelfth Census Enumeration Volume. Persons listing Mexican nativity constituted 8 percent of the census population in 1882, and 11 percent of the census population in 1900.

Persons listing Irish nativity constituted 10.5 percent of the census population of 1882, and, if one includes persons listing both parents native to Ireland, constituted 7 percent of the census population of 1900.

21. Krznarich, oral history tape 862; "Jerome, Arizona," vertical file, Special Collections, Northern Arizona University; McDonald, "Development of Jerome," 86, 92. McDonald records that Jerome school attendance remained between 60 and 70 percent Latino every year from 1931 to 1940, inclusive, with the single exception of 1935, when Latinos constituted 59 percent of Jerome's students. Latinos averaged 65 percent of the student body from 1931 to 1935, and 66 percent from 1936 to 1940, inclusive. Stout estimates in 1951 that "three quarters of the remaining population is Mexican; virtually all of the 150 miners still working in the shaft are of Mexican blood, though Arizona-born and American citizens" ("Want to Buy a Ghost Town?"). *Prescott Courier,* 5 Dec. 1936, 5 May 1937; "Report on Racial Elements of Yavapai County," 27 Mar. 1937, in WPA Writers' Program, Manuscripts, box 8). Ralph Mann calls the depression era of 1856–1863 in the Mother Lode country of California one "of relative ethnic and class tranquility" (*After the Gold Rush,* 3). He has also found evidence of significant ethnic access to business and professional occupations—excepting law and journalism—and believes class and occupation to be better indicators of social status than ethnicity (124, 170).

22. Ronald M. James, Richard D. Adkins, and Rachel T. Hartigan, "Competition and Coexistence in the Laundry: A View of the Comstock," 171, 182–83; Joseph Axford, *Around Western Campfires,* 39; *Daily Tombstone,* 25 Feb., 10 Mar., 9 Apr., 10 Apr., 22 Apr., 30 Apr., 8 May 1886; *Arizona Daily Star,* 11 May, 15 May 1886.

23. *Daily Tombstone,* 12 Apr. 1886; *Arizona Weekly Citizen,* 17 Apr. 1886.

24. D. Smith, *Rocky Mountain Mining Camps,* 30; *Daily Tombstone,* 20 Aug. 1886. In my reading of Tombstone's anti-Chinese newspaper I found its racist arguments and appeals to racial solidarity to be rare and subdued, with the exception of one article—a cartoonish and improbable racial characterization of the Chinese. Most of the anti-Chinese arguments authored by the paper were economic. Rodolfo Acuña believes the Mexican repatriation of the 1930s to be motivated primarily by economic, rather than racial, considerations (*Occupied America: The Chicano's Struggle toward Liberation,* 193). *Verde Copper News,* 30 June 1931.

25. *Arizona Weekly Citizen,* 7 Sept. 1889; *Arizona Daily Star,* 2 Mar. 1886; First National Bank of Arizona, Ledgers and Account Books; Connolly Collection.

26. Cochise County, Ariz., Great Registers, 1882, 1892; Lamb, "Jewish Pioneers in Arizona," 159–60, 162; *Tombstone Prospector,* 26 Sept. 1889.

27. Florence C. Lister and Robert H. Lister, *The Chinese of Early Tucson: Historic Archaeology from the Tucson Urban Renewal Project*, 3–4; Josiah Heyman, "The Oral History of the Mexican-American Community of Douglas, Arizona," 194–95; Hatch, Dahood, and Fernandez, "Clifton," 334; *Prescott Courier*, 30 June 1976; *Verde Copper News*, 22 Dec., 29 Dec. 1933; Jerome Historical Society, *Jerome Chronicle*, spring 1976, in "Jerome," file folder 27, Sharlot Hall Museum and Archives; *Verde Independent*, 15 Apr. 1965.

28. First National Bank of Arizona, Ledgers and Account Books; Muretic, *Heaven in the West*, 25.

29. Cochise County, Ariz., Great Registers, 1882, 1892; *Arizona Weekly Citizen*, 11 July 1879. A comparison of federal census data from 1880 and 1900 indicates that roughly the same percentage of miners listed themselves as Irish, while the percentage of miners who were Mexicans increased from less than 1 percent in 1880 to more than five percent in 1900. Due to the high percentage of those enumerated who did not list any occupation, these data are only suggestive.

30. A comparison of federal census data for Tombstone from 1880 and 1900 shows roughly the same thing, though the figures are too fragmentary to draw a firm conclusion. Of the three principal minorities, Irish, Chinese, and Mexican, the percentage of Irish and Chinese miners showed almost no change in the two enumerations, while the percentage of Mexican miners rose sharply, though their numbers were small in both cases. *Prescott Courier*, 24 Jan 1931.

31. Mrs. Dosa Schuster, "The Last of the Faithful: Society Notes of Old Tombstone Days," Tombstone General History File, Arizona Historical Society, Tucson, 10; Rockfellow Papers, 54, 135; Helen Younge Lind, Tombstone, 1881–1891, Biographical File, Arizona Historical Society, Tucson; Biography-Obituary Files, Hayden Library, Arizona State University, Tempe; Lonnie E. Underhill, ed., *Tombstone Arizona, 1880, Business and Professional Directory*; William C. Disturnell, comp., *Arizona Business Directory and Gazetteer; Colorado, New Mexico, Utah, Nevada, Wyoming, and Arizona Gazetteer and Business Directory*; "Publisher's Department," *Golden Era* (May and June 1889); *Daily Record-Epitaph*, 12 Nov. 1885; *Tombstone Republican*, 23 Feb. 1883.

32. *Arizona Weekly Citizen*, 5 June 1881; Grace McCool, Clip Book, Arizona Historical Society, Tucson; Harriet Rochlin, "The Amazing Adventures of a Good Woman," 286; Cochise County, Index to Deeds of Mines; John Clum, "Nellie Cashman," 13.

33. Hums, "The Attractions of Tombstone." For further discussion on the demographics and opportunities for women on the frontier and elsewhere, see

Mann, *After the Gold Rush*, 202–3; D. Smith, *Silver Saga*, 193; Doyle, *Social Order*, 111–12; Blumin, *Urban Threshold*, 85–95; and Robert L. Griswold, "Apart but Not Adrift: Wives, Divorce, and Independence in California, 1850–1890," 272.

34. U.S. Bureau of the Census, Territory of Arizona, Tenth Census Enumeration Volume; ibid., Twelfth Census Enumeration Volume.

35. First National Bank of Arizona, Ledgers and Account Books; Connolly Collection; *Prescott Courier*, 21 Nov., 29 Nov. 1935; *Verde Copper News*, 25 Nov. 1929, 11 Mar., 15 Apr., 23 Dec. 1930, 10 Mar., 18 Dec. 1931, 5 May, 18 Aug. 1933, 2 Mar., 6 Apr. 1934; William R. Freudenburg, "Women and Men in an Energy Boomtown: Adjustment, Alienation, and Adaptation," 220. Freudenburg has found little support for the idea that women suffered or gained inordinately from boom, and the same observation can be made for bust. On the more general question of whether women found their frontier experience liberating, Zanjani concludes that the answer is "essentially affirmative" (*Goldfield*, 107).

36. Cochise County, Census, 1882; U.S. Bureau of the Census, Territory of Arizona, Twelfth Census Enumeration Volume.

37. The 1930 federal census shows a total population for Jerome of 4,932, and "other" of 2,853, of whom perhaps thirty were Chinese and others. That produces a Latino population of around 2,800, approximately 57 percent of the town's population. This figure may very well be low, as the census tabulations included only incorporated Jerome. Probably more Latinos than persons from other groups lived outside city limits. Rafaila Rodriquez, Personal Recollections, oral history tape 848, Sharlot Hall Museum and Archives, Prescott; Krznarich, oral history tape 862; *Arizona Republic*, 28 June 1957; Stout, "Want to Buy a Ghost Town?"

38. *Prescott Courier*, 24 Jan 1931; Larson to A. H. Whiteley, 6 Apr. 1937, Larson Collection. Thirty-five percent of those enumerated in Tombstone in the 1880 federal census were foreign-born; in 1920 that figure had slipped to 31 percent. In Jerome the decline in the percentage of foreign-born residents was much more significant—from 48 percent in the federal census of 1920 to 21 percent in census of 1950.

39. *Jerome Chamber of Commerce News Bulletin*, 4 May 1936; *Tombstone Epitaph*, 5 June 1930; Dilsaver, "From Boom to Bust," 399–400.

40. *Prescott Courier*, 24 Jan. 1931, 28 Aug. 1933; *Verde Copper News*, 28 Mar. 1930; Atkins, "Everything but the Mortgage," 47. James drew the same conclusions about the Comstock. He took as a sign of bust a decline in the percentage of males aged fifteen to twenty-four in the general population. These men left

the district in search of work. When the district revived somewhat after 1910, the percentage of males in Storey County's population increased (*Roar and the Silence*, 238, 246).

41. Duane Smith has determined the median age in eighteen sampled mining camps. He has discovered that the younger the camp, the younger the median age, although even the population of an older camp was young when compared to eastern towns (*Rocky Mountain Mining Camps*, 28). On the demographics of age, see also D. Smith, *Silver Saga*, 192; Mann, *After the Gold Rush*, 17; Doyle, *Social Order*, 111; Riney-Kehrberg, "In God We Trusted," 103–4; Olien and Olien, *Oil Booms*, 62; and Davenport, "Jerome School Plant," 28. The *Verde Copper News* (5 Sept. 1930) also made the reasonable argument that scarcity of employment during bust kept older children in school. The paper claimed that in better times older students usually dropped out of school to look for work. Other authors have noted that an aging population is a hallmark of declining small towns generally (Davies, *Main Street Blues*, 192–94).

Chapter Eight. The Painful Necessity of Closing the Church

1. *Tombstone Prospector*, 12 May 1890; *Arizona Weekly Citizen*, 17 May 1890.

2. *Arizona Weekly Citizen*, 18 Oct. 1884, 28 Nov. 1885; *Tombstone Prospector*, 29 July 1892. Zanjani reports several suicides associated with the collapse of Goldfield, Nevada's stock market in 1907 (*Goldfield*, 62–63, 176–77).

3. Buss and Redburn, *Shutdown at Youngstown*, 134–37; Lantz, *People of Coaltown*, 194. Page Smith notes the importance of a town's self-image to its citizens (*As a City Upon a Hill*, 200). Given that, one might well wonder if bust affects individual as well as community self-esteem.

4. Jay Dorian Jurie, *Arizona Mining Towns: Public Policy during Boom and Bust*, 9.

5. Buss, *Mass Unemployment*, 19, 28, 30, 33, 38, 72, 86–89, 143, 158, 167; Briar, *Effect of Long-Term Unemployment*, 12, 82, 89–90; Lantz, *People of Coaltown*, 195; Buss and Redburn, *Shutdown at Youngstown*, 35–36, 40–41, 52–53, 59, 60–62, 82, 88, 116.

6. Hayhurst, *Hardpan*, 53. For those writers who believe boomtown violence overrated in Tombstone, and elsewhere, see Boller, Reminiscence of Early 1880s, 4; Traywick, *"Tombstone Epitaph" and Clum*, 34; and Schuster, "Last of the Faithful." Olien and Olien report of the Texas oil towns that "crime was one problem accompanying oil booms that was much less serious than authors have long

described it. . . . Oil boomtowns were always somewhat rowdy, but rarely bloody" (*Oil Booms,* 128, 140, 169).

7. Inquest files for Jerome Precinct, Jerome State Historic Park Collection, Special Collections, Northern Arizona University, Flagstaff; Briar, *Effect of Long-Term Unemployment,* 12; Buss, *Mass Unemployment,* 28–30; *Verde Copper News,* 4 Mar. 1932. A historian of Goldfield, Nevada, has found that town's homicide rate to be lowest in 1906, the best year of its boom. She has determined that both homicide and suicide rates increased as the district declined (Sally S. Zanjani, "To Die in Goldfield: Mortality in the Last Boomtown on the Mining Frontier," 63–64, 66).

8. Kaye Lynn Briegel, "Alianza Hispano-Americana, 1894–1965: A Mexican-American Fraternal Insurance Society," 142; Buss and Redburn, *Shutdown at Youngstown,* 76.

9. *Verde Copper News,* 6 Jan. 1931, 1 Apr. 1932; *Prescott Courier,* 10 Nov. 1933. Zanjani has discovered several contradictory trends in mortality related to bust. She believes that contract mining, with its haphazard safety practices—which as we have seen, returned during bust—might have increased the number of accidental deaths underground. She has found that the overcrowding, transience, poor accommodations, and heavy drinking of the boom days contributed to the spread of infectious diseases and alcohol problems. Once those conditions ended with the bust, deaths from infectious disease and alcohol abuse declined as well ("To Die in Goldfield," 51–53, 62–63).

10. *Tucson and Tombstone Directory,* 110; Town of Tombstone, City Report for the Month of May, 1894, Tombstone Courthouse State Historic Park.

11. Krznarich, oral history tape 862; Pritchard, "Community," 42. Both Zanjani (*Goldfield,* 186, 194) and James (*Roar and the Silence,* 172–74) have found that crime in general declined significantly after the boom ended. James has found increased arrests for some of the lesser crimes on the Comstock, but he speculates that this was because officers could pay more attention to these offenses with fewer major crimes occupying their attention. He also attributes the lessening crime rate to the demographic shift toward a more balanced population, which, as we have seen, is in some ways related to bust and in others is the natural product of the maturation of a mining district.

12. *Verde Copper News,* 17 July 1925, 21 Dec. 1934; *Arizona Weekly Citizen,* 28 Jan. 1888; Town of Tombstone, Minutes of Town Council, 18 Jan. 1888. On crime in general, Duane Smith concludes that "once the boom period passed, the problem of lawlessness declined, becoming similar to that of any country village" (*Rocky Mountain Mining Camps,* 123). Buss and Redburn have discovered a slightly

different situation in Youngstown, Ohio, in the 1970s. They have found that crime rates fluctuated, rather than rising or falling uniformly in the years surrounding the shutdown of the local steel mill (*Shutdown at Youngstown*, 75).

13. Cochise County, Ariz., Superior Court, Civil, Divorces, 1881–1892, Arizona State Department of Library and Archives, Phoenix; Blackburn and Ricards, "Prostitutes and Gamblers," 243.

14. *Verde Copper News*, 8 Jan. 1932; *Jerome Chamber of Commerce News Bulletin*, 7 Jan. 1936; Buss and Redburn, *Shutdown at Youngstown*, 82; Buss, *Mass Unemployment*, 132; Olien and Olien, *Oil Booms*, 101; Briar, *Effect of Long-Term Unemployment*, 13; Riney-Kehrberg, "In God We Trusted," 88–89.

15. Jackson, *Treasure Hill*, 86; Mann, *After the Gold Rush*, 2, 42. Doyle believes that rampant institution-building served to offset rampant mobility in the frontier farm town of Jacksonville, Illinois ("Social Theory," 165).

16. *Tucson and Tombstone Directory*, 110; Hums, "The Attractions of Tombstone"; G. W. Parsons, Diaries, 18 May 1884, 30 June 1886.

17. *Arizona Weekly Citizen*, 17 Jan. 1885; J. Rowland Hill, "Arizona's Developments," 205; Greg, "Religious Growth of Tombstone," 15.

18. *Tombstone Prospector*, 18 Apr., 7 Mar., 22 Mar., 29 Oct., 18 Dec. 1890, 7 Mar. 1891; G. W. Parsons, Diaries, 17 Oct. 1886; Clum, "Nellie Cashman," 23; James W. Byrkit, "The Word on the Frontier: Anglo Protestant Churches in Arizona, 1859–1899," 80; *Prescott Courier*, 4 Mar. 1937; Greg, "Religious Growth of Tombstone," 40; *Tombstone Epitaph*, 5 May 1955.

19. G. W. Parsons, Diaries, 10 Mar., 15 Aug., 10 Oct. 1886; D. Smith, *Silver Saga*, 198; St. Paul's Episcopal Church, Tombstone, Record Book and Minutes, Nov. 1881–Apr. 1896, Special Collections, University of Arizona, Tucson; *Prescott Courier*, 13 Apr. 1937.

20. Nancy R. Smith, "Hispanic Churches in Jerome," Hayden Library, Arizona State University, Tempe, 15–16; *Jerome Tourguide*, 8, 21; "Jerome," file folder 27, Sharlot Hall Museum and Archives; Greg, "Religious Growth of Tombstone," 28, 40; McCool, Clip Book; Rev. Charles R. Nugent, "Closing of Tombstone Church"; Byrkit, "Word on the Frontier," 74; *Tombstone Epitaph*, 5 May 1955.

21. Greg, "Religious Growth of Tombstone," 15; *Tombstone Epitaph*, 5 May 1955; *Verde Copper News*, 23 Nov. 1934; Burns, "Collapse of Small Towns," 17; *Jerome Tourguide*, 7.

22. *Prescott Courier*, 8 Aug. 1980; "Jerome," file folder 27, Sharlot Hall Museum and Archives; *Verde Copper News*, 13 Apr. 1934; Greg, "Religious Growth of Tombstone," 30; Edwards, "Drought and Depression," 80.

23. Ancient Order of United Workmen, Lodge no. 3, Elections and Miscellaneous Papers, Tombstone Courthouse State Historic Park, Tombstone; Baumann, "Tombstone, Arizona," 399; D. D. Martin, *Tombstone's "Epitaph,"* 38; Town of Tombstone, Minutes of Town Council, 22 Apr. 1889; WPA Writers' Program, Manuscripts, box 7; *Jerome Chamber of Commerce News Bulletin,* 24 Feb. 1936; *Verde Copper News,* 31 Mar. 1933, 9 June, 1933. The Special Collections, University of Arizona Library, Tucson, have collections for the Tombstone Improved Order of Red Men, Opata Tribe no. 15, through 1908; Knights of Pythias, Lodge no. 4, to 1906; and the Independent Order of Odd Fellows, Cochise Lodge no. 5, Miscellaneous Papers, 1909–1915.

24. *Prescott Courier,* 5 Dec. 1936, 5 May 1937; *Verde Copper News,* 26 Jan., 20 Apr. 1934; Biography-Obituary Files; *Daily Tombstone,* 2 Feb. 1886; *Arizona Weekly Citizen,* 4 Apr. 1885; *Verde Copper News,* 25 Jan. 1934.

25. *Tombstone Prospector,* 17 May 1887, 17 Apr. 1890; *Prescott Courier,* 29 Aug., 1 Nov. 1935, 13 Aug. 1938, 12 July 1939; *Verde District Shopping News,* 21 July 1938; *Jerome Chamber of Commerce News Bulletin,* 31 Dec. 1935, 6 Apr. 1936; *Verde Copper News,* 9 June 1931, 17 June, 21 June 1932, 5 Jan., 5 Oct., 7 Dec. 1934; Edwards, "Drought and Depression," 80.

26. Olien and Olien, *Oil Booms,* 160. To the Oliens' idea that voluntary associations might be strengthened by the passing of the boom period with its commercial entertainments for transients, we might add that people were less able to afford commercial entertainments during bust. Doyle, *Social Order,* 186–90; Blumin, *Urban Threshold,* 160; Robert V. Hine, *Community on the American Frontier: Separate but not Alone,* 141. Hine calls nineteenth-century fraternal organizations "institutionalized neighborhoods."

27. Briegel, "Alianza Hispano-Americana," 135; Mark C. Carnes, *Secret Ritual and Manhood in Victorian America,* 151; David T. Beito, *From Mutual Aid to the Welfare State: Fraternal Societies and Social Services, 1890–1967,* 142, 161, 153, 213, 231.

28. Alianza Hispano Americana, Libro de Actas de Logia no. 13, 2 Feb. 1930, 17 May, 21 June 1931, Hayden Library, Arizona State University, Tempe. The AHA went from $1,284.90 cash on hand at the end of December 1929 to $618.40 at the end of March 1931. Report on 1938 Meeting of Highway 79 Association, box 12, in Sparkes, Collected Papers; *Verde Copper News,* 8 Nov. 1932; *Tombstone Prospector,* 28 Feb. 1888; Mann, *After the Gold Rush,* 210; *Prescott Courier,* 11 Sept. 1935, 7 Apr. 1937; Edwards, "Drought and Depression," 75. Beito notes that the Moose reported about 30 percent of their membership in arrears because of unemployment by early 1933 (*Mutual Aid,* 223).

29. *Arizona Weekly Citizen*, 10 Oct. 1885, 2 June 1888, 20 July, 26 Oct. 1889, 10 May 1890; *Tombstone Prospector*, 12 Apr., 1 May, 10 May 1890.

30. *Verde Copper News*, 15 Nov. 1932, 17 Nov. 1933, 30 Mar., 18 May 1934; *Prescott Courier*, 22 July 1935, 8 Mar. 1937, 6 Aug. 1938; Krznarich, oral history tape 863. Davies makes the important point that these town and high school teams meant much more than entertainment to their communities. They were an important source of small-town identification and pride. Thus, their loss had psychological as well as recreational import (*Main Street Blues*, 155).

31. *Jerome Chamber of Commerce News Bulletin*, 17 Dec. 1935; *Verde Copper News*, 7 Jan. 1930, 11 Aug. 1931, 12 Jan. 1932, 6 Jan., 3 Feb., 1 Dec. 1933, 5 Jan. 1934; G. W. Parsons, Diaries, 19 Feb. 1886; *Tombstone Prospector*, 21 Nov. 1892.

32. D. Smith, *Silver Saga*, 200; Edwards, "Drought and Depression," 75, 80; Olien and Olien, *Oil Booms*, 156; Elliott, *Growing Up*, 36–74; Krznarich, oral history tape 862; *Verde Copper News*, 26 June, 24 July 1931, 2 Feb. 1934.

33. Pegues, "Recreation in the Verde District"; *Verde Copper News*, 2 Dec. 1932, 4 Jan. 1935; Riney-Kehrberg, "In God We Trusted," 119. John Spratt wrote of the company coal town of Thurber, Texas: "I cannot recall a big July 4 or a Labor Day picnic in Thurber after the mines shut down in 1921. The Sunday evening band concerts ended" (*Thurber, Texas*, 119).

34. *Jerome Chamber of Commerce News Bulletin*, 6 Apr., 27 Apr. 1936; *Prescott Courier*, 11 July, 25 July, 22 Oct. 1936; *Verde Copper News*, 22 Sept. 1931, 1 Apr., 20 May 1932.

35. *Arizona Daily Star*, 27 June 1886; *Tombstone Prospector*, 29 Nov. 1889, 30 June 1890; G. W. Parsons, Diaries, 5 July 1886.

36. *Verde Copper News*, 18 Sept. 1928, 15 Sept. 1931; Ling to Wooldridge, 16 Sept. 1931, Ling Collection.

37. *Verde Copper News*, 3 May, 5 July 1932, 12 May 1933. Edwards has found large community celebrations discontinued in dust-bowl Kansas ("Drought and Depression," 75).

38. *Verde Copper News*, 22 Mar., 27 Sept. 1932, 3 Feb., 17 Feb., 14 July 1933; *Prescott Courier*, 14 Apr. 1936.

Chapter Nine. As Tombstone Has Empty Houses to Burn

1. "Highway 79: The Greatest Scenic Drive in America," in Sparkes, Collected Papers.

2. Prichard, "Community," 32; "The Downfall of Tombstone"; Eric Margolis,

"Life in the Coal Towns," 61. Margolis notes that "shacks and hovels did not disappear with advent of the model [company] town." The boom and bust nature of coal mining required elasticity in housing, so companies permitted these slums to exist during boom periods. Olien and Olien, *Oil Booms*, 45–48.

3. Sanborn Map Company, "Sanborn Fire Insurance Maps," Tombstone, 1882; J. H. Young, "Tombstone, Arizona," 485; J. Myers, *Last Chance*, 37; *Arizona Weekly Citizen*, 23 June 1883; Randall Rohe, "The Geography and Material Culture of the Western Mining Town," 104.

4. J. Myers, *Last Chance*, 48; Sanborn Map Company, "Sanborn Fire Insurance Maps," Tombstone, 1882; James, Adkins, and Hartigan, "Competition and Coexistence," 165; Rohe, "Geography and Culture," 111. Esslinger has found that in South Bend, Indiana, proximity to occupation had more to do with choice of residence than ethnic group, though physical segregation by ethnic group was increasing by 1880 (*Immigrants and the City*, 49, 61). Mann has discovered "a growing specialization of residential areas by occupation" in the aging gold mining towns that he studied. He has also found racial segregation less than absolute, thanks to boarding, domestic servants, and interracial relationships (*After the Gold Rush*, 96–97, 121). Blumin believes segregation of neighborhoods in Kingston, New York, was by class and occupation (*Urban Threshold*, 85). Physical segregation by occupation applied especially to the vices (Elliott West, "The Saloon in Territorial Arizona," 70).

5. Ling to Arizona Industrial Commission, 21 Dec. 1933, Ling Collection; *Jerome Tourguide*, 20; Duffey, "Housing and Service," 98–99; Krznarich, oral history tape 863.

6. Sanborn Map Company, "Sanborn Fire Insurance Maps," Jerome, 1924; H. V. Young, *They Came to Jerome*, 8.

7. WPA Writers' Program, *WPA Guide to Arizona*, 333; *Verde Copper News*, 29 May 1928, 6 May, 30 Sept., 2 Dec., 9 Dec. 1930, 16 June, 3 Nov. 1931, 17 Mar. 1933; Davenport, "Jerome School Plant," 50; *Prescott Courier*, 10 Dec. 1936; First National Bank of Arizona, Ledgers and Account Books.

8. Sanborn Map Company, "Sanborn Fire Insurance Maps," Jerome, 1924, 1938, Tombstone, 1889, 1904.

9. Sanborn Map Company, "Sanborn Fire Insurance Maps," Jerome, 1924, 1938; Davenport, "Jerome School Plant," 54–55.

10. Rodriquez, oral history tape 848; *Verde Copper News*, 7 Mar., 21 Apr. 1933.

11. Davenport, "Jerome School Plant," 87; Shinn, *Story of the Mine*, 262. James's findings on this point disagree with those of this study and with Shinn's

observation. He reports that archaeological research shows that "poorer residents of the Comstock frequently used a depression-era strategy of living on the outskirts of the urban centers. ... As more people became unemployed, many retreated to these humble homes [up the hill] where they could live without paying rent, waiting for the next boom" (*Roar and the Silence*, 76, 219).

12. *Tombstone Prospector*, 18 Feb. 1892; *Verde Copper News*, 2 Dec. 1930.

13. Sanborn Map Company, "Sanborn Fire Insurance Maps," Tombstone, 1882; "Jerome, Arizona," vertical file, Special Collections, Northern Arizona University; *Verde Copper News*, 16 June 1933. Davies finds among the physical signs of a town's decline "peeling paint, grass growing in streets, deteriorating sidewalks, unkempt yards, trash and litter strewn about, and broken windows in abandoned buildings" (*Main Street Blues*, 172). James points out that residents understood the importance of the physical community to a town's image. Residents rebuilt Virginia City, Nevada, quickly and substantially after its great fire of 1875, fearing that to do less would signal to investors the decline of the district (*Roar and the Silence*, 116–17).

14. Limerick, "Haunted by Rhyolite," 38; Olien and Olien, *Oil Booms*, 63. Sometimes these improvements were undertaken during periods of stagnation or decline in an attempt to rally the town. James records that the Comstock's most important improvements to its industrial infrastructure—the Virginia and Truckee Railroad and the Sutro Tunnel—were undertaken during the bust year of 1869. He believes that Comstockers, having weathered bust times before, were simply showing their faith in the district's future. The railroad and tunnel would increase the district's viability by reducing the mines' operating costs (*Roar and the Silence*, 90).

15. *Tucson and Tombstone Directory*, 108–9; Hal LaMar Hayhurst, "On Tombstone Gas Company and Lights," Arizona Historical Society, Tucson; *Verde Copper News*, 16 Nov. 1928, 20 June 1930, 12 Apr. 1932; Town of Jerome, Minutes of Town Council, 11 June, 9 July 1929.

16. *Verde Copper News*, 13 Jan. 1933, 12 Jan., 11 May 1934; George C. Baxley, comp., "Arizona State Fireman's Association Convention, 1930," Hayden Library, Arizona State University, Tempe, 103; *Jerome Chamber of Commerce News Bulletin*, 24 Feb. 1936; Annual Reports of the Jerome Volunteer Fire Department to the Arizona Corporate Commission, 15 Jan. 1935, 15 Jan. 1936, 13 Jan. 1937, 13 Jan. 1938, 8 May 1939, 7 Feb. 1940, Ling Collection; *Tombstone Prospector*, 19 Apr., 20 June 1888; *Tombstone Epitaph*, 5 May 1955; Hattich, *Tombstone*, 17; Town of Jerome, Minutes of Town Council, 20 Apr., 11 June, 1 Aug. 1935, 11 May, 8 June 1937, 15 Mar., 14 June 1938.

17. Russel Wahmann and Robert des Granges, *Verde Valley Railroads: A Search of the Past and Comparison with the Present of the Several Railroads That Operated in the Verde Valley of Arizona*; Thomas Hardin Peterson, "The Tombstone Stagecoach Lines, 1878–1903: A Study in Frontier Transportation," 36–38, 83, 124–27, 134, 138–40, 147; *Arizona Daily Star*, 28 June 1886; Sonnichsen, *Billy King's Tombstone*, 204–5.

18. *Tombstone Republican*, 29 Dec. 1883, 22 Mar. 1884; Town of Tombstone, Minutes of Town Council, 17 Oct., 9 Nov., 14 Dec. 1883; *Arizona Weekly Citizen*, 20 Oct. 1883.

19. Arizona Corporate Commission, "History of Huachuca Water Co."; *Tombstone Republican*, 22 Mar. 1884; Blake, *Tombstone and Its Mines*, 23; Baumann, "Tombstone, Arizona," 399; Hill, "Arizona's Developments," 203; *Arizona Daily Star*, 17 July 1942, 10 Mar. 1969; Hayhurst, "On Tombstone Gas Company and Lights"; Canty and Greeley, *History of Mining in Arizona*, 231.

20. Town of Tombstone, Minutes of Town Council, 11 July, 18 July 1888, 29 Jan., 30 Jan., 7 Feb., 28 Feb., 6 Mar., 12 June, 20 June 1890, 24 May 1891; *Tombstone Prospector*, 31 Jan., 1 Feb., 26 Mar., 5 May, 6 May 1890, 29 Dec. 1891, 21 Nov. 1892; Hattich, *Tombstone*, 17, 31.

21. Hill, "Arizona's Developments," 204; *Tombstone Prospector*, 25 Apr. 1888.

22. *Tombstone Prospector*, 26 Apr. 1888.

23. WPA Writers' Program, Manuscripts, box 7; Pritchard, "Community," 31; Jerome Historical Society, *They Came to Jerome*, 2, 6; Jerome Historical Society, *Bring It All Back to Jerome: Proceedings of the Sixth Annual Historic Symposium . . . 27 Aug. 1983*, 4; *Prescott Courier*, 26 July 1939; Davenport, "Jerome School Plant," 24, 49–50.

24. H. V. Young, *They Came to Jerome*, 8; Jerome Historical Society, *Three Epochs of Jerome's History: Proceedings of the Fifth Annual Historic Symposium, 28 August 1982*, 6; Pritchard, "Community," 31; *Jerome Tourguide*, 11, 20, 22.

25. James, *Roar and the Silence*, 74–75, 244; Limerick, "Haunted by Rhyolite," 24; Brian Levine, *A Guide to the Cripple Creek–Victor Mining District*, 58; D. C. Miller, *Ghost Towns of California*, 45.

26. Ramsey, *Ghost Towns of British Columbia*, 225; D. Fitzgerald, *Ghost Towns of Kansas*, 192, 199; Dallas, *Colorado and Utah Ghost Towns*, 79; Gregory Martin, *Mountain City*, 34.

27. Elliott, *Nevada's Mining Boom*, 44; Dallas, *Colorado and Utah Ghost Towns*, 217; D. C. Miller, *Ghost Towns of Montana*, 56, 72; D. C. Miller, *Ghost Towns of Wyoming*, 86; McCool, *Sunday Trails*, 89; Sherman and Sherman, *Ghost Towns*

of Arizona, 58. McGrath notes that when it declined, most of Aurora, Nevada's mills "were disassembled and shipped off to Bodie or Virginia City to be re-erected" (*Gunfighters, Highwaymen, and Vigilantes,* 3).

28. Olien and Olien, *Oil Booms,* 45, 48. Gregory Martin wrote that "in the boom and bust mining towns of Nevada and the West, where drastic population swings, like the weather, are predictably unpredictable, the trailer home is a symbol of a certain kind of progress. People have learned something. They've realized that the resources their jobs and lives depend on will someday run out, and they're saying to themselves: Chances are, this won't last forever. A few years if we're lucky. Let's get a house that moves so that when the price of gold (or silver or copper or whatever) goes . . . , we can get the hell out" (*Mountain City,* 55).

29. Love, "History of Tombstone," 97; Ethel Macia, "Tombstone in the Early Days," 10; *Tombstone Prospector,* 2 June 1890.

30. Town of Tombstone, Minutes of Town Council, 6 Apr. 1887; *Tombstone Prospector,* 8 Apr. 1887.

31. *Jerome Tourguide,* 2; Krznarich, oral history tape 863; Rodriquez, oral history tape 848.

32. *Jerome Chamber of Commerce News Bulletin,* 3 Feb. 1936; Larson to Nihell, 15 Jan. 1941, Larson Collection; Biography-Obituary Files, Curtis; Dallas, *Colorado and Utah Ghost Towns,* 126; D. C. Miller, *Ghost Towns of Montana,* 123. Hatch, Dahood, and Fernandez record an instance of a Clifton storekeeper trading a guitar for a five-room house. The storekeeper then had the house torn down and sold the lumber for many times the value of the guitar ("Clifton," 334–36). Barton has found that newspaper editors in depressed coal towns in Illinois during the Great Depression regarded the destruction of surplus housing as a good thing, believing that it would increase the value of the remaining structures (*Communities in Disaster,* 320–21).

James reports that residents of the Depression-era Comstock bought adjacent houses at tax sales and reduced them to firewood during the winter, leaving themselves with a larger yard in the process. Modern visitors are thereby tricked into a vision of a frontier Comstock with single-family detached houses on large lots, rather than the nineteenth-century reality of crowded rows of houses (*Roar and the Silence,* 252).

33. Jerome Historical Society, *Three Epochs of Jerome's History,* 6; Pritchard, "Community," 31; *Verde Copper News,* 3 June 1932; Town of Tombstone, Minutes of Town Council, 20 Jan. 1890; Sherman and Sherman, *Ghost Towns of Arizona,* 109.

34. Stout, "Want to Buy a Ghost Town?"; Pritchard, "Community," 42.

35. Weis, *Ghost Towns of the Northwest*, 52; D. C. Miller, *Ghost Towns of Montana*, 15–16, 169; D. Fitzgerald, *Ghost Towns of Kansas*, xiii; McGrath, *Gunfighters, Highwaymen, and Vigilantes*, 8–9; Fatout, *Meadow Lake: Gold Town*, 112.

36. Olien and Olien, *Oil Booms*, 166. Sometimes this process works in reverse, of course. The former supporting town goes on to prosperity, pulling the mining town in its wake. The old gold camps of Jamestown, Gold Hill, and Ward survive today as bedroom communities for or weekend retreats from their former supply town of Boulder, Colorado.

37. *Arizona Daily Star*, 25 June 1884; Faulk, *Tombstone: Myth and Reality*, 169, 172; Sherman and Sherman, *Ghost Towns of Arizona*, 26–27, 40–41.

38. Traywick, *Some Ghosts*, 15, 27, 31; McCool, *Clip Book*; Hill, "Arizona's Developments," 248; Kerry C. S. Tiller, "Charleston Townsite Revisited," 247; Alma Ready, "Charleston, the Town That Never Grew Old," 6; McCool, *Sunday Trails*, 15–16, 30; Dallas, *Colorado and Utah Ghost Towns*, 16; D. C. Miller, *Ghost Towns of California*, 159.

39. Duffey, "Housing and Service," 98–99; *Jerome Tourguide*, 21.

40. Spratt, *Thurber, Texas*, 126; G. Martin, *Mountain City*, 34; Margolis, "Life in the Coal Towns," 60; Sherman and Sherman, *Ghost Towns of New Mexico*, 65. Sometimes company officials were only relocating their town to the next boom when they ordered it dismantled. The town of Lavoye occupied several locations in the Salt Creek Oil Fields of Wyoming (D. C. Miller, *Ghost Towns of Wyoming*, 53). Much mining and milling equipment disappeared in the scrap drives of World War II (D. C. Miller, *Ghost Towns of Idaho*, 100; D. C. Miller, *Ghost Towns of Montana*, 170).

41. D. C. Miller, *Ghost Towns of California*, 44, 84; Ramsey, *Ghost Towns of British Columbia*, 30, 96, 124; Baker, *Ghost Towns of Texas*, 25, 73; D. Fitzgerald, *Ghost Towns of Kansas*, 47; Sherman and Sherman, *Ghost Towns of New Mexico*, 20.

42. Sherman and Sherman, *Ghost Towns of Arizona*, 127; Dallas, *Colorado and Utah Ghost Towns*, 11; D. Smith, *Silver Saga*, 223, 228; D. C. Miller, *Ghost Towns of Nevada*, 63, 68, 111; Jackson, *Treasure Hill*, 148–49, 205; Elliott, *Nevada's Mining Boom*, 95; Zanjani, *Goldfield*, 234; Ramsey, *Ghost Towns of British Columbia*, 184; Weis, *Ghost Towns of the Northwest*, 6–7, 12; Baker, *Ghost Towns of Texas*, 69, 93.

43. Florin, *Colorado and Utah Ghost Towns*, 71; Ramsey, *Ghost Towns of British Columbia*, 222; Dallas, *Colorado and Utah Ghost Towns*, 133; D. C. Miller, *Ghost Towns of Nevada*, 41; D. C. Miller, *Ghost Towns of California*, 132; D. C. Miller, *Ghost Towns of Nevada*, 86; D. Fitzgerald, *Ghost Towns of Kansas*, x; Baker, *Ghost Towns of Texas*, 99; D. C. Miller, *Ghost Towns of Idaho*, 87.

44. D. C. Miller, *Ghost Towns of California*, 6, 67, 93, 133; Greenland, *Hydraulic Mining in California*, 78, 175; D. C. Miller, *Ghost Towns of Nevada*, 77; WPA Writers' Program, *The WPA Guide to 1930s Colorado*, xi.

45. Richard V. Francaviglia, "Copper Mining and Landscape Evolution: A Century of Change in the Warren Mining District, Arizona," 289; D. C. Miller, *Ghost Towns of Nevada*, 123; Elliott, *Growing Up*, 177; Lopez, *Forever*, foreword, 84; Sherman and Sherman, *Ghost Towns of New Mexico*, 189; D. C. Miller, *Ghost Towns of Wyoming*, 60; Sherman and Sherman, *Ghost Towns of Arizona*, 96; Dallas, *Colorado and Utah Ghost Towns*, 221.

46. Brogdon, "History of Jerome, Arizona," 147–52; WPA Writers' Program, Manuscripts, box 8; *Verde Copper News*, 10 Oct., 2 Dec. 1930, 14 July 1931; H. V. Young, *They Came to Jerome*, 50; Mills, "Ground Movement and Subsidence," 167, 169.

47. Town of Jerome, Minutes of Town Council, 13 Oct. 1935, 10 Mar. 1936; Caillou, *Experience Jerome*, 262; Stout, "Want to Buy a Ghost Town?"; Larson to Whiteley, 12 Oct. 1936, Larson Collection.

48. Town of Jerome, Minutes of Town Council, 12 Nov. 1935, 5 Mar., 10 Mar. 1936.

49. Perez to Nihell, 18 Mar. 1937; and Larson to Nihell, 27 Mar. 1937, both in Larson Collection.

50. WPA Writers' Program, Manuscripts, box 8; Larson to Nihell, 27 Oct. 1940, Larson Collection.

51. Larson to Nihell, ? May 1936, ibid.; *Prescott Citizen*, 6 Aug., 10 Aug. 1931.

52. Larson to Nihell, 24 Feb. 1939, Larson Collection.

53. Larson to "Friend Harry," 26 Apr. 1938, ibid.; *Prescott Citizen*, 18 Apr. 1938, 10 Feb., 2 June, 17 June, 3 Nov. 1939.

54. Nihell to Larson, 30 Nov. 1935; and Larson to Nihell, 28 Feb. 1936, 6 Dec. 1938, all in Larson Collection.

55. Larson to Nihell, 19 July 1937, 25 Mar. 1940, ibid.

56. Larson to Nihell, 18 June 1939, ibid.; Krznarich, oral history tape 862; *Verde Copper News*, 23 Mar. 1934.

57. *Verde Copper News*, 21 July, 4 Aug. 1931, 12 Apr. 1932; Citizens' Committee to Jerome Town Council, in re subsidence damage, 13 Aug. 1935, Ling Collection; Town of Jerome, Minutes of Town Council, 13 Aug., 10 Dec. 1935, 13 Apr., 14 Sept. 1937; Jerome Mayor and Clerk to Arizona Governor, 15 Dec. 1936, Governor's Office, Arizona, Governors' Files.

58. Town of Jerome, Minutes of Town Council, 9 Nov., 23 Nov., 14 Dec.

1937, 15 Mar., 12 Apr. 1938; Quigley to Jerome Town Council, 6 July 1938, Ling Collection.

59. United Verde Copper Company, Records, 1900–1940, Special Collections, Northern Arizona University, Flagstaff; *Prescott Courier*, 12 July 1938, 15 Mar. 1939; Town of Jerome, Minutes of Town Council, 13 Dec. 1938, 17 Feb., 14 Mar., 11 Apr. 1939; box 11, Ling Collection.

60. Larson to Nihell, 27 Jan. 1941, 27 Feb. 1941, 29 Mar. 1941, Larson Collection; "Jerome, Arizona," vertical file, Special Collections, Northern Arizona University; D. E. Miller and Sharpless, *Kingdom of Coal*, 321.

Chapter Ten. Due Economy in Town Administration

1. *Verde Copper News*, 12 July 1932.
2. *Verde Copper News*, 29 July 1932.
3. *Tombstone Republican*, 29 Dec. 1883; *Verde Copper News*, 16 Nov. 1928.
4. Town of Tombstone, Minutes of Town Council, 1 Apr., 6 May, 25 July, 15 Aug., 7 Sept. 1885, 7 Jan. 1886, 30 Jan., 9 Feb., 10 Feb., 11 Feb., 10 July, 30 Sept. 1886, 1 Apr. 1891. The minutes for March 1885 show four unsuccessful attempts to achieve a quorum. Peterson reports that by the summer of 1885 Tombstone City Warrants were worth less than half of their face value ("Tombstone Stagecoach Lines," 130–31).
5. D. Smith, *Rocky Mountain Mining Camps*, 90–95, 99.
6. *Verde Copper News*, 16 Mar., 23 Apr. 1926, 27 Apr., 29 May 1928; Jackson, *Treasure Hill*, 86, 93–94, 100–101.
7. *Arizona Weekly Citizen*, 18 Sept. 1880, 25 June 1882; *Verde Copper News*, 25 May, 19 Oct. 1926, 2 Nov. 1928, 9 Sept. 1930.
8. Town of Tombstone, Minutes of Town Council, 14 Dec. 1883; Mann, *After the Gold Rush*, 204.
9. *Verde Copper News*, 5 Mar., 30 Mar. 1926, 11 Jan. 1927.
10. *Verde Copper News*, 26 Oct. 1928, 20 Aug., 29 Oct. 1929.
11. F. E. Doucette gives total tax rates for Jerome as 3.3917 in 1928 and 3.674 in 1929, counting state, county, school district, and town taxes (*Arizona Year Book, 1930–1931*, 41). The comparable figures for Prescott came to 4.2588 and 4.1302. *Verde Copper News*, 19 June 1931.
12. *Prescott Courier*, 4 Dec., 5 Dec. 1931, 4 Jan., 26 Oct. 1934.
13. *Prescott Courier*, 26 Oct. 1934. The previous December, upon news of a presidential proclamation raising the price of silver more than 25 percent, Tally

told the *Prescott Courier* that what Verde Valley mining companies really needed was reduced taxation (22 Dec. 1933). Even when a government came to his aid, Tally could not refrain from throwing his antitax jab.

14. Hunt to Pinchot, 19 Nov. 1931, Governor's Office, Arizona, Governors' Files; *Verde Copper News*, 13 July 1934; "Copper Price vs. Tax Valuations; Yavapai County, Ariz.," graph, Governor's Office, Arizona, Governors' Files; WPA Writers' Program, Manuscripts, box 8. The Oliens report that when the Great Depression hit the oil town of Wink, Texas, property valuations fell by almost half between 1930 and 1931 (*Oil Booms*, 63).

15. Total fines collected in Jerome were tabulated from figures recorded in Town of Jerome, Minutes of Town Council, 1929–1939; Town of Jerome, Police Court Docket, Mar.–May, July–Sept., Nov. 1935; Town of Tombstone, Minutes of Town Council, 30 Jan. 1886, 6 Feb. 1889. At their meeting of 12 Jan. 1932, Jerome's town council discussed reducing the pay of city employees, in part because "of the fact that the Town licenses and Police Court fines have fallen considerably below the amount estimated for the current budget" (Town of Jerome, Minutes of Town Council). Jerome Historical Society, *Bring It All Back to Jerome*, 5; West, "Saloon in Territorial Arizona," 69; D. Smith, *Rocky Mountain Mining Camps*, 146; Riney-Kehrberg, "In God We Trusted," 200.

16. Town of Tombstone, Minutes of Town Council, 20 Oct. 1883, 23 Oct. 1884, 25 May 1888, 19 May 1890, 29 May 1891; *Tombstone Republican*, 29 Dec. 1883; *Tombstone Prospector*, 23 Dec. 1889, 22 Jan. 1890; *Tombstone Epitaph*, 29 Mar. 1890.

17. *Verde Copper News*, 18 Aug. 1931. Budgets, assessed valuations, and town tax rates are recorded in Town of Jerome, Minutes of Town Council, 1929–1939.

18. Town of Jerome, Minutes of Town Council, 1929–1939; Town of Tombstone, Minutes of Town Council, 19 Nov. 1885.

19. G. W. Parsons, Diaries, 5 Feb., 11 Feb., 24 May 1886; *Tombstone Prospector*, 3 June 1892.

20. *Verde Copper News*, 15 Apr., 19 Apr., 26 Apr., 10 May 1932, 28 July 1933; Town of Jerome, Minutes of Town Council, 9 July 1931, 23 May 1932, 9 Aug. 1933, 13 Aug. 1935; *Prescott Courier*, 13 July, 18 Aug. 1935. The Jerome Taxpayers' Association also continued to lobby county officials, although a county taxpayers' association was formed in 1935 to do that.

21. *Prescott Courier*, 5 Nov. 1931, 21 Dec. 1934, 19 Apr., 24 Apr., 6 June, 13 Nov. 1935, 1 Aug. 1938. Arizona's governors issued a series of proclamations authorizing the provision of funds to defend the state against tax suits brought by various

mining companies and other large corporations, including United Verde, United Verde Extension, and Phelps Dodge. See Governor's Office, Arizona, Proclamations, 3 July, 3 Aug., 2 Sept., 19 Oct. 1931, 23 May 1932, 28 Sept., 8 Nov. 1933, 19 Nov. 1934, 31 Jan. 1936.

22. *Prescott Courier*, 4 Nov. 1931; *Verde Copper News*, 29 Apr., 7 June, 19 July, 30 Dec. 1932; Ling to Kingdon, 21 July 1932; and Kingdon to Ling, 22 July 1932, both in Ling Collection.

23. *Prescott Courier*, 24 May 1935.

24. Town of Tombstone, Minutes of Town Council, 4 June, 11 June, 18 Dec. 1884, 4 Oct. 1889. Salary, personnel, and monthly expenses derived from Town of Tombstone, Record of Warrants Issued, 5 Jan. 1882–16 Nov. 1910.

25. Town of Tombstone, Minutes of Town Council, 2 Feb. 1887, 18 Mar. 1891; *Tombstone Prospector*, 21 Jan., 22 Jan. 1890, 19 Mar. 1891; Town of Tombstone, Record of Warrants Issued, 5 Jan. 1882–16 Nov. 1910; Pat M. Ryan, "Trailblazer of Civilization: John P. Clum's Tucson and Tombstone Years," 64 n. 39; Hill, "Arizona's Developments," 201.

26. *Verde Copper News*, 18 Aug. 1931, 19 Jan., 15 Apr. 1932, 27 July 1934; Town of Jerome, Minutes of Town Council, 10 June 1930, 13 Jan., 11 Aug., 10 Nov. 1931, 15 Jan., 15 Apr., 23 May 1932, 14 Feb., 14 Mar. 1933, 12 June 1934, 9 July 1935, 9 June 1936, 9 Mar., 13 July 1937, 18 July 1939; *Prescott Courier*, 9 Aug. 1939. Jerome's town council also ordered a reduction in salaries for town officials during the depression of 1921. Brogdon, "History of Jerome, Arizona," 143.

27. Speaker of Arizona House of Representatives to Governor Hunt, 14 Dec. 1931, Governor's Office, Arizona, Governors' Files. The speaker proposed a 25 percent reduction, voluntary or otherwise, for state employees. Edwards, "Drought and Depression," 72; Riney-Kehrberg, "In God We Trusted," 200; Fatout, *Meadow Lake: Gold Town*, 83; D. Smith, *Silver Saga*, 244. Zanjani records that busted Goldfield, Nevada's methods to reduce expenses and generate income included auctioning properties seized for nonpayment of taxes, seeking bank loans, leaving teachers and firemen unpaid, using prisoners to maintain city streets, and raising the salaries of police officers by levying covert assessments in the red-light district (*Goldfield*, 217).

28. *Arizona Daily Star*, 24 Jan. 1885, 2 Feb., 16 Feb. 1886; *Arizona Weekly Citizen*, 24 Jan. 1885; *Tombstone Prospector*, 8 Mar., 10 Mar., 24 May 1887.

29. *Tombstone Prospector*, 26 Feb. 1890; Town of Tombstone, Minutes of Town Council, 21 Mar. 1890.

30. *Arizona Weekly Citizen*, 12 Apr. 1890; *Tombstone Prospector*, 18 Mar. 1890.

31. *Tombstone Epitaph*, 3 May 1890; *Tombstone Prospector*, 22 Mar., 1 May, 6 May 1890.

32. *Tombstone Prospector*, 1 May 1890, 14 Mar., 6 Aug. 1892; Town of Tombstone, Record of Warrants Issued, 5 Jan. 1882–16 Nov. 1910.

33. *Verde Copper News*, 12 Apr., 19 Apr., 17 May 1932, 28 July 1933.

34. *Verde Copper News*, 6 Nov. 1931; Yavapai County, Ariz., Yavapai County Assessor, Assessment and Tax Rolls, 1919–1959. This register recorded valuations and revenues for Yavapai County School District no. 9, Jerome. Hill, "Arizona's Developments," 204; Davenport, "Jerome School Plant," 25.

35. *Arizona Weekly Citizen*, 23 Jan. 1886; *Arizona Daily Star*, 2 Feb., 16 Feb. 1886; Matia McClelland Burke, "The Beginnings of the Tombstone School," 253.

36. *Verde Copper News*, 12 June, 6 Nov. 1931, 30 June 1933; WPA Writers' Program, Manuscripts, box 7.

37. *Verde Copper News*, 5 June, 6 Nov. 1931, 9 Sept. 1932; *Prescott Courier*, 2 Sept. 1935, 21 Aug. 1936; McDonald, "Development of Jerome," 92; Burke, "Tombstone School," 254.

38. *Verde Copper News*, 12 June 1931, 30 June 1933; Davenport, "Jerome School Plant," 98; Burke, "Tombstone School," 254.

39. Davenport, "Jerome School Plant," 33; *Jerome Tourguide*, 21; H. V. Young, *They Came to Jerome*, 135; Jerome Historical Society, *Three Epochs of Jerome's History*; *Verde Copper News*, 26 Jan. 1934; Edwards, "Drought and Depression," 72, 80. Davies notes how badly the loss of a community's school through such a consolidation movement could hurt the town. In his case example, Camden, Ohio, the town's schools were "always a focus of community life and pride." Like small-town schools all over the United States, Camden's schools "not only provided local students with a basic educational program but also played a central role in the social life of the community.... At times, town and school became virtually indistinguishable." Davies reports that during its decline Camden lost what he considers to be a small town's three most important institutions: its newspaper, school, and bank (*Main Street Blues*, 164–65, 177).

40. *Prescott Courier*, 15 Oct. 1930, 23 Aug. 1932, 29 Nov. 1934; *Verde Copper News*, 11 Nov. 1932, 6 Oct. 1933, 9 Nov. 1934. Briar has found that among her subjects from the 1970s, unemployment did not alter political views (*Effect of Long-Term Unemployment*, 69, 89–91).

41. *Arizona Weekly Citizen*, 6 Oct. 1888; *Tombstone Prospector*, 4 Oct. 1888, 3 Oct. 1890, 25 Oct. 1892.

42. *Prescott Courier*, 6 Apr. 1934, 1 Apr. 1936, 16 Feb., 30 Mar. 1938; *Jerome Chamber of Commerce News Bulletin*, 23 Mar., 6 Apr. 1936.

43. H. V. Young, *They Came to Jerome*, 183–84; Jerome Historical Society, *Bring It All Back to Jerome*, 5; *Verde Copper News*, 5 Dec. 1930.

44. Barton, *Communities in Disaster*, 312.

45. P. Smith, *As a City Upon a Hill*, 123–25; Richard S. Alcorn, "Leadership and Stability in Mid-Nineteenth Century America," 695. Despite the title of their study, Robert S. Lynd and Helen Merrell Lynd discovered of Depression-era Muncie that "leadership in the community has not shifted in kind, but has become more concentrated in the same central group observed in 1925. . . . The city's local government and religion have remained as before [the] most resistant of all its institutions to change" (*Middletown in Transition: A Study in Cultural Conflicts*, 489).

46. *Tombstone Epitaph*, 29 Mar. 1890.

47. *Tombstone Prospector*, 24 May 1887; *Tombstone Epitaph*, 29 Mar. 1890; Town of Jerome, Minutes of Town Council, 15 Apr. 1932, 9 Aug. 1933.

48. *Verde Copper News*, 15 July, 5 Aug. 1932.

49. Barton, *Communities in Disaster*, 290–92; Riney-Kehrberg, "In God We Trusted," 109, 186; Edwards, "Drought and Depression," 79.

50. Lantz, *People of Coaltown*, 206–7; William Simon and John H. Gagnon, "The Decline and Fall of the Small Town," 44, 49, 51.

Chapter Eleven. Our Last Dollar on a Sure Thing

1. Brogdon, "History of Jerome, Arizona," 154–56.

2. *Arizona Weekly Citizen*, 24 May 1884.

3. *Tombstone Republican*, 15 Sept. 1883.

4. Quoted in *Arizona Weekly Citizen*, 24 Apr. 1886; *Daily Tombstone*, 27 May 1886; *Tombstone Prospector*, 16 Oct. 1887, 28 June 1888. Even a general exodus by Tombstone's residents to a new strike in the Centennial District did not excessively agitate the *Prospector*'s editor (6 June 1889). The places of those departing, he assured his readers, would be occupied by fresh faces and new blood, and the publicity that the new discoveries brought to the area could only benefit Tombstone. He concluded that only the narrow-minded could view these latest developments with jealousy.

5. *Tombstone Prospector*, 8 Jan., 14 Mar. 1892; *Verde Copper News*, 7 Oct. 1930; Duane A. Smith, *When Coal Was King: A History of Crested Butte, Colorado, 1880–1952*,

60–61. Zanjani notes local newspapers' practice of suppressing news about epidemics that, like other bad news, discouraged investment (*Goldfield*, 119).

6. *Tombstone Republican*, 8 Sept. 1883.

7. *Daily Tombstone*, 1 July 1886.

8. Lantz, *People of Coaltown*, 197; Riney-Kehrberg, "In God We Trusted," 338–39; Robbins, *Hard Times in Paradise*, 82; *Verde Copper News*, 1 Nov. 1932.

9. Hill, "Arizona's Developments," 202; *Tombstone Prospector*, 25 June 1892. Otis E. Young Jr. observes that even after bust had undeniably come to stay, "virtually every ghost camp had its handful of resident eccentrics who steadfastly held that the ore body had only been scratched and that wealth in immoderate quantities lay only a bit farther down" (*Western Mining*, 9).

10. WPA Writers' Program, Manuscripts, box 8; Stout, "Want to Buy a Ghost Town?" Lantz has discovered that these delusions are not confined to precious-metal mining communities. Informants assured him that coal reserves still existed—or even that they should exist—under their town, and extolled the virtues of these large coal beds, which convinced the informants that mining must surely resume someday (*People of Coaltown*, 198).

11. *Arizona Weekly Citizen*, 24 July 1886; Lantz, *People of Coaltown*, 194. I have had some experience in vicarious living through rumor while serving aboard ships and can report that rumors do serve useful purposes. Pleasant rumors boost morale and gave the mind enjoyable fodder to consume, while unpleasant rumors can simply be dismissed as such. If unpleasant rumors come true, one has already been braced for them. If pleasant rumors fail to come true, an hour's disappointment salves the wound, and then that rumor is dismissed as only rumor anyway, or replaced by a new rumor. One learns to treat anything as suspect until it occurs, which keeps one's mind on the task at hand, and keeps one somewhat insulated from disappointment.

12. *Tombstone Prospector*, 14 Sept. 1889, 18 Feb. 1890; *Verde Copper News*, 31 Oct. 1930, 13 Feb., 3 Nov. 1931.

13. Lantz, *People of Coaltown*, 198; Robbins, *Hard Times in Paradise*, 167. Riney-Kehrberg believes that optimism was realistic in a boom-and-bust business like dryland farming and that it was certainly unrealistic to expect eight straight years of drought and depression in the 1930s. She also notes that the farmers who survived the "Dirty Thirties" prospered in the 1940s ("In God We Trusted," 291).

14. Hums, "The Attractions of Tombstone." In their reexamination of Middletown published in 1937, Lynd and Lynd chide the people of Muncie, Indiana, for treating the Great Depression as an anomaly (*Middletown in Transition*, 494).

But the Great Depression was an anomaly. An economic disaster of that magnitude had never happened before in American history and has never happened since.

James observes that "veterans of the mining world knew that the industry, even when prosperous, rarely provided stability. Those who obtained stocks, claims, or businesses during a depression, when prices were lowest, capture the most return on investments. Similarly, workers had the best chance of good employment if they lingered until the mining camp sprang back to life" (*Roar and the Silence*, 75, 236).

15. *Arizona Weekly Citizen*, 15 May 1881; *Tombstone Prospector*, 25 May, 16 Oct. 1887. The *Arizona Daily Star* admonished readers that "nothing is more cowardly than capital and when threatened [it] will always seek a safe retreat" (13 May 1884). Others have since drawn the same conclusion: Jackson, *Treasure Hill*, 149; Dykstra, *The Cattle Towns*, 115; D. Smith, *Rocky Mountain Mining Camps*, 175. Buss and Redburn believe that "a major closing signals to other investors the need to reconsider plans for local expansion" (*Shutdown at Youngstown*, 161).

16. *Verde Copper News*, 31 Jan. 1930; Larson to Nihell, 28 Jan. 1940, 15 June 1941, Larson Collection.

17. Briar, *Effect of Long-Term Unemployment*, 73; D. Smith, *Silver Saga*, 180; Atkins, "Everything but the Mortgage," 18; Edwards, "Drought and Depression," 1, 93; Lynd and Lynd, *Middletown in Transition*, 494.

18. Buss and Redburn, *Shutdown at Youngstown*, 161; Simon and Gagnon, "Decline and Fall," 50; Pritchard, "Community," 47.

19. D. E. Miller and Sharpless, *Kingdom of Coal*, 314; Robbins, *Hard Times in Paradise*, 72, 74, 79; D. Smith, *Silver Saga*, 195–96; Atkins, "Everything but the Mortgage," 59; Krznarich, oral history tape 863; *Jerome Chamber of Commerce News Bulletin*, 3 Feb. 1936; Robbins, *Hard Times in Paradise*, 15, 87, 91; Spratt, *Thurber, Texas*, 124; Riney-Kehrberg, "In God We Trusted," 135–39, 286; D. E. Miller and Sharpless, *Kingdom of Coal*, 314; D. Smith, *Silver Saga*, 195–98; Heyman, "Oral History of Douglas," 194; Krznarich, oral history tape 862; Governor Hunt to Chairman of Reconstruction Finance Corp., 11 Aug. 1932, Governor's Office, Arizona, Governors' Files; Davies, *Main Street Blues*, 91.

20. D. E. Miller and Sharpless, *Kingdom of Coal*, 314; *Verde Copper News*, 16 Dec. 1930, 20 Mar. 1931, 3 Mar. 1933, 12 Jan., 13 Apr. 1934; D. Smith, *When Coal Was King*, 110; Sparkes, Collected Papers; Sparkes to Governor Stanford, 7 July 1938, Governor's Office, Arizona, Governors' Files; Yavapai Portfolio, 1942, Hayden Library, Arizona State University, Tempe; Robbins, *Hard Times in Paradise*, 72, 87;

Riney-Kehrberg, "In God We Trusted," 135–39; Spence, *Wood River or Bust*, 220; Davies, *Main Street Blues*, 92.

21. Robbins, *Hard Times in Paradise*, 89–90; Atkins, "Everything but the Mortgage," 59; D. Smith, *When Coal Was King*, 110; D. E. Miller and Sharpless, *Kingdom of Coal*, 313; Riney-Kehrberg, "In God We Trusted," 135–39; Krznarich, oral history tape 862.

22. *Verde Copper News*, 12 Dec. 1930, 4 Dec. 1931; Robbins, *Hard Times in Paradise*, 11, 158. While doing research at Bisbee, Arizona, fifteen years after Phelps Dodge suspended major operations there, the author discussed making a living in the local economy with County Recorder Christine Rhodes. She mentioned the common practice of surviving by holding several part-time jobs rather than one full-time job. The author then witnessed an example of this practice at the local Chinese restaurant, eating there on three different nights, served by three different waiters, each of whom was the only waiter on duty that evening.

23. Stewart Udall wrote in the foreword to the 1989 reprint of the *WPA Guide to Arizona* that "one reason the Great Depression did not cut a wide swath in Arizona was that, by national standards, Arizona's economy was chronically depressed" (v). Davies, *Main Street Blues*, 93; Krznarich, oral history tape 862; Robbins, *Hard Times in Paradise*, 72, 88–89.

24. Buss, *Mass Unemployment*, 32, 73, 164–65. Buss and Redburn have found that a "web of personal relationships not only offers emotional comfort but also constitutes an informal system of economic exchange and employment," and that "our studies offer compelling evidence that the informal social support network ... may have lessened the burden of mass unemployment for many workers" (*Shutdown at Youngstown*, 35–36, 116). Miller and Sharpless have discovered that "hard times brought people closer together, often breaking down long-standing ethnic divisions" (*Kingdom of Coal*, 311, 320). Atkins, "Everything but the Mortgage," 63–64; Muretic, *Heaven in the West*, 35; Macia, "Tombstone in the Early Days," 10 (emphasis hers).

25. *Verde Copper News*, 1 Apr., 5 Apr., 8 Apr., 22 Apr., 19 Aug., 23 Aug., 18 Oct. 1932.

26. St. Paul's Episcopal Church, Tombstone, Record Book and Minutes; Willson, *Mimes and Miners*, 59–60, 78; Riney-Kehrberg, "In God We Trusted," 160–62; *Tombstone Prospector*, 21 Aug. 1890; *Verde Copper News*, 26 Dec. 1930, 27 Nov. 1931, 15 Jan. 1932, 3 Feb., 31 Mar., 20 Oct. 1933; *Prescott Courier*, 12 Nov. 1937.

27. *Verde Copper News*, 17 Dec., 21 Dec. 1926, 18 May, 16 Oct. 1928.

28. *Verde Copper News*, 21 Nov., 28 Nov., 5 Dec. 1930, 14 Apr., 1 Dec. 1931. The

district also had a Verde District Welfare Federation, organized in 1929. This umbrella organization for the area's service clubs used local benefits to raise money to provide eyeglasses, school lunches, or medical operations to needy children. *Verde Copper News*, 28 Jan., 31 Jan., 12 Dec. 1930.

29. *Verde Copper News*, 6 Feb., 5 June, 24 July, 28 July, 23 Oct., 27 Nov., 1 Dec. 1931, 12 Apr. 1932; Ling to Watson, 2 Feb. 1932, Ling Collection; Riney-Kehrberg, "In God We Trusted," 149–51, 154–55; Robbins, *Hard Times in Paradise*, 81, 84.

30. *Verde Copper News*, 25 Oct., 1 Nov. 1932, 17 Nov. 1933. In McGill, Nevada, the resident copper company worked through the local Red Cross chapter during the Depression to supply milk, flour, and other essentials to the needy. The company permitted families to remain in company housing while deferring rent, coal, water, and electricity payments (Elliott, *Growing Up*, 138).

31. *Prescott Courier*, 20 June 1935, 16 July 1936, 12 July 1939; *Jerome Chamber of Commerce News Bulletin*, 18 May 1936.

32. *Tombstone Prospector*, 28 Mar. 1890, 14 Jan., 18 Jan. 1892.

33. *Verde Copper News*, 20 Sept. 1927, 4 Jan., 12 Apr. 1929, 29 Jan. 1932; *Tombstone Prospector*, 31 Mar. 1888; *Jerome Chamber of Commerce News Bulletin*, 10 Dec., 31 Dec. 1935; *Verde District Shopping News*, 21 July 1938; D. Smith, *When Coal Was King*, 113. Davies reports that the business districts of Camden, Ohio, and other communities used lotteries to arrest their declines. In Camden's case the idea was used in 1951 to lure people away from their television sets and downtown on what had previously been the town's busiest commercial evening. Receipts from local stores were used as tickets in a cash lottery held on Saturday nights. The idea—which produced some interest initially among Camden's townspeople but not much revenue for its businesses—was scrapped within a year (*Main Street Blues*, 161–62).

34. *Prescott Courier*, 9 Sept., 10 Dec. 1930, 21 Dec. 1931, 28 Mar. 1934, 26 July 1939; Governor's Office, Arizona, Proclamations, 23 Jan., 14 Mar. 1930, 8 June 1931, 26 Mar. 1932; Riney-Kehrberg, "In God We Trusted," 128–31, 292–93; Robbins, *Hard Times in Paradise*, 81.

35. *Verde Copper News*, 12 June 1931, 15 Nov., 18 Nov., 25 Nov. 1932, 11 Jan. 1935.

36. Governor Hunt to Chairman of Reconstruction Finance Corp., 11 Aug. 1932, Governor's Office, Arizona, Governors' Files; Prescott Salvation Army Circular, 20 Feb. 1936, ibid; Riney-Kehrberg, "In God We Trusted," 157–59.

37. *Tombstone Prospector*, 20 July, 17 Aug. 1889.

38. *Tombstone Prospector*, 15 Oct., 26 Oct. 1889, 4 May 1892.

39. *Prescott Courier*, 9 Dec. 1930. Governor Hunt claimed in the spring of 1931 that U.S. overproduction was the least of the problem and that a tariff would give domestic copper "a fair chance" for a market in the United States. By the end of that year, Hunt saw "little hope for revival ... until Congress accords us tariff protection against the cheap-labor metal of Africa, Mexico and South America" ("What Is the Matter with Copper Mining?"; Gov. Hunt to Gov. Pinchot of Pennsylvania, 19 Nov. 1931, Governor's Office, Arizona, Governors' Files).

40. *Prescott Courier*, 5 Apr. 1932, 7 Feb. 1933. The *Courier* also asked pointedly "if the federal government can enter the field of farming by buying up commodities and can loan China $50,000,000 with which to purchase farm products in this country, why not do the same with mining?" (6 June 1933).

41. *Verde Copper News*, 18 Aug. 1933; *Prescott Courier*, 23 Jan. 1935; United Verde Mine and Smelter Committees, Petition to President Franklin D. Roosevelt, 7 Sept. 1933, box 6, Governor's Office, Arizona, Governors' Files.

42. Governor Moeur telegram to General Johnson, 26 Dec. 1933; Moeur to President Roosevelt, 30 Mar. 1934; Chairman of Copper Tariff Board to Governor Stanford, 20 June 1939, all in Governor's Office, Arizona, Governors' Files; *Prescott Courier*, 12 June 1937, 22 Nov., 16 Dec., 20 Dec. 1939.

43. Yavapai Chamber of Commerce, Resolution, 5 Sept. 1933, in Sparkes, Collected Papers; Governor Stanford telegram to President Roosevelt, 7 Dec. 1938; and Governor of Nevada's Telegram to Western Governors, 14 June 1939, both in Governor's Office, Arizona, Governors' Files; *Prescott Courier*, 10 Mar. 1934; *Verde Copper News*, 29 Dec. 1933; *Prescott Courier*, 22 Dec. 1933, 22 May 1934, 11 Apr., 3 May 1935; WPA Writers' Program, Manuscripts, box 8.

44. *Verde Copper News*, 1 Sept. 1933; *Prescott Courier*, 26 Aug., 14 Oct. 1932, 21 May 1935, 28 Aug. 1936.

45. *Prescott Courier*, 2 July, 5 Apr. 1932, 20 Feb., 30 Mar., 1 Oct. 1934.

46. Dembo, "Pacific Northwest Lumber Industry," 51; Lowitt, *New Deal and the West*, 115, 117, 121.

47. *Verde Copper News*, 3 Mar. 1931, 22 July 1932; Maxwell, "Depression in Yavapai County," 212, 219. The *Verde Copper News* printed the text of a bulletin from the Arizona director of President Hoover's committee on unemployment relief. The director concluded his bulletin by writing something of an epitaph for the Hoover administration: "The county directors," he admonished, "are warned to be very careful about approving or recommending wild or untried schemes for relief. ... It is much better to move slowly in this problem than it is to dissipate funds on unwise plans. Every attempt should be made to provide work instead of

338 *Notes to Pages 242–248*

the dole. I feel that we should move slowly and surely in work of this character and stick to policies that are fundamentally sound" (25 Sept. 1931).

48. *Prescott Courier,* 5 Aug., 11 Aug., 1 Sept., 1 Dec. 1933, 9 Feb. 1934, 21 Aug. 1935; N. Smith, "Jerome," 14; Sparkes, Collected Papers, box 2; Maxwell, "Depression in Yavapai County," 221–23; box 8, Governor's Office, Arizona, Governors' Files; *Verde Copper News,* 2 June, 27 Oct., 29 Dec. 1933, 9 Feb. 1934; Sparkes, Collected Papers.

49. *Prescott Courier,* 14 Nov. 1932, 3 May, 7 Dec., 15 Dec. 1933, 6 June 1936, 11 May 1938, 25 July 1939; N. Smith, "Jerome," 14; Leonard Arrington, "Arizona in the Great Depression Years," vertical file, Special Collections, Northern Arizona University, Flagstaff, 17; "Review of Activities Under Federal Civil Works Administration in the State of Arizona," and FERA Data, in Sparkes, Collected Papers.

50. *Verde Copper News,* 21 Nov., 25 Nov. 1930, 2 Aug., 20 Sept. 1932; Connolly Collection.

51. David F. Myrick, "The Railroads of Southern Arizona: An Approach to Tombstone," 165; *Arizona Weekly Citizen,* 17 Apr. 1886.

52. *Tombstone Prospector,* 13 Apr., 17 Apr., 6 June, 14 June, 15 June 1888.

53. *Tombstone Prospector,* 7 June, 17 June 1888; Town of Tombstone, Minutes of Town Council, 6 June, 15 June 1888; *Arizona Weekly Citizen,* 16 June 1888.

54. Myrick, "Railroads of Southern Arizona," 166; *Tombstone Prospector,* 23 June, 24 June, 28 June, 29 June 1888.

55. *Tombstone Prospector,* 20 Oct., 24 Oct., 10 Nov. 1891.

56. Town of Tombstone, Minutes of Town Council, 8 Oct., 15 Oct. 1891; *Tombstone Prospector,* 10 Nov., 11 Nov., 24 Nov. 1891.

57. *Verde Copper News,* 15 Nov. 1927, 26 Apr., 3 Dec., 6 Dec. 1929.

58. "Review of Activities under Federal Civil Works Administration in the State of Arizona," in Sparkes, Collected Papers, 24; *Verde Copper News,* 9 May, 28 Nov. 1930, 13 Jan., 6 Mar., 5 June 1931, 5 Jan., 2 Feb. 1932.

59. Yavapai County Chamber of Commerce and Flagstaff Chamber of Commerce to Regional Director, U.S. Forest Service, Albuquerque, 15 June 1933; Verde District Kiwanis Club Telegrams to Chairman of Arizona Highway Commission and Highway Commissioner Ray N. Vyne, 15 June 1933, both in Sparkes, Collected Papers.

60. London to Sparkes, 20 July 1933; "State Highway 79 Association, Minute Book, 1933–1939; meetings of 16 Aug. 1933, 19 Mar. 1935, undated membership list, and membership listed dated 17 Mar. 1937, all in Sparkes, Collected Papers.

61. *Prescott Courier*, 21 Feb., 20 Sept. 1935, 28 May, 11 July, 12 Aug., 5 Sept. 1936, 16 Mar., 17 Mar., 21 July 1937, 4 Sept., 7 Oct. 1939; Highway 79 Association Minute Book, 30 Mar. 1938, and pamphlet "Highway 79 'The Greatest Scenic Drive in America,'" in Sparkes, Collected Papers; *Verde Copper News*, 17 Nov. 1933, 22 June 1934; *Jerome Chamber of Commerce News Bulletin*, 3 Dec. 1935; "Jerome, Arizona," vertical file, Special Collections, Northern Arizona University.

62. Dilsaver, "From Boom to Bust," 4; Francaviglia, "Copper Mining and Landscape Evolution," 277; D. Smith, *Rocky Mountain Mining Camps*, 7–8.

63. *Verde Copper News*, 22 July 1932; *Prescott Courier*, 16 May 1935; telegrams to the governor from all over Arizona describing conditions, Oct. 1937; Gov. Stanford to Sens. Hayden and Ashurst and Rep. Murdock, 10 Feb. 1938, both in Governor's Office, Arizona, Governors' Files. Booster organizations such as the Highway 79 Association besieged en masse the state agencies responsible for dispensing funds in bare-knuckled efforts to land as much money for local projects as possible. Governor Hunt wrote to the Arizona Good Roads Association on 12 May 1932 noting "with regret what I consider to be an over-assertion of local highway interest. This has found expression in open and violent hostility to the construction or improvement of any state highways which do not serve a particular region" (Governor's Office, Arizona, Governors' Files).

64. *Tombstone Republican* 15 Dec. 1883; *Tombstone Prospector*, 10 Nov. 1891; Hamilton, *The Resources of Arizona* (1884), 79.

65. Dilsaver, "From Boom to Bust," 259–62, 339, 392–97, 402–3; D. Smith, *Rocky Mountain Mining Camps*, 8–10, 45; Richard V. Francaviglia, "Bisbee, Arizona: A Mining Town Survives a Decade of Closure," 6–8; Spratt, *Thurber, Texas*, 125; Olien and Olien, *Oil Booms*, 168; Mann, *After the Gold Rush*, 39; D. Smith, *When Coal Was King*, 85; D. E. Miller and Sharpless, *Kingdom of Coal*, 323; Burke, "Tombstone School," 256; *Verde Copper News*, 21 May 1929; Faulk, *Tombstone: Myth and Reality*, 176; *Tombstone Prospector*, 17 May 1887, 31 Mar. 1888, 24 Sept., 17 Oct. 1889. The *Arizona Daily Star* informed its readers on 2 Mar. 1886 that the cattle trade seemed to be reviving Tombstone. The paper held that "the grazing interest of Cochise county, which is considerable, finds a natural center in Tombstone, and the agricultural interest of the upper San Pedro valley adds much to the future of the county seat of Cochise. The business men of Tombstone, while not doing so large a business as in the past, are doing a safer and more legitimate business. . . . Tombstone has got down to a solid conservative basis, and her future is as well fixed and determined as any town in Arizona."

66. *Verde Copper News*, 24 May 1932, 9 Nov. 1934; *Jerome Chamber of Commerce*

News Bulletin, 24 Dec. 1935, 11 May 1936; WPA Writers' Program, *WPA Guide to Arizona*, 87.

67. *Arizona Republic*, 28 June 1953; Caillou, *Experience Jerome*, 263–65; Pritchard, "Community," 34–36, 38–39.

68. *Tombstone Courthouse State Historic Park*; Richard Francaviglia, *Hard Places: Reading the Landscape of America's Historic Mining Districts*, 167, 177–80, 188, 198. Francaviglia notes that "even townscapes such as Tombstone, Arizona, came to confuse popular imagery of television westerns with reality" (198). Tombstone's merchants, once they had discovered their secret of success, were loath to risk interrupting the flow of wealth. When preservationists suggested restoring the town to a more historically accurate image, away from its movie-set appearance, they received a hostile reception from the business community. Comstock merchants went even further, covering the facades of their brick buildings with rough cedar boards during the 1960s to make the town look more like the Virginia City of the television western *Bonanza*. With that series long out of production, only the Bonanza Saloon still retains its wooden false facade. James, *Roar and the Silence*, 262, 268.

I term the most extreme manifestation of this denial of history "mining camp chic." These are towns—like Aspen and Telluride, Colorado, or Park City, Utah—so utterly transformed that they scarcely acknowledge their origins, now inhabited by wealthy celebrities. These people are usually horrified when someone suggests reopening a mine in the area.

69. *Arizona Daily Star*, 26 June 1891; *Tombstone Prospector*, 11 Jan. 1882; Stout, "Want to Buy a Ghost Town?"; Brogdon, "History of Jerome, Arizona," 154–56; Robbins, *Hard Times in Paradise*, 154; *Verde Independent*, 15 Apr. 1965.

70. D. Smith, *Rocky Mountain Mining Camps*, 8–10; *Tombstone Epitaph*, 5 May 1955. McGill, Nevada, survived the final curtailment of its mining operations in the 1970s by being named by the state legislature as the site for a new state prison, becoming the headquarters for the newly created Great Basin National Park nearby, and transforming the local mining railroad into a tourist railroad and museum (Elliott, *Growing Up*, 184–85, 189). While many of the towns in Idaho's Wood River District disappeared, Ketchum, Hailey, and Belleview survived by tapping the economic potential of stock raising, lumbering, tourism, skiing, fishing and hunting, hot springs resorts, and, lately, celebrities (Spence, *For Wood River or Bust*, 231).

71. *Tombstone Prospector*, 25 Apr. 1891, 4 May, 15 June 1892.

72. Barton has found that a busted community regained its equilibrium either through diversification of its economy, or through its residents' willingness to

accept a reduced standard of living. He concludes that if the decline was gradual, collective decision-making could help readjust the system (*Communities in Disaster*, 65).

Chapter Twelve. But a Monument to the Glories of the Past?

1. D. Smith, *Silver Saga*, 218; Spence, *Wood River or Bust*, 214; Traywick, *The Mines of Tombstone*, 17; "Tombstone Mines, Arizona," 919; R. B. Brinsmade, "Tombstone Arizona, Restored," 371.

2. D. D. Martin, *Tombstone's "Epitaph*," 260–61; Blake, *Tombstone and Its Mines*, 22.

3. "Mining Reports, Arizona, 1901–1930," 283–84; D. D. Martin, *Tombstone's "Epitaph*," 262; Blake, *Tombstone and Its Mines*, 19; "The Mining Revival at Tombstone," 314–15.

4. Blake, *Tombstone and Its Mines*, 19–20; "The Resurrection of Tombstone"; "Mining Revival at Tombstone," 315.

5. D. D. Martin, *Tombstone's "Epitaph*," 262–64, 267; "Mining Reports, Arizona, 1901–1930," 285; Traywick, *"Tombstone Epitaph" and Clum*, 4; Brinsmade, "Tombstone Arizona, Restored," 371–72; Sharlot M. Hall, "The Re-Making of an Old Bonanza," 243, 245–46; O. E. Young Jr., *Western Mining*, 278.

6. "Mining Reports, Arizona, 1901–1930," 321; D. D. Martin, *Tombstone's "Epitaph*," 267–68.

7. "Mining Reports, Arizona, 1901–1930," 357–59, 362, 378; D. D. Martin, *Tombstone's "Epitaph*," 270–72; Traywick, *The Mines of Tombstone*, 19.

8. Traywick, *The Mines of Tombstone*, 19, 21, 23, 25, 28, 36; Wayne Winters, "Tombstone Rises Again."

9. Mining Properties Folders, "Contention Consolidated," Arizona State Mining Museum, Phoenix; *Arizona Weekly Citizen*, 29 June 1984.

10. *Prescott Courier*, 20 Nov. 1929.

11. Andy Anderson, "Somewhere in Arizona, 1930," Biographical File, Arizona Historical Society, Tucson; U.S. Bureau of Census, *Sixteenth Census of the United States*, 91.

12. Traywick, *"Tombstone Epitaph" and Clum*, 5–6; *Prescott Courier*, 26 Oct. 1929; Faulk, *Tombstone: Myth and Reality*, 200, 203.

13. "Newsclippings—Tombstone," Arizona Ephemera Collection, Hayden Library, Arizona State University, Tempe; Paul L. Greer, "Tombstone Promotes Its Past to Ensure Its Future," 8; U.S. Bureau of Census, *1990 Census of Population*

and Housing . . . Arizona, 8–12; *Arizona Republic*, 12 Dec. 1993; Winters, "Tombstone Rises Again," 42; personal observations by the author. James reports that when the Comstock briefly resumed mining in 1979, there was considerable division of opinion about the wisdom of doing so. Locals saw it either as a resumption of the district's purpose or as a threat to their quality of life and their newer industry, tourism (*Roar and the Silence*, 269).

14. Alenius, "Brief History"; Yavapai Board of Supervisors to Yavapai County Attorney, 22 Apr. 1941; Larson to Nihell, 29 Feb., 28 June 1940, both in Larson Collection; "United Verde Extension Cavein, Property Damage and Lawsuits," 1, 8; and "Chronological History of Mining in Arizona," both in WPA Writers' Program, Manuscripts, box 8.

15. Larson to Nihell, 15 June, 4 Oct. 1941, 4 Nov., ? Nov. 1942; and Hulda Larson to Nihell, 6 Feb. 1942, all in Larson Collection. Hulda Larson noticed the same transience as had her husband, Joe. "Jerome is a pretty busy place," she wrote to Nihell, "miners coming and going. Most of them don't stay very long, as now they know they can get work in any camp they go to. Makes it hard on the Company though." The same labor shortage plagued the coal mines of Crested Butte, Colorado, and Madrid, New Mexico, throughout World War II. D. Smith, *When Coal Was King*, 115; Melzer, *Madrid Revisited*, 40.

16. McDonald, *Jerome, Arizona*, 21; Krznarich, oral history tape 862; Larson to Nihell, 9 Sept., ? Nov. 1942, 5 May 1943, 18 Jan. 1944, Larson Collection; Stout, "Want to Buy a Ghost Town?"; Brogdon, "History of Jerome, Arizona," 76–77; Muretic, *Heaven in the West*, 92; Ling to Turner, 29 Apr. 1944, Ling Collection.

17. Mining Properties Folders, "United Verde," Arizona State Mining Museum, Phoenix; "Jerome, Arizona," vertical file, Special Collections, Northern Arizona University.

18. Krznarich, oral history tape 862; Stout, "Want to Buy a Ghost Town?"; Jerome Historical Society, *Three Epochs of Jerome's History*, 1–2; "Story of the United Verde," United Verde Mine, Miscellaneous News Clippings, file folder "United Verde 2," Sharlot Hall Museum and Archives; *Prescott Courier*, 30 June 1976; Brewer, *Jerome*, 11.

19. Brewer, *Jerome*, 9–10; Stout, "Want to Buy a Ghost Town?"; *Verde Independent*, 15 Apr. 1965; Mining Properties Folders, "United Verde" and "United Verde Extension," Arizona State Mining Museum, Phoenix.

20. Jerome Historical Society, *Three Epochs of Jerome's History*, 3–8; *Arizona Republic*, 28 June 1953; "Jerome, Arizona," vertical file, Special Collections, Northern Arizona University; *Los Angeles Times*, 2 Feb. 1953; Evans, *Two Generations*, 89–90; *Verde Independent*, 15 Apr. 1965.

21. *Verde Independent,* 15 Apr. 1965; "Jerome, Arizona," vertical file, Special Collections, Northern Arizona University; John Fitzgerald and John Barton, *Jerome, Arizona: Yesterday and Today,* 2.

22. "Jerome," file folder 27A, Sharlot Hall Museum and Archives; "Jerome, Arizona," vertical file, Special Collections, Northern Arizona University.

23. Pritchard, "Community," 70–71; "Jerome," file folder 27, Sharlot Hall Museum and Archives.

24. Cochise County, Ariz., Great Registers, 1882, 1892; *Tombstone Prospector,* 7 June 1888, 12 May 1890, 13 Mar. 1920; Mann, *After the Gold Rush,* 89–91; D. Smith, *Silver Saga,* 194. Simon and Gagnon note in their study of three busted Illinois coal towns "a deep-seated resistance to social change of any real significance." They have found that only those children who stood to inherit businesses stayed behind, which meant that residents lacked real commitment to the future, and had no reason to rock the boat. The authors believe that these people might regard too much prosperity as a bad thing under those circumstances, as it might restore real competition against established businesses and elites ("Decline and Fall," 50).

25. D. Smith, *Mining America,* 21; Mark Twain, *Roughing It,* 309; Powell Greenland, *Hydraulic Mining in California: A Tarnished Legacy,* 54.

26. Bernard DeVoto, "The West: A Plundered Province," 358, 360.

27. Ibid., 360.

28. Bernard DeVoto, "Our Great West: Boom or Bust?" 47.

29. William G. Robbins, "The 'Plundered Province' Thesis and the Recent Historiography of the American West," 578–80, 582.

30. Richard D. Lamm and Michael McCarthy, *The Angry West: A Vulnerable Land and Its Future,* 5–6, 99.

31. Patricia Nelson Limerick, *The Legacy of Conquest: The Unbroken Past of the American West,* 87; Limerick, "Haunted by Rhyolite," 22–23.

32. Limerick, "Haunted by Rhyolite," 22–23, 34, 39.

33. My ideas about the meaning of the ghost town in the West have been developed from considering the statements of participants in nineteenth-century mining rushes, some of which have been included in this study; from Gerald Thompson, "The New Western History: A Critical Analysis"; and from conversations with fellow historian Richard D. Adkins.

As we have seen in numerous examples, it is also possible to declare failure too early, as did the eminent historian Michael Malone ("Collapse of Western Metal Mining"). He believed in 1986 that "with prices hovering at sixty cents per pound, copper is down, and down to stay." By the early 1990s the price of copper had rebounded to $1.15, and copper mining had resumed. Malone was certainly correct in his assertion that things would never be the same in western metal mining. They never are.

34. R. Myers, "Boom and Bust," 9; Robbins, *Hard Times in Paradise*, 5, 7, 168. Some authors have seen a more complex problem than East versus West, identifying intraregional colonialisms originating in San Francisco or Denver. Robbins holds the timber harvesting along the Oregon coast in the nineteenth century to be largely an intraregional colonialism by the lumber barons of San Francisco. In a brief sojourn in the Alaskan fishing industry, I learned about Alaskans' resentment of Seattle's heavy hand.

35. Lamm and McCarthy, *Angry West*, 5, 7, 99.

36. *Tucson and Tombstone Directory*, 114.

37. O. E. Young Jr., *Western Mining*, 288; Spence, "I Was a Stranger"; Zanjani, *Goldfield*, 62–63, 238; "Story of the United Verde," United Verde Mine, file folder "United Verde 2," Sharlot Hall Museum and Archives; R. Myers, "Boom and Bust," 9. For some famously first-class swindling, see also Rice, *My Adventures*.

38. Olien and Olien, *Oil Booms*, 170–71; D. Smith, *Silver Saga*, 248–49; O. E. Young Jr., *Western Mining*, 288; Faulk, *Tombstone: Myth and Reality*, 185.

39. DeVoto, "Our Great West: Boom or Bust?" 47.

40. Miller, Manuscript, 4, 8.

41. Suckertown, California, is mentioned in D. C. Miller, *Ghost Towns of California*, 161. Small Hopes, Colorado, is mentioned in Dallas, *Colorado and Utah Ghost Towns*, 4. Fatout has compiled a delightful list of names of California placer camps that includes disappointments like Humbug, Sucker Flat, Bunkumville, Dead Broke, Bummerville, and Boneyard (*Meadow Lake: Gold Town*, 4,9). Greenland records nine different Mother Lode placer camps with the word *humbug* in their names (*Hydraulic Mining in California*, 198).

42. Fatout, *Meadow Lake: Gold Town*, 4; Spence, *Wood River or Bust*, 2; Zanjani, *Goldfield*, 73; Elliott, *Growing Up*, 3; Hums, "The Attractions of Tombstone." Hums's observations were not unique. Orrin Hilton wrote that in Goldfield, Nevada, a man should "stand the inconveniences for awhile, make his wad, and then duck out" (Zanjani, "To Die in Goldfield," 66). George Featherstonhaugh wrote of the Mineral Point, Wisconsin, lead district in 1837 that "men do not always seem to select situations in that country with a view to living tranquilly and happily, but to try to find ready money by digging for it, or to live upon others; the moment they find there is no likelihood of success, they go to another place" (D. Smith, *Mining America*, 20). James Fergus wrote some years later of

Bannack City, Montana, that "Bannack was not supposed to be a settlement, but simply a mining camp where everyone was trying to get what he could, then go home" (38). Duane Smith argues that this transience contributed to the environmental carnage inflicted upon the American mining landscape.

43. *Arizona Weekly Citizen*, 16 June 1883; Miller, Manuscript, 1.

44. Clum, "Nellie Cashman," 23, 26–27; G. W. Parsons, Diaries, 19 July 1886; *Arizona Daily Star*, 18 Aug. 1892; *Tombstone Prospector*, 29 Apr., 6 May 1890; Rochlin, "Amazing Adventures," 284–86, 292–94. There are legions of sources about boomers upon which to draw; see *Tombstone Prospector*, 12 Jan., 23 Jan. 1889; *Arizona Weekly Citizen*, 9 Nov. 1887, 24 Sept., 8 Nov., 25 Nov. 1889, 4 Jan., 25 Jan., 25 Mar. 1890, 19 Dec. 1891; and *Arizona Weekly Citizen*, 24 July 1886, 15 Oct. 1887. For discussions of mobility, see Mann, *After the Gold Rush*, 89; Dilsaver, "From Boom to Bust," 399–400; D. Smith, *Silver Saga*, 246; D. E. Miller and Sharpless, *Kingdom of Coal*, 311; William S. Greever, *The Bonanza West: The Story of the Western Mining Rushes, 1848–1900*, 367; and Robbins, *Hard Times in Paradise*, 76. Edith Dorsey, born in Nevada and the daughter of a mining engineer, wrote to lifelong Tombstone resident Ethel Macia in March 1939: "It seems so strange that you have lived in Tombstone all this time—We have been the rolling stones and have lived in Mexico, Mont., Wyo., Wash., ... Idaho ... [and] California" (Reminiscence of Life in the 1880s, Biographical File, Arizona Historical Society, Tucson).

45. Some variation of this story is told in most books about Tombstone (Underhill, *Silver Tombstone*, 14, 52, 83). The quote is from Love, "The History of Tombstone," 5. Zanjani and her informants make similar observations (*Goldfield*, 182). The poet Robert Service celebrated this wanderlust in his poetry. See particularly "The Men That Don't Fit In," "The Wanderlust," and "The Spell of the Yukon," which contains the line: "Yet it isn't the gold that I'm wanting; so much as just finding the gold" (*The Best of Robert Service*, 1, 23, 84).

46. Elliott, *Nevada's Mining Boom*, 159.

47. Krznarich, oral history tape 862; *Los Angeles Times*, 2 Feb. 1953; United Verde Mine and Smelter Committees, Petition to President Roosevelt, 7 Sept. 1933, Governor's Office, Arizona, Governors' Files; Muretic, *Heaven in the West*, 95.

48. For a good account of tramp mining in the twentieth century, see Voynick, *Hardrock Miner*. Robbins, *Hard Times in Paradise*, 164. Buss and Redburn report unemployed steel workers "only slightly more likely" than the general population to express an intention to migrate. The authors believe that an elaborate social support network might be the reason for workers' reluctance to trade their

present locale, however depressed, for an unknown chance in an unfamiliar place (*Shutdown at Youngstown*, 103, 107).

49. *Verde Copper News*, 25 Sept., 28 Sept. 1928, 7 May, 17 Sept. 1929, 4 Mar. 1932, 11 Jan. 1935; Rodriquez, oral history tape 848. A number of Latinos—Jesus Perez and Octavio Placios among them—resided in the district for over thirty years.

50. Lopez, *Forever*, 17. James records of the Comstock's residents at the end of that district's last boom that "this was not a fickle boomtown population, ever ready to lose faith and switch attention to the next strike. It was a stable community supporting an industrial giant. Considerable investment, personal and corporate, emotional and economic, inspired most to cling to hope" (*Roar and the Silence*, 237).

51. Olien and Olien, *Oil Booms*, 100; D. Smith, *When Coal Was King*, 111; Miller and Sharpless, *Kingdom of Coal*, 311; D. Smith, *Silver Saga*, 246; Dan Lee, "Ghost Town Class Reunion."

Manuscript Sources

Alianza Hispano Americana. Acta de Instalacion de la Logia no. 13, 18 Apr. 1899. Hayden Library, Arizona State University, Tempe.

——. Libro de Actas de Logia 13, Aug. 1926–Dec. 1927, 5 Jan. 1930–9 Dec. 1932. Hayden Library, Arizona State University, Tempe.

——. Logia no. 13. Membership List, 1939. Hayden Library, Arizona State University, Tempe.

Ancient Order of United Workmen, Lodge no. 3. Elections and Miscellaneous Papers. Tombstone Courthouse State Historic Park, Tombstone, Ariz.

Anderson, Andy. "Somewhere in Arizona, 1930." Biographical File. Arizona Historical Society, Tucson.

"Arizona, Early Description." File Folder. Sharlot Hall Museum and Archives, Prescott.

Arrington, Leonard. "Arizona in the Great Depression Years." Vertical File. Special Collections, Northern Arizona University, Flagstaff.

Ashburn, Frank D. Biography of Endicott Peabody, Religion in Tombstone. Biographical File. Arizona Historical Society, Tucson.

Bacon, Terry. "Steins, New Mexico: A Railhead Town Abandoned by Progress." Vertical File. Special Collections, Northern Arizona University, Flagstaff.

Baxley, George C., comp. "Arizona State Firemen's Association Convention, 1930." Hayden Library, Arizona State University, Tempe.

Biographies in *Tombstone Prospector*. Hayden Library, Arizona State University, Tempe.

Biography-Obituary Files. Hayden Library, Arizona State University, Tempe.

Blackburn, C. W. Biographical File. Arizona Historical Society, Tucson.

Boller, Robert M. Reminiscence of Early 1880s. Biographical File. Arizona Historical Society, Tucson.

Boyer, Glenn N. "A Different Look at Some Pioneers." MS 87. Arizona Historical Society, Tucson.

Comstock, L. B. Correspondence. Special Collections, University of Arizona, Tucson.

Connolly, John P. Collection. MS 137. Special Collections, Northern Arizona University, Flagstaff.

Dorsey, Mrs. Edith Johnson. Reminiscence of Life in the 1880s. Biographical File. Arizona Historical Society, Tucson.

Drachman, Mary A. Collection. MS 225. Arizona Historical Society, Tucson.

First National Bank of Arizona. Ledgers and Account Books, 1860s–1948. First Interstate Bank Collection, MS COL 263. Special Collections, Northern Arizona University, Flagstaff.

Governor's Office, Arizona. Governors' Files. Boxes 4–13, 13A, 69, 116A, 120. Arizona State Department of Library and Archives, Phoenix.

——. Proclamations, 1912–1939. Arizona State Department of Library and Archives, Phoenix.

Gray, John Pleasant. "When All Roads Led to Tombstone." 1940. MS 312. Arizona Historical Society, Tucson.

Greg, Mike. "History of the Religious Growth of Tombstone, Arizona." 7 June 1960. MS 231. Arizona Historical Society, Tucson.

Haven Methodist Visitor 1 (Mar. 1945). "Jerome, Arizona" File. Special Collections, Northern Arizona University, Flagstaff.

Hayhurst, Hal LaMar. "On Tombstone Gas Company and Lights." MS 343. Arizona Historical Society, Tucson.

Historical Folder of Tombstone, Ariz. History, maps, historic buildings, and sites. Hayden Library, Arizona State University, Tempe.

Hums [pseud.]. "The Attractions of Tombstone." 1882. Arizona State Department of Library and Archives, Phoenix.

——. "Religious Observances in the West, Tombstone, Arizona." 1881. Arizona State Department of Library and Archives, Phoenix.

"Jerome." File folder 27. Sharlot Hall Museum and Archives, Prescott.

"Jerome, Arizona," "Jerome Historical Society, Inc.," "Jerome, History," and "Jerome State Historic Park." Vertical Files. Special Collections, Northern Arizona University, Flagstaff.

Jerome Historical Society. *Bring It All Back to Jerome: Proceedings of the Sixth Annual Historic Symposium . . . 27 Aug. 1983.* Arizona State Department of Library and Archives, Phoenix.

———. *They Came to Jerome: Proceedings of the Seventh Annual Historic Symposium, 25 Aug. 1984.* Arizona State Department of Library and Archives, Phoenix.

———. *Three Epochs of Jerome's History: Proceedings of the Fifth Annual Historic Symposium, 28 August 1982.* Arizona State Department of Library and Archives, Phoenix.

Jerome State Historic Park Collection. Phelps Dodge Corp., United Verde Branch Records, 1900–1952. Boxes 7–11. Box 13 contains inquest files for Jerome Precinct, 1932–1947. MS 198. Special Collections, Northern Arizona University, Flagstaff.

Johnson, Albert Jacob. Hayden File. Arizona Historical Society, Tucson.

Knights of Columbus. Verde Council, Jerome, Ariz., 1920s–1940s. MS COL 109. Special Collections, Northern Arizona University, Flagstaff.

Knights of Pythias. Miscellaneous Papers of Lodge no. 4, Tombstone, Arizona, 1884–1997. AZ 241. Special Collections, University of Arizona, Tucson.

Krznarich, John. Personal Recollections of Jerome, Ariz., 1930s and 1940s. Oral history tapes 862 and 863, doc. box 330. Sharlot Hall Museum and Archives, Prescott.

Lamb, Walter A. Biographical File. Arizona Historical Society, Tucson.

Larson, Joe. Collection. MS 129. Special Collections, Northern Arizona University, Flagstaff.

Lind, Helen Younge. Tombstone, 1881–1891. Biographical File. Arizona Historical Society, Tucson.

Ling, Perry. Collection, 1920s–1940s. MS 93. Special Collections, Northern Arizona University, Flagstaff.

MacNeil, Donald, and Bessie MacNeil. Personal and Business Correspondence, and Records of Cochise Hardware and Trading Co. MS 1039. Arizona Historical Society, Tucson.

McCool, Grace. Clip Book, articles by McCool on history of Tombstone and the like. Arizona Historical Society, Tucson.

McDonald, Lewis. Collection. Jerome Slide Series, Arizona State Parks. Videotape 73. Special Collections, Northern Arizona University, Flagstaff.

Miller, William N. Manuscript. MS 498. Arizona Historical Society, Tucson.

"Mine Production of Gold, Silver, Lead, and Zinc, Yavapai County, Arizona, 1880–1936." Ephemera Collection. Hayden Library, Arizona State University, Tempe.

"The Mining Manual." Mining Companies Operating in Arizona, 1936–1937. Produced by the Statistical Research Bureau, San Francisco. Excerpts at Arizona State Department of Library and Archives, Phoenix.

Mining Properties Folders. "Contention Consolidated," "Grand Central," "Tombstone Consolidated," "Tombstone Mill and Mining Company," "United Verde," "United Verde Extension," and "Verde Central." Arizona State Mining Museum, Phoenix.

"Mining Reports, Arizona, 1901–1930." Hayden Library, Arizona State University, Tempe.

"Miscellaneous News Articles Plus Photos." File folder 27A. Sharlot Hall Museum and Archives, Prescott.

Monmonier Family. Papers, 1878–1968. AZ 371. Special Collections, University of Arizona, Tucson.

Mountain States Telephone and Telegraph Company. Telephone Directories, 1930–1940. Microfilm. Arizona State Department of Library and Archives, Phoenix.

"Newsclippings–Tombstone." Arizona Ephemera Collection. Hayden Library, Arizona State University, Tempe.

Newspaper and Ephemera File. Arizona Historical Society, Tucson.

Noon, Adolphus. "Visit to Tombstone City." Manuscript copy of letter to *Chicago Tribune*. MS 590. Arizona Historical Society, Tucson.

Palmquist, Robert F. "The Fight for the Tombstone Townsite." Palmquist Ephemera File. Arizona Historical Society, Tucson.

Parsons, George W. Diaries. MS 645. Arizona Historical Society, Tucson.

"Peabody Tells of Tombstone." 26 Jan. 1941. Tombstone General History File. Arizona Historical Society, Tucson.

Peterson, Andrew Dominic. "A Tribute to Father John." Hayden Library, Arizona State University, Tempe.

Refermat, Thomas. "Transportation in Jerome." Hayden Library, Arizona State University, Tempe.

Rockfellow, John A. Papers, 1879–1935. MS 694. Arizona Historical Society, Tucson.

Rodriques, Rafaila. Personal Recollections. Oral history tapes 848 and 849, Doc. box 330. Sharlot Hall Museum and Archives, Prescott.

Schuster, Mrs. Dosa. "The Last of the Faithful: Society Notes of Old Tombstone Days." Oral history TS, 8–9 Mar. 1937. Tombstone General History File. Arizona Historical Society, Tucson.

Smith, Nancy R. "Hispanic Churches in Jerome." Hayden Library, Arizona State University, Tempe.

Sparkes, Grace M. Collected Papers, 1904–1953. MS S67. Hayden Library, Arizona State University, Tempe.

——. Correspondence, Papers, etc., Related to Mining and the Reconstruction Finance Corp. MS 752. Arizona Historical Society, Tucson.

St. Paul's Episcopal Church, Tombstone. Record Book and Minutes, Nov. 1881–Apr. 1896. AZ 358. Special Collections, University of Arizona, Tucson.

Sullivan Hotel. Sullivan Hotel of Jerome Collection. MS COL 108. Special Collections, Northern Arizona University, Flagstaff.

Talbot, Rev. Mr. "St. Paul's Episcopal Church, Tombstone, Arizona." Hayden Library, Arizona State University, Tempe.

Tombstone Consolidated Mines. Bankruptcy Papers, 1911. MS 506. Hayden Library, Arizona State University, Tempe.

United Verde Copper Company and United Verde Extension Copper Company. Records, 1900–1940. MS COL 199. Special Collections, Northern Arizona University, Flagstaff.

United Verde Extension. Employee Records. (As of this research, these were incorrectly indexed as the United Verde Collection; they are UVX records.) Doc. boxes 271–72. Sharlot Hall Museum and Archives, Prescott.

United Verde Mine. Miscellaneous News Clippings. File folders, "United Verde 1" and "United Verde 2." Sharlot Hall Museum and Archives, Prescott.

Warnekros, Paul B. Biographical File. Arizona Historical Society, Tucson.

Whitmore, William Vincent. Biographical Sketch of Dr. Goodfellow of Tombstone, 1880–91. MS 858. Arizona Historical Society, Tucson.

Willson, C. E. "From Variety Theater to Coffee Shoppe: History of the Bird Cage Theater, Tombstone, to 1934." Biographical File. Arizona Historical Society, Tucson.

Wolf, James G. "Story of James G. Wolf." Interview with Edward J. Kelley, July–Nov. 1937. MS 880. Arizona Historical Society, Tucson.

Wood, R. W. Collection. Tombstone Courthouse State Historic Park, Tombstone, Ariz.

WPA Writers' Program. Manuscripts. Boxes 6–8. Arizona State Department of Library and Archives, Phoenix.

"Yavapai County, Description of, c. 1940." Doc. box 14. Sharlot Hall Museum and Archives, Prescott.

Yavapai County Chamber of Commerce. "Mine Production . . . Yavapai County, 1880–1936." Hayden Library, Arizona State University, Tempe.

Yavapai Portfolio, 1942. Pamphlet, *Desert Magazine* (Feb. 1942). Ephemera Collection. Hayden Library, Arizona State University, Tempe.

Government Documents

Arizona Corporate Commission. "History of Huachuca Water Co." Dockett no. 1331-E-104, decision no. 1418, 23 Dec. 1921. Tombstone Courthouse State Historic Park, Tombstone, Ariz.

Arizona State Board of Health. *Annual Report of Vital Statistics, 1933.* Phoenix: n.p., 1933.

——. *Arizona Public Health News* 35-147 (Oct. 1925–June 1938).

——. *Biennial Report of Vital Statistics, 1930–1931.* Phoenix: Sims Printing, 1932.

Arizona State Mine Inspector. *Annual Reports.* [Phoenix]: Board of Control, 1921–1954.

Cochise County, Ariz. Census, 1882. Arizona State Department of Library and Archives, Phoenix.

——. Delinquent Tax Roll, 1882–1886, inclusive. Cochise County Recorder's Office, Bisbee.

——. District Court, Criminal Register of Actions. Vol. 1 (1881–1907). Arizona State Department of Library and Archives, Phoenix.

——. District Court, Judgements. Vol. 1 (18 Apr. 1881–8 June 1885). Arizona State Department of Library and Archives, Phoenix.

——. Great Register of Cochise County, 1900. Special Collections, University of Arizona.

——. Great Registers of Cochise County, 1882, 1884. Arizona State Department of Library and Archives, Phoenix.

——. Great Registers of Cochise County, 1884, 1886, 1890, 1892. Arizona Historical Society, Tucson.

——. Index to Deeds of Mines, Cochise County. Vols. 1–2 (1881–1904). Cochise County Recorder's Office, Bisbee.

——. Index to Leases, Cochise County. Vol. 1 (1881–1931). Cochise County Recorder's Office, Bisbee.

——. Index to Maps, Cochise County. Vol. 1 (1881–). Cochise County Recorder's Office, Bisbee.

———. Superior Court, Civil, Divorces, no. 2-6522, 1881–1892. Box 92. Arizona State Department of Library and Archives, Phoenix.

———. Superior Court, Civil, Divorces Not Granted, no. 90-6509, 1886–1912. Box 127. Arizona State Department of Library and Archives, Phoenix.

———. Superior Court, Marriage Records, 1881–1978. Arizona State Department of Library and Archives, Phoenix.

———. Tax Collector, Licence Fees Collected, 1883–1888. Arizona State Department of Library and Archives, Phoenix.

Jerome, Town of. Minutes of Town Council, 1929–1939. City Hall, Jerome, Ariz.

———. Police Court Dockets, 1932–1937. City Hall, Jerome, Ariz.

Pima County, Ariz. Great Register of Pima County, 1880. Arizona State Department of Library and Archives, Phoenix.

Tombstone, Town of. City Report. Report of Chief of Police, License and Tax Collection for the Month of May 1894. Tombstone Courthouse State Historic Park, Tombstone, Ariz.

———. City Report. Summary of Business Transactions during the Month of March 1882. Tombstone Courthouse State Historic Park, Tombstone, Ariz.

———. Minutes of Town Council, 1881–1909. Arizona State Department of Library and Archives, Phoenix.

———. Record of Warrants Issued, 5 Jan. 1882–16 Nov. 1910. Tombstone Courthouse State Historic Park, Tombstone, Ariz.

U.S. Bureau of the Census. *Eighteenth Decennial Census of the United States: Census of Population, 1960*. Vol. 1. Washington D.C.: United States Government Printing Office, 1961.

———. *Fifteenth Census of the United States*. Vol. 3. Washington D.C.: United States Government Printing Office, 1932.

———. *Fourteenth Census of the United States*. Vol. 3. Washington D.C.: United States Government Printing Office, 1922.

———. *1950 Census of Population*. Vol. 2. Washington D.C.: United States Government Printing Office, 1953.

———. *1990 Census of Population and Housing . . . Arizona*. Washington D.C.: United States Government Printing Office, 1993.

———. *Sixteenth Census of the United States*. Vol. 1. Washington D.C.: United States Government Printing Office, 1942.

———. *Statistics of the Population of the United States at the Tenth Census*. Vol. 1. Washington D.C.: United States Government Printing Office, 1883.

———. Territory of Arizona. Tenth Census Enumeration Volume [1880]. Arizona State Department of Library and Archives, Phoenix.

——. Territory of Arizona. Twelfth Census Enumeration Volume [1900]. Arizona State Department of Library and Archives, Phoenix.

——. *Thirteenth Census of the United States: Abstract of the Census . . . with Supplement for Arizona.* Washington D.C.: United States Government Printing Office, 1913.

U.S. Strategic Bombing Survey. *The Effects of Strategic Bombing on German Morale.* Washington D.C.: United States Government Printing Office, 1947.

——. *The Effects of Strategic Bombing on Japanese Morale.* Washington D.C.: United States Government Printing Office, 1947.

Yavapai County, Ariz. General Index to Mines, 14 and 15 (1924–1938). Yavapai County Recorder's Office, Prescott.

——. Justice of the Peace, Monthly Records, 1928–1934. Doc. box 113B. Sharlot Hall Museum and Archives, Prescott.

——. Yavapai County, Superior Court, Civil, 1864–1972. Arizona State Department of Library and Archives, Phoenix.

——. Yavapai County, Tax Collector, Cash Books, 1922–1937. Arizona State Department of Library and Archives, Phoenix.

——. Yavapai County, Taxes, Miscellaneous Letters, 1895–1935. Doc. box 190. Sharlot Hall Museum and Archives, Prescott.

——. Yavapai County Assessor, Assessment and Tax Rolls, 1919–1959. Arizona State Department of Library and Archives, Phoenix.

——. Yavapai County Assessor's Office, Automobile Registrations, 1920–1932. Arizona State Department of Library and Archives, Phoenix.

——. Yavapai County Superior Court, 1865–1939. Arizona State Department of Library and Archives, Phoenix.

——. Yavapai County Treasurer. Cash Books, 1927–1934. Arizona State Department of Library and Archives, Phoenix.

——. Yavapai County Treasurer. Property Taxes Paid, 1924–1932. Delinquent Tax Records, Tax Bills, Tax Suits, 1923–1933. SB32, box 5. Arizona State Department of Library and Archives, Phoenix.

Maps

Bufkin, Don. "Tombstone, Arizona, and Vicinity, and the San Pedro River Mills." 1:84,480 and 1:63,360, 1968.

Edwards, W. S. "Tombstone Mining District and Richmond Townsite and Waterworks." 1:79,200, 1879.

Gilchriese, John D. "Tombstone, Arizona, 1881–82." 1:1500, 1971.

Howe, E. G. "Cochise County, Arizona." 1:380,160, 1902.

"Jerome, Arizona." One inch equals fifty feet, 1927.

Sanborn Map Company. "Sanborn Fire Insurance Map." Tombstone, Aug. 1882. Photocopy at Arizona Historical Society, Tucson.

——. "Sanborn Fire Insurance Maps." Teaneck, N.J.: Chadwyck Healy Co. 1983. Microfilm. Map of Tombstone, July 1886, Jan. 1889, May 1904, and Jan. 1909. Map of Jerome, July 1924 and Dec. 1938.

U.S. Geological Survey. "Tombstone Mining Map, Cochise County, Arizona." 1:6,000, 1908.

Newspapers

Arizona Daily Star (Tucson), 1879–1892, 1910, 1911, 1938, 1941, 1942, 1969.

Arizona Journal Miner (Prescott), 1880.

Arizona Mining Index (Tucson), 1883–1886.

Arizona Republic[an] (Phoenix), 1938–1993.

Arizona Weekly Citizen (Tucson), 1878–1984.

Arizona Weekly Gazette (Tucson?), 1882.

Bisbee Gazette, 1987.

Daily Record-Epitaph (Tombstone), 1885.

Daily Tombstone, 1885–1886.

Holbrook Tribune, 1921.

Jerome Chamber of Commerce News Bulletin, 1935–1936.

Los Angeles Times, 1917, 1953.

Prescott Courier, 1930–1939, 1976, 1980, 1985.

Tombstone Epitaph, 1881–1955.

Tombstone Nugget, 1879–1882.

Tombstone Prospector, 1887–1920.

Tombstone Republican, 1883–1884.

Verde Copper News (Jerome), 1926–1935.

Verde District Shopping News (Jerome), 1938–1942.

Verde Independent (Cottonwood), 1965.

Secondary Sources

Aaron, Samuel. "An Arizona Pioneer: Memoirs of Sam Aaron." Ed. Jacob R. Marcus. *American Jewish Archives* 10 (Oct. 1958): 95–120.

Acuña, Rudolfo. *Occupied America: The Chicano's Struggle Toward Liberation.* San Francisco: Canfield Press, 1972.

Alcorn, Richard S. "Leadership and Stability in Mid-Nineteenth Century America." *Journal of American History* 61 (1974): 685–702.

Alenius, E. M. J. "A Brief History of the United Verde Open Pit, Jerome, Arizona." *Arizona Bureau of Mines, Bulletin 178* (1968): 1–34.

"Arizona: Greatest Producer of Copper in the United States." *Yavapai Magazine* 3, no. 2 (Mar. 1924): 3–8.

"An Arizona Pioneer: Teamster, Miner, and Mayor of Tombstone." *Western States Jewish Historical Quarterly* 21 (July 1989): 307–9.

Atherton, Lewis E. *Main Street on the Middle Border.* Bloomington: Indiana University Press, 1954.

Atkins, Annette Marie. "Everything but the Mortgage: The Response to Agricultural Disaster in Minnesota, 1873–1878." Ph.D. diss., Indiana University, 1981.

Axford, Joseph. *Around Western Campfires.* Tucson: University of Arizona Press, 1969.

Baker, T. Lindsay. *Ghost Towns of Texas.* Norman: University of Oklahoma Press, 1986.

Bamford, Lawrence. "Streets from Silver: Leadville's Early History through Its Built Environment." *Colorado Heritage,* no. 4 (1987): 2–11.

Barkdull, Tom. "Nellie Cashman." *Old West* (summer 1980): 26–28.

Barton, Allen H. *Communities in Disaster: A Sociological Analysis of Collective Stress Situations.* Garden City, N.Y.: Doubleday, 1969.

Baskin, John. *New Burlington: The Life and Death of an American Village.* New York: W. W. Norton, 1976.

Baumann, Jules. "Tombstone, Arizona." *Mining and Scientific Press* 59 (23 Nov. 1889): 399.

Beito, David T. *From Mutual Aid to the Welfare State: Fraternal Societies and Social Services, 1890–1967.* Chapel Hill: University of North Carolina Press, 2000.

Bird, Allan G. *Bordellos of Blair Street: The Story of Silverton, Colorado's Notorious Red Light District.* Grand Rapids, Mich.: Other Shop, 1987.

Bishop, William Henry. *Across Arizona in 1883: Including Glimpses of Yuma, Tombstone, Tucson.* Olympic Valley, Calif.: Outbooks, 1977.

Blackburn, George M., and Sherman L. Ricards. "The Chinese of Virginia City, Nevada, 1870." *Amerasia Journal* 7 (spring 1980): 51–68.

———. "The Prostitutes and Gamblers of Virginia City, Nevada: 1870." *Pacific Historical Review* 48 (1979): 235–59.

Blake, William P. *Tombstone and Its Mines: A Report upon the Past and Present Condition of the Mines of Tombstone, Cochise County, Arizona.* New York: Cheltenham Press, 1902.

"Blinn's Lumber Yard." *Arizona Quarterly Illustrated* 1 (Jan. 1881): 23.

Blumin, Stuart M. *The Urban Threshold: Growth and Change in a Nineteenth-Century American Community.* Chicago: University of Chicago Press, 1976.

Bowden, Martin John. "The Dynamics of City Growth: An Historical Geography of the San Francisco Central District, 1850–1913." Ph.D. diss., University of California at Berkeley, 1967.

Brewer, James. *Jerome: A Story of Mines, Men, and Money.* Globe, Ariz.: Southwest Monuments Association, 1967.

Briar, Katharine H. *The Effect of Long-Term Unemployment on Workers and Their Families.* San Francisco: R and E Research Associates, 1978.

Briegel, Kaye Lynn. "Alianza Hispano-Americana, 1894–1965: A Mexican-American Fraternal Insurance Society." Ph.D. diss., University of Southern California, 1974.

Brinsmade, R. B. "Tombstone Arizona, Restored." *Mines and Minerals* 28 (1907): 371–74.

Brogdon, John Carl. "History of Jerome, Arizona." Master's thesis, University of Arizona, 1952.

Burgess, Opie Rundle. "Quong Kee: Pioneer of Tombstone." *Arizona Highways* 25 (July 1949): 14–17.

Burke, Matia McClelland. "The Beginnings of the Tombstone School." *Arizona and the West* 1 (fall 1959): 248–57.

Burkhardt, C. J. "A Mining Town's Ups and Downs." *Desert* (July 1979): 24.

Burns, Nancy. "The Collapse of Small Towns on the Great Plains: A Bibliography." *Emporia State Research Studies* 31 (summer 1982): 1–36.

Buss, Terry F. *Mass Unemployment: Plant Closings and Community Mental Health.* Beverly Hills: Sage Publications, 1983.

Buss, Terry F., and F. Stevens Redburn. *Shutdown at Youngstown.* Albany: State University of New York Press, 1983.

Butler, Anne M. *Daughters of Joy, Sisters of Misery: Prostitutes in the American West.* Urbana: University of Illinois Press, 1985.

Byington, Margaret. "The Family in a Typical Mill Town." *American Journal of Sociology* 14 (1909): 648–59.

Byrkit, James W. "The Word on the Frontier: Anglo Protestant Churches in Arizona, 1859–1899." *Journal of Arizona History* 21 (spring 1980): 63–86.

Caillou, Aliza, ed. *Experience Jerome and the Verde Valley.* Sedona, Ariz.: Thorne Enterprises, 1990.

Canty, J. Michael, and Michael N. Greeley. *History of Mining in Arizona.* Tucson: Mining Club of the Southwest Foundation, 1987.

Carnes, Mark C. *Secret Ritual and Manhood in Victorian America.* New Haven: Yale University Press, 1989.

Cartmell, Windy. "A Breezy Letter on the Verde District." *Yavapai Magazine* 20 (Mar. 1930): 5–8.

Cassaday, L. W. "The Economics of a One-Industry Town." *Arizona Business and Economic Review* 3 (Dec. 1954): 1–5.

Chamberlain, D. S. "Tombstone in 1879: The Lighter Side." *Journal of Arizona History* 13 (winter 1972): 229–34.

Chapman, Mary Margaret. "Boomtown in Transition: Changing Patterns of Local Decision Making." Ph.D. diss., University of Colorado at Denver, 1982.

Church, John A. "Concentration and Smelting at Tombstone." *Transactions American Institute of Mining Engineers* 15 (1887): 601–13.

——. "The Tombstone, Arizona, Mining District." *Engineering and Mining Journal* 73 (26 Apr. 1902): 584–85.

——. "The Tombstone, Arizona, Mining District." *Transactions American Institute of Mining Engineers* 33 (1903): 3–37.

Clum, John. "Nellie Cashman." *Arizona Historical Review* 3 (Jan. 1931): 9–34.

Colorado, New Mexico, Utah, Nevada, Wyoming, and Arizona Gazetteer and Business Directory. Chicago: R. L. Polk, 1884; micropaque, Louisville, Ky.: Lost Cause Press, 1968.

Cook, Nancy. "Cleary." *Alaska Journal* 6 (spring 1976): 106–12.

Coonfield, Ed. "Bullfrog." *Nevada Magazine,* no. 2 (1976): 20–22.

Cooper, Evelyn. "C. S. Fly of Arizona: The Life and Times of a Frontier Photographer." *History of Photography* 13 (Jan.–Mar. 1989): 31–47.

Cortese, Charles F., and Bernie Jones. "The Sociological Analysis of Boomtowns." *Western Sociological Review* 8 (1977): 76–90.

Cosulich, Bernice. "Mr. Douglas of Arizona: Friend of Cowboys and Kings." *Arizona Highways* 29 (Sept. 1953): 2–11.

Cottrell, W. F. "Death by Dieselization." *American Sociological Review* 16 (June 1951): 358–65.

Covington, E. Gorton. "Safford Flared but Briefly." *Nevada Magazine,* no. 4 (1976): 36–37.

Crane, James M. "Analysis of the Great Register of Cochise County, Arizona Territory, 1884." *Cochise Quarterly* (fall 1988): 3–9.

Dallas, Sandra. *Colorado and Utah Ghost Towns and Mining Camps.* Norman: University of Oklahoma Press, 1985.

Davenport, Maynard. "An Analysis of the Capacity and Utilization of the Jerome School Plant." Master's thesis, Northern Arizona University at Flagstaff, 1941.

Davies, Richard O. *Main Street Blues: The Decline of Small Town America.* Columbus: Ohio State University Press, 1998.

Davis, J. F. "A Formal Interpretation of the Theory of Relative Deprivation." *Sociometry* 22 (1959): 280–96.

Davis, Ronald L., and Harry D. Holmes, eds. "Studies in Western Urbanization." *Journal of the West* 13 (July 1974): 61–73.

Dembo, Jonathan. "The Pacific Northwest Lumber Industry during the Great Depression." *Journal of the West* 24 (Oct. 1985): 51–61.

Devere, Jeanne. "The Tombstone Bonanza, 1878–1886." *Arizoniana* 1 (fall 1960): 16–20.

DeVoto, Bernard. "Our Great West: Boom or Bust?" *Collier's* (25 Dec. 1953): 46–63.

——. "The West: A Plundered Province." *Harper's* 169 (Aug. 1934): 355–64.

Dickson, Christina Ellen, and Robert Henry Dickson. *Dickson Saga: Story of Our Married Life.* Sun City, Ariz.: n.p., 1970.

Dilsaver, Lary Michael. "From Boom to Bust: Post Gold-Rush Patterns of Adjustment in a California Mining Region." Ph.D. diss., Louisiana State University, 1982.

Disturnell, William C., comp. *Arizona Business Directory and Gazetteer.* San Francisco: W. C. Disturnell, 1881.

"Dividends Paid in the Tombstone District." *Arizona Quarterly Illustrated* 1 (Jan. 1881): 12.

Doucette, F. E., comp. *Arizona Year Book, 1930–1931.* Phoenix: n.p., 1930.

"The Downfall of Tombstone." *Arizona Graphic* (10 Feb. 1900).

Doyle, Don Harrison. *The Social Order of a Frontier Community: Jacksonville, Illinois, 1825–70.* Urbana: University of Illinois Press, 1978.

——. "Social Theory and New Communities in Nineteenth-Century America." *Western Historical Quarterly* 8 (1977): 151–65.

Duffey, R. K. "Housing and Service for Employees." *Mining Congress Journal* 16 (Apr. 1930): 98–99.

Dykstra, Robert R. *The Cattle Towns.* New York: Atheneum, 1968.

Eblin, Jack. "An Analysis of Nineteenth-Century Frontier Populations." *Demography* 2 (1965): 399–413.

Edwards, Allen D. "Influence of Drought and Depression on a Rural Community: A Case Study of Haskell County, Kansas." Ph.D. diss., Duke University, 1939.

Elliott, Russell. *Growing Up in a Company Town: A Family in the Copper Camp of McGill, Nevada*. Reno: Nevada Historical Society, 1990.

——. *Nevada's Twentieth Century Mining Boom: Tonopah, Goldfield, Ely*. Reno: University of Nevada Press, 1966.

Esslinger, Dean. *Immigrants and the City: Ethnicity and Mobility in a Nineteenth Century City*. Port Washington, N.Y.: Kennikat Press, 1975.

Evans, William A. *Two Generations in the Southwest*. Phoenix: Sims Printing, 1971.

Fahys-Smith, Virginia Ellen. "The Migration of Boomtown Construction Workers: Wanderlust or Adaptation?" Ph.D. diss., University of Colorado, 1982.

Fatout, Paul. *Meadow Lake: Gold Town*. Bloomington: Indiana University Press, 1969.

Faulk, Odie B. *Arizona: A Short History*. Norman: University of Oklahoma Press, 1970.

——. "Life in Tombstone." *Journal of the West* 11 (July 1972): 495–512.

——. *Tombstone: Myth and Reality*. New York: Oxford University Press, 1972.

Finsterbusch, Kurt, ed. *Social Impact Assessment Methods*. Beverly Hills: Sage Publications, 1983.

Fitzgerald, Daniel. *Ghost Towns of Kansas: A Traveler's Guide*. Lawrence: University Press of Kansas, 1988.

Fitzgerald, John, and John Barton. *Jerome, Arizona: Yesterday and Today*. N.p., 1975.

Florin, Lambert. *Colorado and Utah Ghost Towns*. Seattle: Superior Publishing, 1971.

Fong, Lawrence Michael. "Sojourners and Settlers: The Chinese Experience in Arizona." *Journal of Arizona History* 21 (fall 1980): 227–56.

Foster, James C. "Western Miners and Silicosis: The Scourge of the Underground Toiler, 1890–1943." *Industrial and Labor Relations Review* 37 (1984): 371–85.

Francaviglia, Richard V. "Bisbee, Arizona: A Mining Town Survives a Decade of Closure." *Small Town* 13 (Jan.–Feb. 1983): 4–8.

——. "Copper Mining and Landscape Evolution: A Century of Change in the Warren Mining District, Arizona." *Journal of Arizona History* 23 (fall 1982): 267–98.

——. *Hard Places: Reading the Landscape of America's Historic Mining Districts*. Iowa City: University of Iowa Press, 1991.

Freeman, Joshua, Nelson Lichtenstein, Stephen Brier, David Bensman, Susan Porter Benson, David Brondage, Bret Eynow, Bruce Levine, and Bryan Palmer. *Who Built America? Working People and the Nation's Economy, Politics, Culture, and Society*. Vol. 2. New York: Pantheon Books, 1992.

Freudenburg, William R. "Women and Men in an Energy Boomtown: Adjustment, Alienation, and Adaptation." *Rural Sociology* 46 (summer 1981): 220–44.

Goldman, Marion. *Gold Diggers and Silver Miners: Prostitution and Social Life on the Comstock Lode.* Ann Arbor: University of Michigan Press, 1981.

Goodale, G. W. "Reminiscences of Early Days in Tombstone, Arizona." *Arizona Mining Journal* 10 (1927): 60–62.

Greenland, Powell. *Hydraulic Mining in California: A Tarnished Legacy.* Spokane: Arthur H. Clark, 2001.

Greer, Paul L. "Tombstone Promotes Its Past to Ensure Its Future." *Tempo* (28 June 1979): 8–10.

Greever, William S. *The Bonanza West: The Story of the Western Mining Rushes, 1848–1900.* 1963. Reprint, Moscow, Idaho: University of Idaho Press, 1993.

Griswold, Robert L. "Apart but Not Adrift: Wives, Divorce, and Independence in California, 1850–1890." *Pacific Historical Review* 49 (1980): 265–83.

Gulliford, Andrew. *Boomtown Blues: Colorado Oil Shale, 1885–1985.* Niwot: University Press of Colorado, 1989.

Haeger, John D. "The Abandoned Townsite on the Midwestern Frontier: A Case Study of Rockwell, Illinois." *Journal of the Early Republic* 3 (summer 1983): 165–83.

Halaas, David F., and Gerald C. Morton. "Boom and Bust." *Colorado Heritage News,* no. 1–2 (1983): 9–24.

Hall, Sharlot M. "The Re-Making of an Old Bonanza." *Out West* 25 (Sept. 1906): 235–46.

Hamilton, Patrick. *The Resources of Arizona.* Prescott: n.p., 1881.

———. *The Resources of Arizona.* San Francisco: A. L. Bancroft, 1883, 1884.

Harding, J. C., and C. L. Guynn. "Employment and Welfare at the United Verde." *Mining Congress Journal* 16 (Apr. 1930): 97, 103.

Hart, Mary Nicklanovich. "Merchant and Miner: Two Serbs in Early Bisbee." *Journal of Arizona History* 21 (fall 1980): 313–34.

Hatch, Heather, Karen Dahood, and Al V. Fernandez. "Clifton: Photography and Folk History in a Mining Town." *Journal of Arizona History* 23 (fall 1982): 317–40.

Hattich, William. *Tombstone.* 1903. Reprint, Norman: University of Oklahoma Press, 1981.

Hayhurst, Hal LaMar. *Hardpan: A Story of Early Arizona.* N.p., 1938.

Heyman, Josiah. "The Oral History of the Mexican-American Community of Douglas, Arizona." *Journal of the Southwest* 35 (summer 1993): 186–206.

Hill, J. Rowland. "Arizona's Developments." *Golden Era* 38 (May 1889): 195–207, 243–51.

Hine, Robert V. *Community on the American Frontier: Separate but Not Alone*. Norman: University of Oklahoma Press, 1985.

Hoover, Dwight W. "Changing Views of Community Studies: Middletown As a Case Study." *Journal of the History of the Behavioral Sciences* 25 (Apr. 1989): 111–24.

Hubner, John. "Just One Goodfellow in Tombstone." *Arizona Highways* 52 (Sept. 1976): 8–14.

Huginnie, Andrea Yvette. "Strikitos: Race, Class, and Work in the Arizona Copper Industry, 1870–1920." Ph.D. diss., Yale University, 1991.

Hunt, George W. P. "What Is the Matter with Copper Mining?" *Yavapai Magazine* 21 (May 1931): 18.

Huntoon, Peter W. "Mary M. Costello: Tombstone Banker." *Journal of Arizona History* 29 (winter 1988): 413–25.

"Intensive Development of Yavapai's Mineral Resources." *Yavapai Magazine* 18 (Apr. 1928): 16–18.

Jackson, W. Turrentine. *Treasure Hill: Portrait of a Silver Mining Camp*. Tucson: University of Arizona Press, 1963.

James, Ronald M. *The Roar and the Silence: A History of Virginia City and the Comstock Lode*. Reno: University of Nevada Press, 1998.

James, Ronald M., Richard D. Adkins, and Rachel T. Hartigan. "Competition and Coexistence in the Laundry: A View of the Comstock." *Western Historical Quarterly* 25 (summer 1994): 164–84.

Jerome Tourguide. Jerome, Ariz.: Jerome Community Service Organization, n.d.

Johnson, Wayne. "Jerome, Arizona: The Billion Dollar Copper, Mineral, and Silver Camp." *California Mining Journal* (Mar.–Apr. 1985): 3, 58–63.

Jurie, Jay Dorian. *Arizona Mining Towns: Public Policy during Boom and Bust*. Tempe: Arizona State University, Morrison Institute for Public Policy, 1984.

Kammen, Michael. *The Past before Us: Contemporary Historical Writing in the United States*. Ithaca: Cornell University Press, 1980.

Killian, L. M. "The Significance of Multiple-Group Membership in Disaster." *American Journal of Sociology* 57 (1952): 309–14.

Knight, Oliver. "Toward an Understanding of the Western Town." *Western Historical Quarterly* 4 (Jan. 1973): 27–42.

Lamb, Blaine Peterson. "Jewish Pioneers in Arizona, 1850–1920." Ph.D. diss., Arizona State University, 1982.

Lamm, Richard D., and Michael McCarthy. *The Angry West: A Vulnerable Land and Its Future.* Boston: Houghton-Mifflin, 1982.

Lantz, Herman R. *The People of Coaltown.* New York: Columbia University Press, 1958.

Larson, T. A. *History of Wyoming.* Lincoln: University of Nebraska Press, 1978.

Lee, Dan. "Ghost Town Class Reunion." *Arizona* (24 Nov. 1974): 23.

Levine, Brian. *A Guide to the Cripple Creek–Victor Mining District.* Colorado Springs: Century Press, 1978.

Lewis, David Rich. "La Plata, 1891–93: Boom, Bust, and Controversy." *Utah Historical Quarterly* 50 (winter 1982): 4–21.

Limerick, Patricia Nelson. "Haunted by Rhyolite: Learning from the Landscape of Failure." *American Art* 6 (fall 1992): 18–39.

———. *The Legacy of Conquest: The Unbroken Past of the American West.* New York: W. W. Norton, 1987.

Lister, Florence C., and Robert H. Lister. *The Chinese of Early Tucson: Historic Archaeology from the Tucson Urban Renewal Project.* Tucson: University of Arizona Press, 1989.

———. "Chinese Sojourners in Territorial Prescott." *Journal of the Southwest* 31 (spring 1989): 1–111.

Lopez, Leonor. *Forever, Sonora, Ray, Barcelona: A Labor of Love.* N.p.: L. Lopez, 1984.

Love, Alice Emily. "The History of Tombstone to 1887." Master's thesis, University of Arizona, 1933.

Lowitt, Richard. *The New Deal and the West.* Bloomington: Indiana University Press, 1984.

Lynd, Robert S., and Helen Merrell Lynd. *Middletown in Transition: A Study in Cultural Conflicts.* New York: Harcourt, Brace, 1937.

Machlis, Gary E., and Jo Ellen Force. "Community Stability and Timber-Dependent Communities." *Rural Sociology* 53 (summer 1988): 220–34.

Macia, Ethel. "Tombstone in the Early Days." *Arizona Cattlelog* (Nov. 1949): 8–11.

Malone, Michael. "The Collapse of Western Metal Mining: An Historical Epitaph." *Pacific Historical Review* 55 (Aug. 1986): 455–64.

Mann, Ralph. *After the Gold Rush: Society in Grass Valley and Nevada City, California, 1849–1870.* Stanford: Stanford University Press, 1982.

———. "Decade after the Gold Rush: Social Structure in Grass Valley and Nevada City, California, 1850–1860." *Pacific Historical Review* 41 (Nov. 1972): 484–504.

Margolis, Eric. "Life in the Coal Towns." *Journal of the West* 24 (1985): 46–69.

Martin, Douglas D. *Tombstone's "Epitaph."* Albuquerque: University of New Mexico Press, 1951.

Martin, Gregory. *Mountain City.* New York: North Point Press, 2000.

Maxwell, Margaret. "The Depression in Yavapai County." *Journal of Arizona History* 25 (summer 1982): 209–28.

McCool, Grace. *Sunday Trails in Old Cochise: A Guide to Ghost Towns, Lost Mines, and Buried Treasure in Cochise County, Arizona.* Tombstone: Epitaph, 1967.

McDonald, Lewis J. "The Development of Jerome: An Arizona Mining Town." Master's thesis, Northern Arizona University, 1941.

———. *Jerome, Arizona: The Unique Town of America.* Jerome, Ariz.: L. J. McDonald, 1948.

McGrath, Roger D. *Gunfighters, Highwaymen, and Vigilantes: Violence on the Frontier.* Berkeley and Los Angeles: University of California Press, 1984.

Melzer, Richard. *Madrid Revisited: Life and Labor in a New Mexican Mining Camp in the Years of the Great Depression.* Santa Fe: Lightning Tree, 1976.

Miller, Donald C. *Ghost Towns of California.* Boulder: Pruett Publishing, 1978.

———. *Ghost Towns of Idaho.* Boulder: Pruett Publishing, 1976

———. *Ghost Towns of Montana.* Boulder, Colorado: Pruett Publishing, 1974.

———. *Ghost Towns of Nevada.* Boulder: Pruett Publishing, 1979.

———. *Ghost Towns of Wyoming.* Boulder: Pruett Publishing, 1977.

Miller, Donald E., and Richard Sharpless. *The Kingdom of Coal: Work, Enterprise, and Ethnic Communities in the Mine Fields.* Philadelphia: University of Pennsylvania Press, 1985.

Mills, C. E. "Ground Movement and Subsidence at the United Verde Mine." *Transactions American Institute of Mining Engineers* 109 (1934): 153–72.

Mining Congress Journal. "The United Verde Copper Company." *Mining Congress Journal* 16 (Apr. 1930): 301–411.

"Mining Resource Edition." *Yavapai Magazine* 19 (Aug. 1929): 2, 9–13, 19.

"The Mining Revival at Tombstone." *Engineering and Mining Journal* 73 (1 Mar. 1902): 314–35.

Morris, Lorine. "Jerome Revives Growth, Heritage." *Westward* (30 June 1976): 13.

———. "Mexican-Spanish Influence Goes Back Centuries." *Westward* (30 June 1976): 19.

———. "United Verde Mine." *Westward* (30 June 1976): 13.

Muretic, John. *To Heaven in the West.* Sedona, Ariz.: Oak Creek Press, 1970.

Myers, Frank D. *Cochise County, Arizona.* N.p.: F. D. Myers, 1910.

Myers, John. *The Last Chance: Tombstone's Early Years.* New York: Dutton, 1950.

Myers, Rex. "Boom and Bust: Montana's Legacy of High Hopes ... and Lost Dreams." *Western Wildlands* 8 (spring–summer 1982): 6–10.

Myrick, David F. "A Chapter in the Life of Raso: A Railroad Ghost Town." *Journal of Arizona History* 17 (winter 1976): 363–74.

———. "The Railroads of Southern Arizona: An Approach to Tombstone." *Journal of Arizona History* 8 (autumn 1967): 155–70.

Noel, Thomas J., Paul F. Mahoney, and Richard E. Stevens. *Historical Atlas of Colorado.* Norman: University of Oklahoma Press, 1993.

Nugent, Charles R. "Closing of Tombstone Church." *Church at Home and Abroad* 17 (Mar. 1895): 198.

Olien, Roger M., and Diana D. Olien. *Oil Booms: Social Changes in Five Texas Towns.* Lincoln: University of Nebraska Press, 1982.

Parker, Charles Franklin, and Jeanne S. Humburg. "Yavapai County: The Mother of Counties and Land of Enchantment." *Arizona Highways* 36 (May 1960): 10–31.

Parsons, George W. "Early Days in Tombstone." *Los Angeles Mining Review* (23 Mar. 1901): 13.

Parsons, L. A. "The New Surface-Plant for the United Verde Copper Company." *Mining and Scientific Press* 122 (25 June 1921): 873–78.

Pegues, Noel. "Recreation in the Verde District." *Mining Congress Journal* 16 (Apr. 1930): 107–9.

Peterson, Richard H. "The Frontier Thesis and Social Mobility on the Mining Frontier." *Californians* 6 (Mar.–Apr. 1988).

Peterson, Thomas Hardin. "The Tombstone Stagecoach Lines, 1878–1903: A Study in Frontier Transportation." Master's thesis, University of Arizona, 1968.

Phelps Dodge: A Copper Centennial, 1881–1981. Supp. to *Arizona Paydirt.* Bisbee, Ariz.: Copper Queen Publishing, 1981.

Pratt, Florence E. "An Arizona Home." *Homemaker* 4 (Sept. 1890): 487.

Prescott City Directory. Prescott: Prescott Courier, 1922, 1937, 1939.

Prescott City Directory. Loveland, Colo.: Johnson Publishing, 1928, 1935, 1939.

"Price of Labor, Provisions, Etc., at Tombstone." *Arizona Quarterly Illustrated* 1 (July 1880): 17.

Pritchard, Nancy Lee. "Community: A Combination of Tradition and the Individual." Master's thesis, Northern Arizona University, 1984.

"Publisher's Department." *Golden Era* 38 (May 1889): 239–41.

"Publisher's Department." *Golden Era* 38 (June 1889): 288–89.

Raht, Carlysle Graham. *Confessions of a Fiddlefoot.* Odessa, Tex.: Rahtbooks, 1967.

Raines, J. C., Lenora E. Berson, and David M. Gracie. *Community and Capital in Conflict: Plant Closings and Job Loss.* Philadelphia: Temple University Press, 1982.

Ramsey, Bruce. *Ghost Towns of British Columbia.* Vancouver, B.C.: Mitchell Press, 1963.

Ray, Scott. "The Depressed Industrial Society: Occupational Movement, Out-Migration, and Residential Mobility in the Industrial-Urbanization of Middle-town, 1880–1925." Ph.D. diss., Ball State University, 1981.

Ready, Alma. "Charleston, the Town That Never Grew Old." *Arizona Highways* 38 (Nov. 1962): 2–7.

"The Resurrection of Tombstone." *Harper's Weekly* (17 May 1902): 626.

Rice, George Graham. *My Adventures with Your Money.* Boston: R. G. Badger, 1913.

Riney-Kehrberg, Pamela Lynn. "In God We Trusted, in Kansas We Busted . . . Again: A Social History of Dust Bowl Kansas." Ph.D. diss., University of Wisconsin at Madison, 1991.

———. *Rooted in Dust: Surviving Drought and Depression in Southwestern Kansas.* Lawrence: University Press of Kansas, 1994.

Ringholtz, Raye Carleson. *Uranium Frenzy: Boom and Bust on the Colorado Plateau.* New York: W. W. Norton, 1989.

Robbins, William G. *Hard Times in Paradise: Coos Bay, Oregon, 1850–1986.* Seattle: University of Washington Press, 1988.

———. "The 'Plundered Province' Thesis and the Recent Historiography of the American West." *Pacific Historical Review* 55 (Nov. 1986): 577–97.

Rochlin, Harriet. "The Amazing Adventures of a Good Woman." *Journal of the West* 12 (Apr. 1973): 281–95.

Rohe, Randall. "The Geography and Material Culture of the Western Mining Town." *Material Culture* 16 (fall 1984): 99–120.

Rohrbough, Malcolm J. *Aspen: The History of a Silver Mining Town, 1879–1893.* New York: Oxford University Press, 1986.

Rusco, Mary K. "Counting the Lovelock Chinese." *Nevada Historical Society Quarterly* 23 (winter 1981): 319–28.

Ryan, Pat M. "Tombstone Theater Tonight! A Chronicle of Entertainment on the Southwestern Mining Frontier." *Smoke Signal* 13 (1966): 49–76.

———. "Trailblazer of Civilization: John P. Clum's Tucson and Tombstone Years." *Journal of Arizona History* 6 (1965): 53–70.

Sayre, John W. *The Santa Fe, Prescott, and Phoenix Railway: The Scenic Line of Arizona.* Boulder: Pruett Publishing, 1990.

Serven, J. E. "C. S. Fly, Tombstone, A.T." *Arizona Highways* 3 (Feb. 1970): 3.

Service, Robert. *The Best of Robert Service.* New York: Penguin Putnam, 1989.

Servin, Manuel P., and Robert L. Spude. "Historical Conditions of Early Mexican Labor in the United States: Arizona–a Neglected Story." *Journal of Mexican American History* 5 (1975): 43–56.

Shaw, S. F. "Mining and Milling in Tombstone District." *Mining World* 30 (1909): 550–89.

Sherman, James E., and Barbara H. Sherman. *Ghost Towns and Mining Camps of New Mexico.* Norman: University of Oklahoma Press, 1975.

———. *Ghost Towns of Arizona.* Norman: University of Oklahoma Press, 1969.

Shinn, Charles H. *The Story of the Mine.* 1896. Reprint, Reno: University of Nevada Press, 1980.

"Silent Movies Were Its Bill: Liberty Theater Served Jerome." *Jerome Traveler* (May 1990): 4.

Simon, William, and John H. Gagnon. "The Decline and Fall of the Small Town." *Trans-Action* (Apr. 1967): 42–51.

Smith, Duane A. "Boom to Bust and Back Again: Mining in the Central Rockies, 1920–1981." *Journal of the West* 21 (Oct. 1982): 3–10.

———. *Mining America: The Industry and the Environment, 1800–1980.* Lawrence: University Press of Kansas, 1987.

———. *Rocky Mountain Mining Camps: The Urban Frontier.* 1967. Reprint, Lincoln: University of Nebraska Press, 1974.

———. "The San Juaner: A Computerized Portrait." *Colorado Magazine* 52 (1975): 137–52.

———. *Silver Saga: The Story of Caribou, Colorado.* Boulder: Pruett Publishing, 1974.

———. *When Coal Was King: A History of Crested Butte, Colorado, 1880–1952.* Golden: Colorado School of Mines Press, 1984.

Smith, Nancy R. "Jerome: Man's Changes to One Mountain." *Smoke Signal* 51 (1989): 14.

———. "Some Early History of the Verde Central Property." *Jerome Chronicle* (spring 1991): 1–5.

Smith, Page. *As a City upon a Hill: The Town in American History.* New York: Knopf, 1966.

Snodgrass, Richard, and Art Clark. *Ballad of a Laughing Mountain.* Tempe: Counterpoint Productions, 1957.

Sonnichsen, Charles Leland. *Billy King's Tombstone: The Private Life of an American Boom Town.* Caldwell, Idaho: Caxton Printers, 1942.

Spence, Clark. *For Wood River or Bust: Idaho's Silver Boom of the 1880s.* Moscow: University of Idaho Press, 1999.

——. "I Was a Stranger and Ye Took Me In." *Montana Magazine* 44 (winter 1994): 42–53.

Spratt, John S., Sr. *Thurber, Texas: The Life and Death of a Company Coal Town.* Austin: University of Texas Press, 1986.

Sproul, Irene. "Other Tombstone Churches." *Cochise Quarterly* (summer and fall 1975): 59.

——. "Sacred Heart Roman Catholic Church." *Cochise Quarterly* (summer and fall 1975): 56.

——. "St. Paul's Episcopal Church." *Cochise Quarterly* (summer and fall 1975): 57–59.

Spude, Robert L. "Swansea, Arizona: The Fortunes and Misfortunes of a Copper Camp." *Journal of Arizona History* 17 (winter 1976): 375–96.

Steffen, Jerome O. *The American West.* Norman: University of Oklahoma Press, 1979.

Stewart, Kenneth. "The Boom and Bust of Central City." *South Dakota History* 2 (summer 1972): 230–60.

Stout, Wesley W. "Want to Buy a Ghost Town?" *Saturday Evening Post* (26 May 1951): 144.

Tally, Robert. "Relations with Employees." *Mining Congress Journal* 16 (Apr. 1930): 96.

Thernstrom, Stephen, and Peter R. Knights. "Men in Motion: Some Data and Speculations about Urban Population Mobility in Nineteenth-Century America." *Journal of Interdisciplinary History* 1 (1970): 2–25.

Thompson, Gerald. "The New Western History: A Critical Analysis." *Continuity: A Journal of History* 17 (fall 1993): 6–24.

Thure, Karen. "Tombstone 1881." *Arizona Highways* 55 (Mar. 1979): 3–9.

Tiller, Kerry C. S. "Charleston Townsite Revisited." *Journal of Arizona History* 23 (fall 1982): 241–48.

"Tombstone." *Arizona Quarterly Illustrated* 1 (Jan. 1881): 22.

Tombstone Courthouse State Historic Park. N.p.: Arizona State Parks, n.d.

"The Tombstone District." *Arizona Quarterly Illustrated* 1 (Jan. 1881): 20–21.

Tombstone Mill and Mining Company. *Annual Report on Mines and Mills, with Production and Expenses for the Year Ending 31 March 1883.* Philadelphia: McCalla and Stavely, 1883.

"The Tombstone Mill and Mining Company, Arizona." *Engineering and Mining Journal* 42 (11 Sept. 1886): 186.

"Tombstone Mines, Arizona." *Engineering and Mining Journal* 77 (9 June 1904): 919–20.

Traywick, Ben T. *The Chinese Dragon in Tombstone.* N.p., 1989.

———. *The Chronicles of Tombstone.* Tombstone: Red Marie's Bookstore, 1990.

———. *Eleanora Dumont, Alias Madam Moustache.* Tombstone: Red Marie's Bookstore, 1990.

———. *The Mines of Tombstone.* Tombstone: Red Marie's Bookstore, 1983.

———. *Some Ghosts along the San Pedro.* Tombstone: Red Marie's Bookstore, 1987.

———. *The "Tombstone Epitaph" and John Philip Clum.* Tombstone: Red Marie's Bookstore, 1985.

———. *A Town Called Tombstone.* Tombstone: B. F. Traywick, 1982.

Tsai, Shih-Shan Henry. "American Exclusion against the Chinese." *Californians* 6 (Mar.–Apr. 1988): 36–41.

Tucson and Tombstone General and Business Directory, for 1883 and 1884. Tucson: Daily Citizen Steam Printing, 1883.

Twain, Mark. *Gold Miners and Guttersnipes.* San Francisco: Chronicle Books, 1991.

———. *Roughing It.* New York: New American Library, Signet, 1962.

Underhill, Lonnie E. *The Silver Tombstone of Edward Schieffelin.* Tucson: Roan Horse Press, 1979.

———, ed. *Tombstone, Arizona, 1880, Business and Professional Directory.* Tucson: Roan Horse Press, 1982.

Van Wormer, Stephen. "The Wetherbee Planning Mill: A Case History of the 1880s Boom and Bust." *Journal of San Diego History* 29 (winter 1983): 20–28.

"Verde Copper King Company." *Los Angeles Mining Review* (23 Mar. 1901): 9.

"Verde District, in the Heart of Yavapai's Wonderland." *Yavapai Magazine* 12 (Mar. 1924): 7–8.

"Verde District, Special Edition." *Arizona Mining Journal* 6 (1 Aug. 1922): 9, 47–50.

"Verde Mining District." *Los Angeles Mining Review* (23 Mar. 1901): 5.

Voynick, Stephen M. *The Making of a Hardrock Miner.* Berkeley: Howell-North Books, 1978.

Wahmann, Russel. "A Centennial Commemorative: United Verde Copper Company, 1882–1982." *Journal of Arizona History* 23 (fall 1982): 249–66.

Wahmann, Russel, and Robert des Granges. *Verde Valley Railroads: A Search of the Past and Comparison with the Present of the Several Railroads That Operated in the Verde Valley of Arizona.* Jerome, Ariz.: Wahmann, 1983.

Walker, Henry P. "Arizona Land Fraud: Model 1880, the Tombstone Townsite Company." *Arizona and the West* 21 (spring 1979): 5–36.

Walters, Lorenzo D. *Tombstone's Yesterday*. Tucson: Acme Printing, 1928.

Weis, Norman D. *Ghost Towns of the Northwest*. Caldwell, Idaho: Caxton Printers, 1971.

West, Elliott. "Heathens and Angels: Childhood in the Rocky Mountain Mining Towns." *Western Historical Quarterly* 14 (Apr. 1983): 145–64.

———. "The Saloon in Territorial Arizona." *Journal of the West* 13 (1974): 61–73.

———. *The Saloon on the Rocky Mountain Mining Frontier*. Lincoln: University of Nebraska Press, 1979.

Wiebe, Robert. *The Search for Order, 1877–1920*. New York: Hill and Wang, 1967.

Willard, Don. *An Old-Timer's Scrapbook: A Roundup from the Range of His Own Recollections and Observations*. Mesa, Ariz.: Marker Graphics, 1984.

Williams, David C. "Boom Then Bust: Managing Rapid Growth Cities." *Management Information Service Project* 9 (Mar. 1977): 1–13.

Willis, Charles F. "Jerome Business Houses." *Arizona Mining Journal* 6 (1 Aug. 1922): 48–50.

———. "Jerome Grand Copper between Verde Central and United Verde." *Arizona Mining Journal* 6 (1 Aug. 1922): 46.

———. "Special Edition on the Verde District." *Arizona Mining Journal* 6 (1 Aug. 1922): 9.

———. "T. F. Miller Company Store Started in Year 1890." *Arizona Mining Journal* 6 (1 Aug. 1922): 47.

———. "Verde Chief Property in South Section of District." *Arizona Mining Journal* 6 (1 Aug. 1922): 45.

Willson, Clair Eugene. *Mimes and Miners: A Historical Study of the Theater in Tombstone*. Tucson: University of Arizona, 1935.

Windham, Joey Samuel. "Grand Encampment Mining District: A Case Study of the Life Cycle of a Typical Western Frontier Mining District." Ed.D. diss., Ball State University, 1981.

Winters, Wayne. "Tombstone Rises Again." *Desert* (May 1980): 42–43.

Wood, Homer R. *The History of Mining in Yavapai County*. Prescott: Yavapai County Chamber of Commerce, 1935.

Worster, Donald. *Dust Bowl: The Southern Plains in the 1930s*. New York: Oxford University Press, 1979.

———. *Under Western Skies: Nature and History in the American West*. New York: Oxford University Press, 1992.

Woyski, Margaret S. "Women and Mining in the Old West." *Journal of the West* 20 (Apr. 1981): 38–47.

WPA Writers' Program. *The WPA Guide to 1930s Arizona.* 1940. Reprint, Tucson: University Press of Arizona, 1989.

——. *The WPA Guide to 1930s Colorado.* 1941. Reprint, Lawrence: University Press of Kansas, 1987.

Wyllys, Rufus Kay. *Arizona: The History of a Frontier State.* Phoenix: Hobson and Herr, 1950.

——. *Men and Women of Arizona.* Phoenix: Pioneer Publishing, 1940.

Yavapai County Chamber of Commerce. "Annual Report." *Yavapai Magazine* 21 (Feb. 1931): 2–7, 14–18.

"Yavapai Mines Have Added $450,000,000 to the World's Wealth since 1888." *Yavapai Magazine* 18 (Apr. 1928): 8–12.

Young, Herbert V. *Ghosts of Cleopatra Hill: Men and Legends of Old Jerome.* Jerome, Ariz.: Jerome Historical Society, 1964.

——. *They Came to Jerome: The Billion Dollar Copper Camp.* Jerome, Ariz.: Jerome Historical Society, 1972.

Young, J. H. "Tombstone, Arizona." *Overland Monthly* 8 (Nov. 1886): 483–87.

Young, Michael. "The Role of the Extended Family in a Disaster." *Human Relations* 7 (1954): 383–91.

Young, Otis E., Jr. *Western Mining: An Informal Account of Precious-Metals Prospecting, Placering, Lode Mining, and Milling on the American Frontier from Spanish Times to 1893.* Norman: University of Oklahoma Press, 1970.

Zanjani, Sally S. *Goldfield: The Last Gold Rush on the Western Frontier.* Athens: Ohio University Press, Swallow Press, 1992.

——. "To Die in Goldfield: Mortality in the Last Boomtown on the Mining Frontier." *Western Historical Quarterly* 21 (1990): 47–69.

Note: Italic page numbers refer to illustrations.